Current trends in Bioinformatics: An Insight

Gulshan Wadhwa • P. Shanmughavel
Atul Kumar Singh • Jayesh R. Bellare
Editors

Current trends in Bioinformatics: An Insight

Springer

Editors
Gulshan Wadhwa
Department of Biotechnology
Apex Bioinformatics Centre
Ministry of Science & Technology
New Delhi, India

P. Shanmughavel
Department of Bioinformatics
Bharathiar University
Coimbatore, Tamil Nadu, India

Atul Kumar Singh
Central Research Facility
Indian Institute of Technology Delhi
New Delhi, India

Centre for Research in Nanotechnology and Sciences
Indian Institute of Technology Bombay
Mumbai, India

Jayesh R. Bellare
Department of Chemical Engineering
Indian Institute of Technology Bombay
Mumbai, India

ISBN 978-981-10-7481-3 ISBN 978-981-10-7483-7 (eBook)
https://doi.org/10.1007/978-981-10-7483-7

Library of Congress Control Number: 2018943304

Printed on acid-free paper

This Springer imprint is published by the registered company Springer Nature Singapore Pte Ltd.
The registered company address is: 152 Beach Road, #21-01/04 Gateway East, Singapore 189721, Singapore

Foreword

It gives us immense pleasure to present this edited book to the biotechnological research communities. Bioinformatics and computational biology – the science of using biological data to develop algorithms and relations among various biological systems – are the cutting edge areas for research. Computational sciences have their roots in the development of increasingly powerful computers over the last few decades. Rather rapidly, the instrumentation and the newly developed methodology with the underlying algorithms became widely appreciated and used as novel research strategies serving in many different fields of academic investigation, particularly in natural, engineering, social sciences, and humanities. Computational sciences have been recognized for their invaluable contributions to data collection, storage, handling, and analysis, thus leading to efficient strategies of modeling, prediction, and design of molecular structures and of their functional properties that are often of immediate relevance for the medical sciences. Computational comparisons of DNA sequences from different organisms provide invaluable insights into past evolutionary developments, and this has become a powerful new tool in the systematics of living organisms.

The growth in high-throughput full genomic sequencing, structural genomics, proteomics, epigenetics, etc., would be rather limited without bioinformatics. In order to give concise information on basic concept and advances in bioinformatics, authors have thought of bringing out an edited volume "Current Trends in Bioinformatics" for the benefit of the students and researchers working in the field of life science, medicine, and pharmaceutical science. It also focuses on reviews on advances in computational molecular/structural biology, encompassing areas such as computing in biomedicine and genomics, computational proteomics and systems biology, and metabolic pathway engineering. Developments in these fields have direct impact on key issues related to health care, medicine, genetic disorders, development of agricultural products, renewable energy, environmental protection, etc. The book has 18 chapters, divided into two sections.

The overview of important aspects of bioinformatics would further contribute to strengthen international contacts and serve as a testament to such a fruitful development for the basic as well as applied sciences. The Department of Biotechnology considering the great significance of this field established a countrywide network of bioinformatics centers in academic institutions. These have paid rich dividends.

We hope that scientific community especially students, in particular, would enjoy reading, learn and make best use of this book.

(Dr. Manju Sharma)

Former Secretary, Department of Biotechnology
New Delhi, India

Manju Sharma

Principal Advisor to the Deptartment of Science and Technology
Gandhinagar, Gujarat, India

Distinguished Women Scientist Chair, NASI
Allahabad, India

Preface

Bioinformatics has become a frontline applied science and is of vital importance to study new biology, which is widely recognized as the new scientific endeavor of the twenty-first century. The growth in full genomic sequencing, structural genomics, proteomics, and microarray will be very slow without application of bioinformatics. In fact the very high importance of bioinformatics comes from its usefulness in these areas to solve complex biological problems. So up-to-date information in the field of bioinformatics is the most needed one. The proposed book *Current Trends in Bioinformatics* fulfills these requirements.

Current Trends in Bioinformatics aims to publish all the latest and outstanding developments in bioinformatics. The book contains a series of timely, in-depth reviews, drug clinical trial studies, and biodiversity informatics and thematic issues written by leaders in the field, covering a wide range of the integration of biology with computer and information science.

It also focuses on reviews on advances in computational molecular/structural biology, encompassing areas such as computing in biomedicine and genomics, computational proteomics and systems biology, and metabolic pathway engineering. Developments in these fields have direct implications on key issues related to health care, medicine, genetic disorders, development of agricultural products, renewable energy, and environmental protection.

This book is an ideal foundation for teaching at the undergraduate and graduate levels. It is also highly suited for self-instruction by research investigators interested in applying bioinformatics methods of analysis and information technologists associated with academic and industrial laboratories.

It is supposed that the nonspecialists would be the principal readers of the book. So, before embarking on the bioinformatics, some fundamental aspects of molecular evolution, taxa-related studies, some core concepts of genomics and some of the important genomic techniques were discussed in this book, to make the readers conceptualize the bioinformatics analysis.

The author would also like to thank colleagues for their encouragement, enthusiasm, and support for the success of this project.

Last but not the least, the author is grateful to the Staff of Springer for making this project a reality, helping to bring it to successful completion, and always being available whenever help and advice were needed.

New Delhi, India Gulshan Wadhwa
Mumbai, India Jayesh R. Bellare
Coimbatore, India P. Shanmughavel
New Delhi, India Atul Kumar Singh
Mumbai, India

Contents

Contributors

C. Awasthi Department of Biotechnology, Gobind Ballabh Pant Engineering College, Pauri Garhwal, Uttarakhand, India

Jayesh R. Bellare Department of Chemical Engineering, Indian Institute of Technology Bombay, Mumbai, India

Vaibhav D. Bhatt Department of Pharmaceutical Sciences, Saurashtra University, Rajkot, Gujarat, India

Chandan Chakraborty School of Medical Science & Technology, Indian Institute of Technology Kharagpur, Kharagpur, West Bengal, India

Dev Kumar Das School of Medical Science & Technology, Indian Institute of Technology Kharagpur, Kharagpur, West Bengal, India

Abhijit Datta Department of Botany, Jhargram Raj College, Medinipur, West Bengal, India

Rajat K. De Machine Intelligence Unit, Indian Statistical Institute, Kolkata, India

Rohit Farmer School of Biosciences, University of Birmingham, Birmingham, UK

Department of Computational Biology and Bioinformatics, Jacob Institute of Biotechnology and Bioengineering, Sam Higginbottom University of Agriculture, Technology and Sciences, Allahabad, Uttar Pradesh, India

J. Fermin Angelo Selvin Department of Physics, Nadar Mahajana Sangam S. Vellaichamy Nadar College, Madurai, Tamil Nadu, India

Sayak Ganguli Theoretical and Computational Biology Division, Amplicon Institute of Interdisciplinary Science and Technology, Palta, West Bengal, India

Budhayash Gautam Department of Computational Biology and Bioinformatics, Jacob Institute of Biotechnology and Bioengineering, Sam Higginbottom University of Agriculture, Technology and Sciences, Allahabad, Uttar Pradesh, India

Kavita Goswami Plant RNAi Biology Group, International Center for Genetic Engineering and Biotechnology, New Delhi, India

T. Hema Thanka Christlet Department of Physics, Dr. Ambedkar Government Arts College, Chennai, Tamil Nadu, India

Chakresh Kumar Jain Department of Biotechnology, Jaypee Institute of Information Technology, Noida, Uttar Pradesh, India

Yogesh K. Jaiswal School of Studies in Biochemistry, Jiwaji University, Gwalior, India

Qazi Mohammad Sajid Jamal Department of Health Information Management, College of Applied Medical Sciences, East Qassim University, Al Qassim-Buraydah, Kingdom of Saudi Arabia

A. Jasmine Department of Physics, Manonmaniam Sundaranar University, Tirunelveli, Tamil Nadu, India

Chaitanya G. Joshi Department of Animal Biotechnology, College of Veterinary Science and Animal Husbandry, Anand Agricultural University, Anand, Gujarat, India

Sukhdeep Kaur Department of Computational Biology and Bioinformatics, Jacob Institute of Biotechnology and Bioengineering, Sam Higginbottom University of Agriculture, Technology and Sciences, Allahabad, Uttar Pradesh, India

Kavindra Kumar Kesari Department of Applied Physics, and Department of Bioproduct & Biosystem, Aalto University, Espoo, Finland

K. Konda Reddy Department of Pharmacy, National University of Singapore, Singapore, Singapore

Ajit Kumar Centre for Bioinformatics, Maharshi Dayanand University, Rohtak, India

Sunil Kumar Bioinformatics Centre, Institute of Life Sciences, Bhubaneswar, Odisha, India

ICAR-NBAIM, Mau, Uttar Pradesh, India

Sharad Singh Lodhi School of Studies in Biochemistry, Jiwaji University, Gwalior, India

Ashok K. Maiti Medipath Clinic (P) Ltd, West Medinipur, West Bengal, India

Jayakanthan Mannu Department of Plant Molecular Biology and Bioinformatics, Centre for Plant Molecular Biology and Biotechnology, Tamil Nadu Agricultural University, Coimbatore, Tamil Nadu, India

Soma S. Marla Indian Council of Agricultural Research, National Bureau of Plant Genetic Resources, New Delhi, India

Premendu P. Mathur Department of Biochemistry and Molecular Biology, School of Life Sciences, Pondicherry University, Pondicherry, India

Ranjeet Maurya Department of Computational Biology and Bioinformatics, Jacob Institute of Biotechnology and Bioengineering, Sam Higginbottom University of Agriculture, Technology and Sciences, Allahabad, Uttar Pradesh, India

Neelofar Mirza Indian Council of Agricultural Research, National Bureau of Plant Genetic Resources, New Delhi, India

Neeti Sanan Mishra Plant RNAi Biology Group, International Center for Genetic Engineering and Biotechnology, New Delhi, India

Pallavi Mishra Department of Computational Biology and Bioinformatics, Jacob Institute of Biotechnology and Bioengineering, Sam Higginbottom University of Agriculture, Technology and Sciences, Allahabad, Uttar Pradesh, India

Rashmi Mukherjee RNLKWC, Vidyasagar University, Midnapur, West Bengal, India

K. D. Nadella Directorate of Knowledge Management Units (DKMU), ICAR, New Delhi, India

Genetics division, ICAR, IARI, New Delhi, India

H. P. Pandey Department of Biochemistry, Nepalgunj Medical College, Chisapani Campus, Kathmandu University, Nepalgunj, Nepal

Monika Patel Department of Animal Biotechnology, College of Veterinary Science and Animal Husbandry, Anand Agricultural University, Anand, Gujarat, India

Anjali Rao Department of Computational Biology and Bioinformatics, Jacob Institute of Biotechnology and Bioengineering, Sam Higginbottom University of Agriculture, Technology and Sciences, Allahabad, Uttar Pradesh, India

P. Shanmughavel Department of Bioinformatics, Bharathiar University, Coimbatore, Tamil Nadu, India

Mohd. Haris Siddiqui Department of Bioengineering, Faculty of Engineering, Integral University, Lucknow, Uttar Pradesh, India

Atul Kumar Singh Central Research Facility, Indian Institute of Technology Delhi, New Delhi, India

Centre for Research in Nanotechnology and Sciences, Indian Institute of Technology Bombay, Mumbai, India

Satendra Singh Department of Computational Biology and Bioinformatics, Jacob Institute of Biotechnology and Bioengineering, Sam Higginbottom University of Agriculture, Technology and Sciences, Allahabad, Uttar Pradesh, India

Manish Sinha Laureate Institute of Pharmacy, Kangra, Himachal Pradesh, India

Gitanjali Tandon Department of Computational Biology and Bioinformatics, Jacob Institute of Biotechnology and Bioengineering, Sam Higginbottom University of Agriculture, Technology and Sciences, Allahabad, Uttar Pradesh, India

Christopher M. Thomas School of Biosciences, University of Birmingham, Birmingham, UK

Namrata Tomar Department of BioMedical Engineering, Medical College of Wisconsin, Milwaukee, WI, USA

Machine Intelligence Unit, Indian Statistical Institute, Kolkata, India

Mobashar Hussain Urf Turabe Fazil Lee Kong Chian School of Medicine, Nanyang Technological University, Singapore, Singapore

K. Veluraja Department of Physics, School of Advanced Sciences, VIT University, Vellore, Tamil Nadu, India

Jyoti Verma Department of Biotechnology, Gobind Ballabh Pant Engineering College, Pauri Garhwal, Uttarakhand, India

Gulshan Wadhwa Department of Biotechnology, Apex Bioinformatics Centre, Ministry of Science & Technology, New Delhi, India

Peter J. Winn School of Biosciences, University of Birmingham, Birmingham, UK

Arvind Kumar Yadav Department of Computational Biology and Bioinformatics, Jacob Institute of Biotechnology and Bioengineering, Sam Higginbottom University of Agriculture, Technology and Sciences, Allahabad, Uttar Pradesh, India

About the Editors

Gulshan Wadhwa is currently the Joint Director in the Department of Biotechnology, Ministry of Science & Technology, Government of India. He has developed the Bioinformatics program in India (BTISnet) and established teaching and training programs and a super computing facility in Bioinformatics in India. He has undergone trainings from WIPO, Geneva, and NUS, Singapore.

Jayesh R. Bellare is currently Institute Chair Professor and Professor with Chemical Engineering Department & Center for Research in NanoTechnology and Science, Indian Institute of Technology-Bombay, Mumbai, India. He was a Post Doctoral Fellow, from M.I.T., Cambridge, USA, and has over 33 years of experience.

P. Shanmughavel is currently working as Associate Professor at Bharathiar University, Coimbatore, India. He has published 7 books in bioinformatics (two books published in Germany), 6 book chapters, 5 conference proceedings, and 31 research papers in reputed national and international journals in Bioinformatics. He has established the Centre of Bioinformatics-"DBT-BIF" in 2006.

Atul Kumar Singh is currently working as Senior Scientist at Indian Institute of Technology-Delhi. He has completed his Doctorate from Indian Institute of Technology-Bombay, Mumbai, in Nanotechnology. He has published 11 research papers in reputed international journals and has 6 patents.

Part I
Overview

1 An Insight of Biological Databases Used in Bioinformatics

Vaibhav D. Bhatt, Monika Patel, and Chaitanya G. Joshi

Abstract

Collections of life sciences information from scientific investigations, high-throughput experiment technology, available literature, and computational analysis are called biological databases. It contains information from research areas comprising genomics, microarray gene expression, proteomics, phylogenetics, metabolomics, gene function, structure, localization and similarities of biological sequences. In a nutshell, databases are libraries for storage and representation of biological data obtained from the scientific community which converts data into knowledge. Utmost biological databases are available from websites that categorize data which operators can browse through the data online. Due to the vast amount of data generated by high-throughput DNA sequencers in the investigation of genome, transcriptome, and exome sequences of various organisms in current times, the biological data has stored with an exponential rate. The availability of enormous amount of biological data (sequences as well as structural) has generated a need for managing, storing, and retrieving this huge data. This chapter reviews current knowledge of the different types of databases available with examples of their file formats.

Keywords

Biological sequences · High-throughput DNA sequencers · Transcriptome and exome sequences

V. D. Bhatt (✉)
Department of Pharmaceutical Sciences, Saurashtra University, Rajkot, Gujarat, India

M. Patel · C. G. Joshi
Department of Animal Biotechnology, College of Veterinary Science and Animal Husbandry, Anand Agricultural University, Anand, Gujarat, India

G. Wadhwa et al. (eds.), *Current trends in Bioinformatics: An Insight*,
https://doi.org/10.1007/978-981-10-7483-7_1

1.1 Introduction

Databases are the convenient system to properly store, search, and recover several types of data. A database helps to easily handle and share large amount of data and supports large-scale analysis by easy access and data update (Liu and Özsu 2009).

Due to the vast amount of data generated in experiments of genome, transcriptome, and exome sequences of various organisms in current times, the biological data has stored with an exponential rate. The availability of enormous amount of biological data (sequences as well as structural data) has generated a need for managing, storing, and retrieving this huge data.

Therefore the biological databases have come into existence as invaluable sources for the biological community. In a nutshell, databases are libraries for storage and representation of biological data obtained from the scientific community which converts data into knowledge.

1.2 History

A book published in 1965, *Atlas of Protein Sequences and Structures, was the first biological database* by Margaret Dayhoff and colleagues, and further they have published other editions of the book in the 1970s; however the first edition was limited to 65 sequences only (Dayhoff and Foundation 1973, 1976; Foundation 1972).

With the discovery of the integrated circuit, the powerful and reliable third generation computers are became the choice of storage of biological databases for scientists. An English scientist Tim Berners-Lee in 1989 invented the "World Wide Web" (WWW) which is the primary tool people use to interact on the Internet and is the way to access all biological databases. Production of high throughput sequencing machines leads production of data rich science, needs an interdisciplinary arena to develop software tools which is used to understand biological data. The field of science with the involvement of computer, statistics and engineering to study biological data is called Bioinformatics.

1.3 Classification of Biological Databases

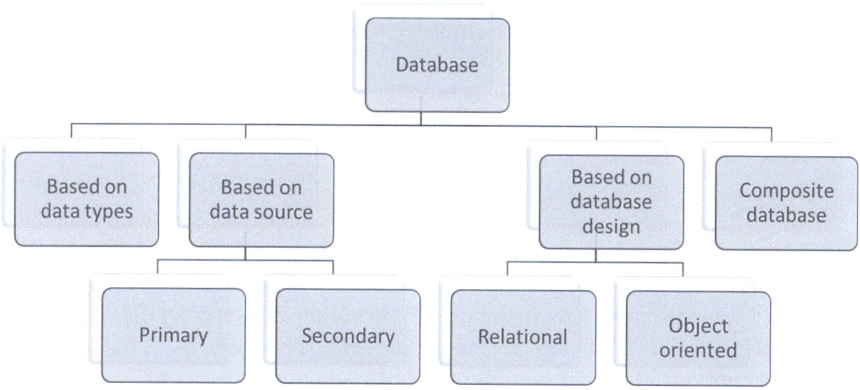

1.3.1 Databases Based on Data Types

This database was divided into several databases; some of the databases were discussed below in detail.

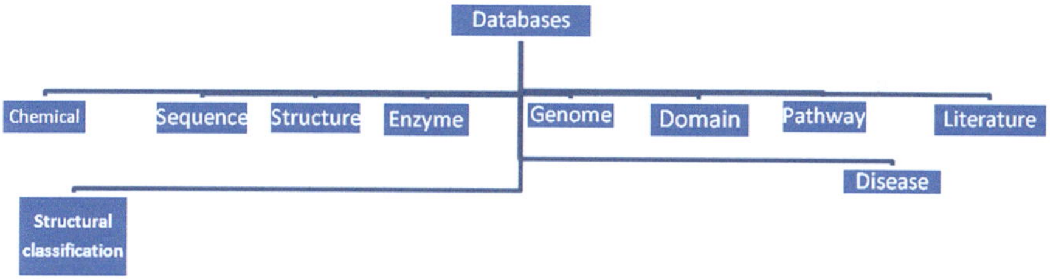

1.3.1.1 Sequence Databases

Sequence databases contain both nucleic acid and protein sequences. First we will discuss about nucleotide sequence repositories.

(I) Nucleic Acid Sequence Database

There are three main nucleotide sequence repositories:

(A) GenBank
(B) European Molecular Biology Laboratory (EMBL)
(C) DNA Data Bank of Japan (DDBJ)

Raw nucleic acid sequences are stored in these databases and make available through Internet sources. Initially, these databases worked independently, but later the *International Nucleotide Sequence Database Collaboration* (INSDC, http://insdc.org) was developed to maintain collaboration between DDBJ, GenBank, and EMBL (Fig. 1.1). These databases started exchanging their data through constant communication between the team at each collaborating organizations in order to access the sequences present in all three different formats.

(A) *GenBank*

GenBank is a collection of raw and annotated nucleotide as well as protein information. GenBank is maintained and accessed through the National Center for Biotechnology Information (NCBI). Every 2 months a new release is made. It is maintained by NCBI as part of the INSDC (Benton 1990). There are approximately 137384889783 bases, from 149819246 sequence records in the GenBank release 188.0 on February 15, 2012. Type "insulin" in the search tab on the GenBank home page to view list of sequences of insulin gene, partial or complete from different organisms (Fig. 1.2).

Example of GenBank Format

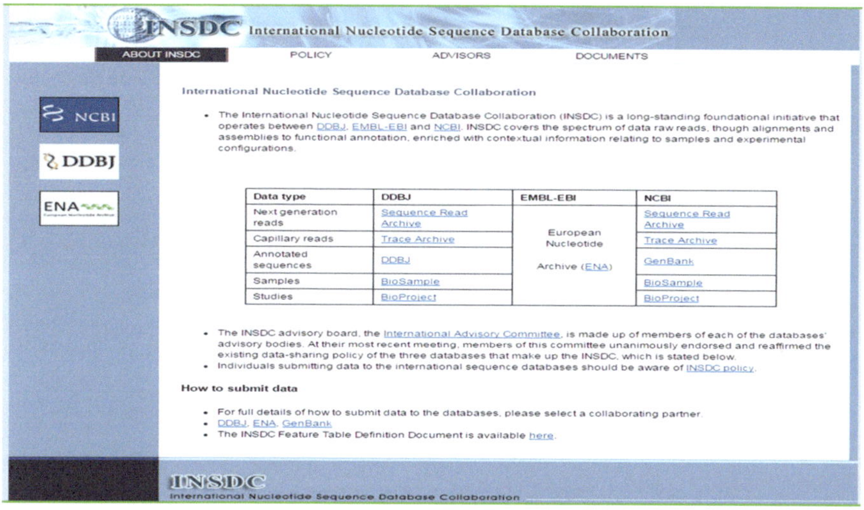

Fig. 1.1 The home page of International Nucleotide Sequence Database Collaboration (INSDC) (http://insdc.org)

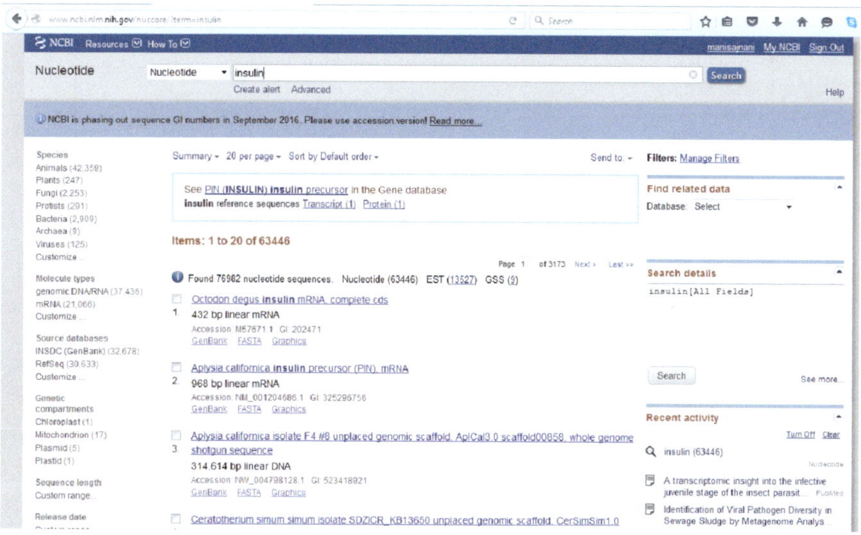

Fig. 1.2 Using GenBank to query insulin sequences (http://www.ncbi.nlm.nih.gov/nuccore/?term=insulin)

Octodon degus insulin mRNA, complete cds

GenBank: M57671.1

FASTA Graphics

Go to:

```
LOCUS       OCOINS                   432 bp    mRNA    linear   ROD 27-APR-1993
DEFINITION  Octodon degus insulin mRNA, complete cds.
ACCESSION   M57671
VERSION     M57671.1
KEYWORDS    insulin; insulin alpha-chain; insulin beta-chain; insulin
            connecting peptide.
SOURCE      Octodon degus (degu)
  ORGANISM  Octodon degus
            Eukaryota; Metazoa; Chordata; Craniata; Vertebrata; Euteleostomi;
            Mammalia; Eutheria; Euarchontoglires; Glires; Rodentia;
            Hystricognathi; Octodontidae; Octodon.
REFERENCE   1  (bases 1 to 432)
  AUTHORS   Nishi,M. and Steiner,D.F.
  TITLE     Cloning of complementary DNAs encoding islet amyloid polypeptide,
            insulin, and glucagon precursors from a New World rodent, the degu,
            Octodon degus
  JOURNAL   Mol. Endocrinol. 4 (8), 1192-1198 (1990)
   PUBMED   2293024
COMMENT     Original source text: Octodon degus pancreas, cDNA to mRNA.
FEATURES             Location/Qualifiers
     source          1..432
                     /organism="Octodon degus"
                     /mol_type="mRNA"
                     /db_xref="taxon:10160"
                     /tissue_type="pancreas"
     gene            1..432
                     /gene="insulin"
     CDS             42..371
                     /gene="insulin"
                     /codon_start=1
                     /product="insulin"
                     /protein_id="AAA40590.1"
                     /translation="MAPWMHLLTVLALLALWGPNSVQAYSSQHLCGSNLVEALYMTCG
                     RSGFYRPHDRRELEDLQVEQAELGLEAGGLQPSALEMILQKRGIVDQCCNNICTFNQL
                     QNYCNVP"
     sig_peptide     42..113
                     /gene="insulin"
     mat_peptide     114..200
                     /gene="insulin"
                     /product="insulin B-chain"
     mat_peptide     207..293
                     /gene="insulin"
                     /product="insulin C-peptide"
     mat_peptide     300..368
                     /gene="insulin"
                     /product="insulin A-chain"
     regulatory      414..419
                     /regulatory_class="polyA_signal_sequence"
                     /gene="insulin"
     polyA_site      432
                     /gene="insulin"
ORIGIN
        1 gcattctgag gcattctcta acaggttctc gaccctccgc catggccccg tggatgcatc
       61 tcctcaccgt gctggccctg ctggccctct ggggacccaa ctctgttcag gcctattcca
      121 gccagcacct gtgcggctcc aacctagtgg aggcactgta catgacatgt ggacggagtg
      181 gcttctatag accccacgac cgccgagagc tggaggacct ccaggtggag caggcagaac
      241 tgggtctgga ggcaggcggc ctgcagcctt cggccctgga gatgattctg cagaagcgcg
      301 gcattgtgga tcagtgctgt aataacattt gcacatttaa ccagctgcag aactactgca
      361 atgtccctta gacacctgcc ttgggcctgg cctgctgctc tgccctggca accaataaac
      421 cccttgaatg ag
//
```

Format Explanation

GenBank format includes *locus name* which is similar to the accession number and unique to the entry, and it is followed by sequence length. In our example sequence length is 587 bp. Definition includes description of source organism, gene/protein name, and other details about sequence.

- *Accession number* is the unique identifier of the sequence (NM_013564).
- *Version* is similar to accession number, but whenever a change occurs in sequence data, the version increases by 1. In our example, version is NM_013564.7; this indicates that sequence has been changed seven times.
- *GI (GenInfo Identifier)* number also runs parallel to the accession number and version system. A new GI is allotted, if the sequence has been changed and the version has increased by unity. In our example, GI is 365192585.
- *Keywords* are words or expressions about sequence. The keyword field contains a dot if nothing is provided.
- *Source* contains name of the organism from which the sequence has been derived.
- *Organism* is a related sub-keyword of source and contains the scientific name of the organism along with the lineage as described in NCBI taxonomy database.
- *Reference* contains the publication by the authors of the sequence.
- *Authors* contain list of authors in the same order as appears in publication.
- *Title* shows the title of published/unpublished work.
- *Journal* contains MEDLINE abbreviations of the journal name where the work is published.
- *PubMed* field provides the PubMed identifier (PMID) of that article.
- *Comment* points out the change occurred in the submitted sequence.
- *Features* provide information about genes and their products, segment of biological significance in the submitted sequence, as well as other characteristics.
- *Gene* provides gene length and gene name and its function and synonyms. CDS represents coding sequence which codes for protein sequence.
- *Origin* contains the sequence data. Finally, GenBank record ends with // sign.

Sequence Submission to GenBank

Sequence submission is done by using different tools available at NCBI. Few of them are:

BankIt: direct submissions are made to GenBank using it (www.ncbi.nlm.nih.gov/WebSub/?tool=genbank).
Sequin: it is a stand-alone submission platform (www.ncbi.nlm.nih.gov/Sequin/).
tbl2asn: it is a command-line program, used for submission of large batches of sequences and complete genomes (www.ncbi.nlm.nih.gov/genbank/tbl2asn2).

Table 1.1 Various databases and software tools of NCBI for sequence analysis

NCBI			
Tools			Databases
Sequence Submission	Sequence	Data mining	Literature
	Analysis		Nucleotide
Sequin	BLAST	Entrez	Protein
BankIt	Blink	My NCBI	Structure
tbl2asn	Stand-alone BLAST	LinkOut	Genome
			OMIM
			SNP
Barcode Submission Tool	e-PCR	Citation	Books
		Matcher	Domain
	ORF Finder		Chemical
			Expression
	Map viewer		Other databases
	Tax plot		
	Trace archive		

Barcode Submission Tool: it is a WWW-based tool for the submission of sequences and trace read data (http://www.ncbi.nlm.nih.gov/WebSub/?tool=barcode).

National Center for Biotechnology Information (NCBI)

NCBI was started in 1988, as a part of the US National Library of Medicine (NLM) located at Bethesda, Maryland. It is a division of the National Institutes of Health and is directed by David Lipman. The responsibility of NCBI is to make available the GenBank nucleotide sequence database since 1992. NCBI is playing a very remarkable role for biological scientists by making available various public databases and software tools for sequence analysis (Table 1.1). GenBank manages with individual laboratories and other sequence databases like those of the EMBL and the DDBJ. Meanwhile in 1992, NCBI has developed to run other databases in addition to GenBank ((US) 2013). The home page of NCBI is shown in Fig. 1.3.

Databases and Tools of NCBI

Database Retrieval Tool

Entrez (www.ncbi.nlm.nih.gov/Entrez/) in Fig. 1.4 is a primary text search engine which comprises of 40 molecular and literature databases. It extracts huge information from the PubMed database, such as DNA and protein sequences and structure, gene, genome, genetic variation, and gene expression.

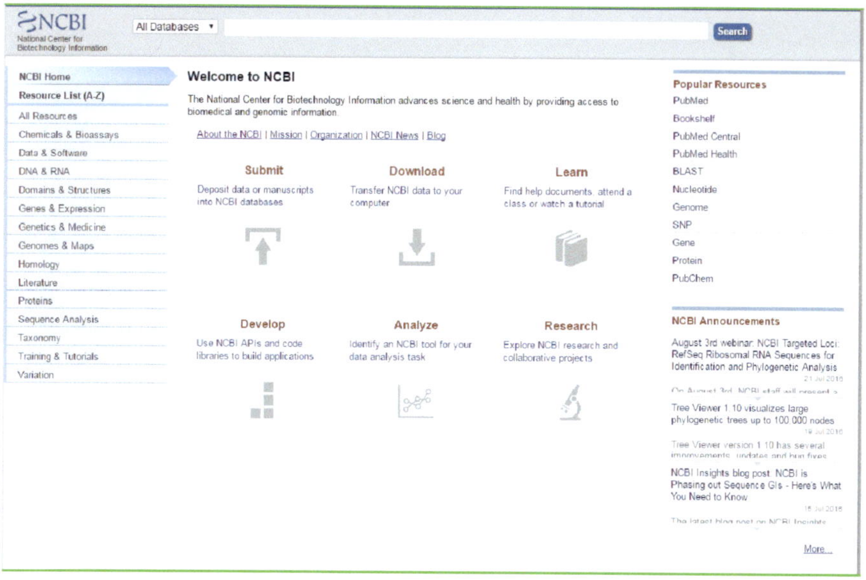

Fig. 1.3 The home page of National Center for Biotechnology Information (http://www.ncbi.nlm.nih.gov/)

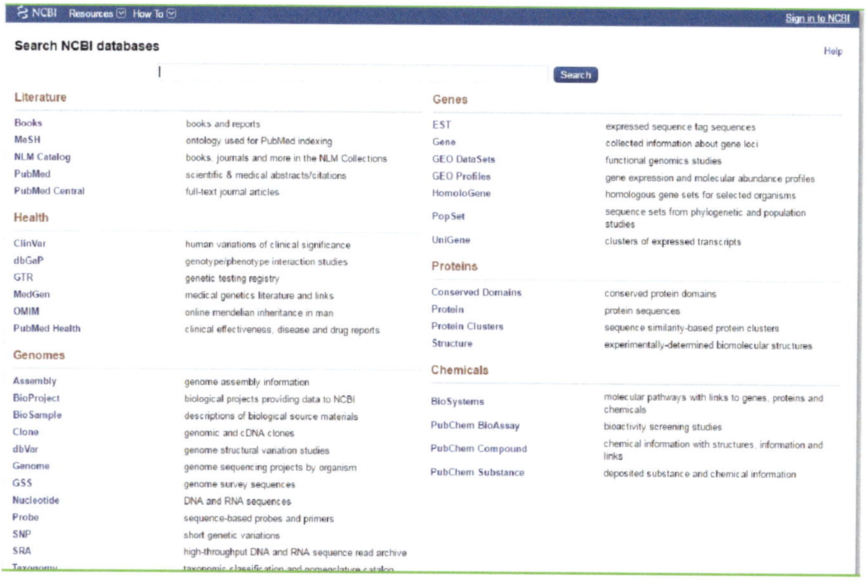

Fig. 1.4 The home page of Entrez (www.ncbi.nlm.nih.gov/Entrez/)

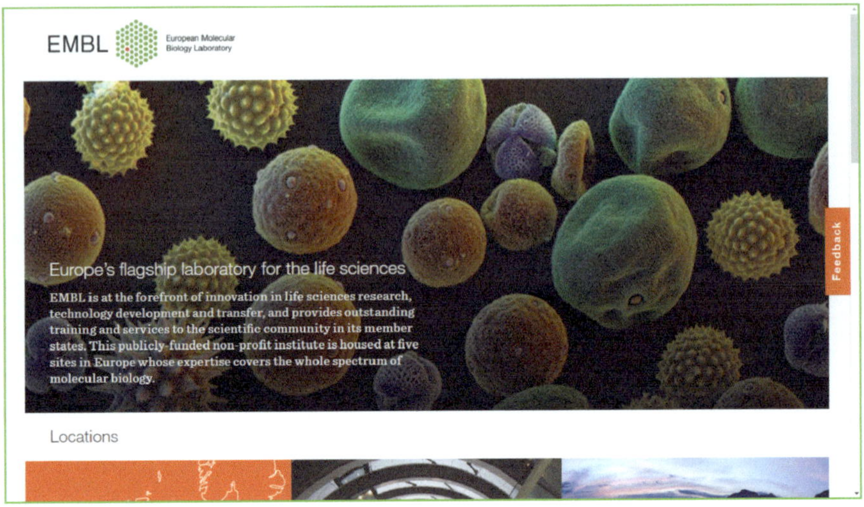

Fig. 1.5 The home page of European molecular biology laboratory (http://www.embl.org/)

(B) *European Molecular Biology Laboratory (EMBL)*

The *European Molecular Biology Laboratory* (*EMBL*) (http://www.embl.org/) in Fig. 1.5 is a molecular biology organization which is maintained by 20 European countries, with Australia as associate member state. It is an intergovernmental organization created in 1974. It develops and maintains a large number of databases, and scientists can access the data free of cost. This research laboratory functions from five different locations, the main laboratory, the European Bioinformatics Institute (EBI), Heidelberg, Germany, is a hub for bioinformatics research and services, directed by Dr. Rolf Apweiler and Dr. Ewan Birney. It is a part of INSDC, which includes DDBJ and GenBank. Typing insulin gene at EMBL search engine produced a result in Fig. 1.6.

EMBL File Format

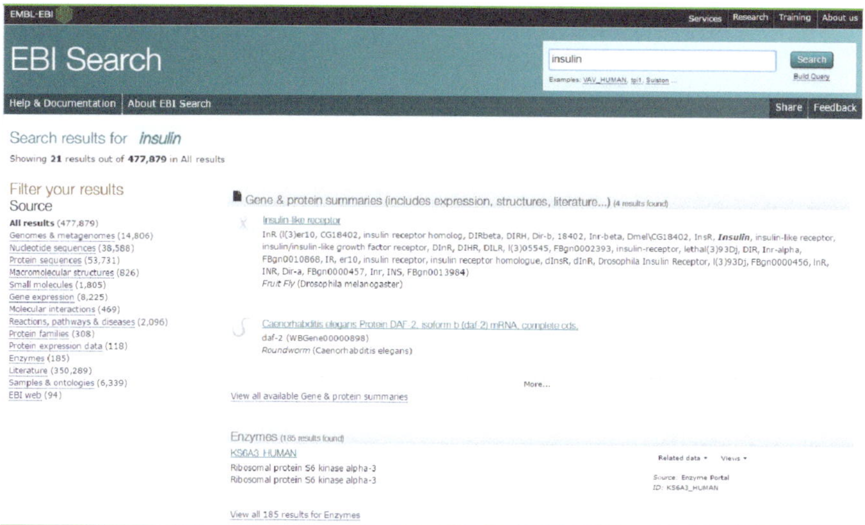

Fig. 1.6 Insulin gene search at European molecular biology laboratory website (https://www.ebi.ac.uk/ebisearch/search.ebi?query=insulin&db=allebi&requestFrom=searchBox)

```
ID   AH002190; SV 2; linear; genomic DNA; STD; ROD; 782 BP.
XX
AC   AH002190; M25583; M25585;
XX
DT   13-JUN-2016 (Rel. 129, Created)
DT   13-JUN-2016 (Rel. 129, Last updated, Version 1)
XX
DE   Rattus norvegicus insulin 2 (INS2) gene, complete cds.
XX
KW   insulin.
XX
OS   Rattus norvegicus (Norway rat)
OC   Eukaryota; Metazoa; Chordata; Craniata; Vertebrata; Euteleostomi; Mammalia;
OC   Eutheria; Euarchontoglires; Glires; Rodentia; Sciurognathi; Muroidea;
OC   Muridae; Murinae; Rattus.
XX
RN   [1]
RP   1-782
RX   DOI; 10.1111/j.1749-6632.1980.tb47271.x.
RX   PUBMED; 6249167.
RA   Lomedico P.T., Rosenthal N., Kolodner R., Efstratiadis A., Gilbert W.;
RT   "The structure of rat preproinsulin genes";
RL   Ann. N. Y. Acad. Sci. 343:425-432(1980).
XX
DR   MD5; 2b03b65970e00e00d50a5054fad8125c.
XX
CC   On or before Jun 10, 2016 this sequence version replaced gi:204949,
CC   gi:204950, gi:204951.
XX
FH   Key             Location/Qualifiers
FH
FT   source          1..782
FT                   /organism="Rattus norvegicus"
FT                   /mol_type="genomic DNA"
FT                   /db_xref="taxon:10116"
FT   gene            1..739
FT                   /gene="INS2"
FT   exon            <1..46
FT                   /gene="INS2"
FT                   /number=1
FT   intron          47..165
FT                   /gene="INS2"
FT                   /number=1
FT   CDS             join(180..366,541..686)
FT                   /codon_start=1
FT                   /gene="INS2"
FT                   /product="insulin 2"
FT                   /note="precursor"
FT                   /protein_id="AAA41440.1"
FT                   /translation="MALWIRFLPLLALLILWEPRPAQAFVKQHLCGSHLVEALYLVCGE
FT                   RGFFYTPMSRREVEDPQVAQLELGGGPGAGDLQTLALEVARQKRGIVDQCCTSICSLYQ
FT                   LENYCN"
FT   sig_peptide     180..251
FT                   /gene="INS2"
FT   exon            180..366
FT                   /gene="INS2"
FT                   /number=2
FT                   /note="first expressed exon"
FT   mat_peptide     252..341
FT                   /gene="INS2"
FT                   /product="beta chain"
FT   mat_peptide     join(348..366,541..614)
FT                   /gene="INS2"
FT                   /product="insulin 2 connecting peptide"
FT   intron          367..>410
FT                   /gene="INS2"
FT                   /number=2
FT   gap             411..510
FT                   /estimated_length=unknown
```

```
FT   intron          <511..540
FT                   /gene="INS2"
FT                   /number=2
FT   exon            541..739
FT                   /gene="INS2"
FT                   /number=3
FT   exon            541..>686
FT                   /gene="INS2"
FT                   /number=3
FT                   /note="preproinsulin 2"
FT   mat_peptide     621..683
FT                   /gene="INS2"
FT                   /product="insulin 2"
FT                   /note="alpha chain"
XX
SQ   Sequence 782 BP; 136 A; 212 C; 173 G; 161 T; 100 other;
     cccagcccta agtgaccagc tacagtcgga aaccatcagc aagcaggtat gtactctcca        60
     aggtgggcct agcttcccca gtcaagactc caaggatttg agggacgctg tgggctcttc       120
     tcttacatgt accttttgct agcctcaacc ctgactatct tccaggtcat tgttccaaca       180
     tggccctgtg gatccgcttc ctgcccctgc tggccctgct catcctctgg gagccccgcc       240
     ctgcccaggc ttttgtcaaa cagcaccttt gtggttctca cttggtggaa gctctctacc       300
     tggtgtgtgg ggagcgtgga ttcttctaca cacccatgtc ccgccgcgaa gtggaggacc       360
     cacaaggtaa gctctgctcc tgaattctat cccaagtgct aactaccctg nnnnnnnnnn       420
     nnnnnnnnnn nnnnnnnnnn nnnnnnnnnn nnnnnnnnnn nnnnnnnnnn nnnnnnnnnn       480
     nnnnnnnnnn nnnnnnnnnn nnnnnnnnnn tgccctgtgc tgacatgacc tccctggcag       540
     tggcacaact ggagctgggt ggaggcccgg gggccggtga ccttcagacc ttggcactgg       600
     aggtggcccg gcagaagcgc ggcatcgtgg atcagtgctg caccagcatc tgctctctct       660
     accaactgga gaactactgc aactaggccc accactaccc tgtccacccc tctgcaatga       720
     ataaaacctt tgaaagagca ctacaagttg tgtgtacatg cgtgcatgtg catatgtggt       780
     gc                                                                       782
//
```

Sequence Retrieval System (SRS)

SRS (http://srs.ebi.ac.uk/) (Fig. 1.7) is a powerful searching tool to retrieve sequences (and other types of data) and also to perform various operations on retrieved information for EMBL. It is similar to Entrez of NCBI, a search engine for extracting all sort of information available at EMBL.

Sequence Submission at EMBL

There are mainly three tools available for submitting data at EMBL.

1. Webin: for nucleotide sequence submission
2. Sequin: a stand-alone tool for submitting nucleotide sequences to GenBank, EMBL, and DDBJ developed by NCBI
3. Webin-Align: a tool for sequence alignment submission

(C) *DNA Data Bank of Japan (DDBJ)*

DDBJ, (http://ddbj.sakura.ne.jp/) (Fig. 1.8) part of *INSDC*, was established at the National Institute of Genetics (NIG), Japan, in 1986 with the support of the Ministry of Education, Culture, Sports, Science and Technology, Japan.

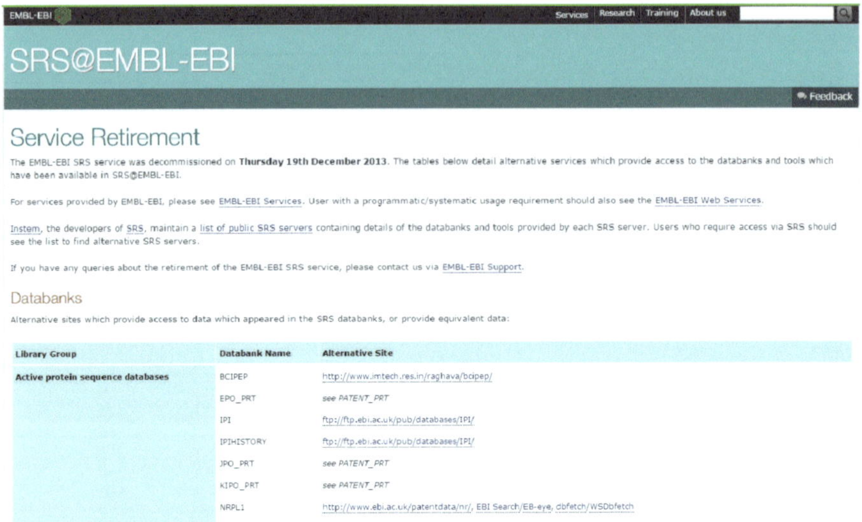

Fig. 1.7 The home page of Sequence Retrieval System (http://srs.ebi.ac.uk/)

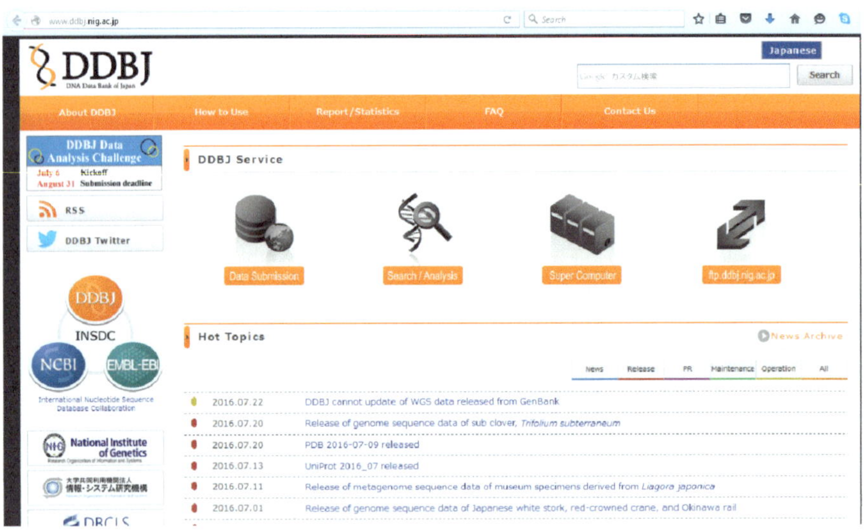

Fig. 1.8 The home page of DNA Data Bank of Japan (http://ddbj.sakura.ne.jp/)

SAKURA

SAKURA (http://sakura.ddbj.nig.ac.jp/top-e.html) is a source for data (nucleotide sequence) submission system through the WWW-based server where one can enter and submit nucleotide sequences and translated amino acid sequences. Since 1995 it is open to the public and scientists community.

DDBJ Format

```
LOCUS       OCOINS                   432 bp    mRNA    linear   ROD 27-APR-1993
DEFINITION  Octodon degus insulin mRNA, complete cds.
ACCESSION   M57671
VERSION     M57671.1
KEYWORDS    insulin; insulin alpha-chain; insulin beta-chain; insulin
            connecting peptide.
SOURCE      Octodon degus (degu)
  ORGANISM  Octodon degus
            Eukaryota; Metazoa; Chordata; Craniata; Vertebrata; Euteleostomi;
            Mammalia; Eutheria; Euarchontoglires; Glires; Rodentia;
            Hystricognathi; Octodontidae; Octodon.
REFERENCE   1  (bases 1 to 432)
  AUTHORS   Nishi,M. and Steiner,D.F.
  TITLE     Cloning of complementary DNAs encoding islet amyloid polypeptide,
            insulin, and glucagon precursors from a New World rodent, the degu,
            Octodon degus
  JOURNAL   Mol. Endocrinol. 4 (8), 1192-1198 (1990)
   PUBMED   2293024
COMMENT     Original source text: Octodon degus pancreas, cDNA to mRNA.
FEATURES             Location/Qualifiers
     source          1..432
                     /organism="Octodon degus"
                     /mol_type="mRNA"
                     /db_xref="taxon:10160"
                     /tissue_type="pancreas"
     gene            1..432
                     /gene="insulin"
     CDS             42..371
                     /gene="insulin"
                     /codon_start=1
                     /product="insulin"
                     /protein_id="AAA40590.1"
                     /db_xref="GI:202472"
                     /translation="MAPWMHLLTVLALLALWGPNSVQAYSSQHLCGSNLVEALYMTCG
                     RSGFYRPHDRRELEDLQVEQAELGLEAGGLQPSALEMILQKRGIVDQCCNNICTFNQL
                     QNYCNVP"
     sig_peptide     42..113
                     /gene="insulin"
     mat_peptide     114..200
                     /gene="insulin"
                     /product="insulin B-chain"
     mat_peptide     207..293
                     /gene="insulin"
                     /product="insulin C-peptide"
     mat_peptide     300..368
                     /gene="insulin"
                     /product="insulin A-chain"
     regulatory      414..419
                     /regulatory_class="polyA_signal_sequence"
                     /gene="insulin"
     polyA_site      432
                     /gene="insulin"
BASE COUNT           86 a          134 c          119 g          93 t
ORIGIN
        1 gcattctgag gcattctcta acaggttctc gaccctccgc catggccccg tggatgcatc
       61 tcctcaccgt gctggccctg ctggccctct ggggacccaa ctctgttcag gcctattcca
      121 gccagcacct gtgcggctcc aacctagtgg aggcactgta catgacatgt ggacggagtg
      181 gcttctatag accccacgac cgccgagagc tggaggacct ccaggtggag caggcagaac
      241 tgggtctgga ggcaggcggc ctgcagcctt cggccctgga gatgattctg cagaagcgcg
      301 gcattgtgga tcagtgctgt aataacattt gcacatttaa ccagctgcag aactactgca
      361 atgtccctta gacacctgcc ttgggcctgg cctgctgctc tgccctggca accaataaac
      421 cccttgaatg ag
//
```

(II) Protein Sequence Databases

The different protein sequence databases available are the following:

(A) Protein Information Resource
(B) UniProt

(A) *Protein Information Resource (PIR)*

Margaret Dayhoff was the inventor of Protein Information Resource (PIR) in the 1960s at the National Biomedical Research Foundation (NBRF) for investigation of evolutionary relationships among proteins. Analysis tools for protein database are provided by PIR which are freely available to the scientists (George et al. 1997).

In 2002 Protein Information Resource and its worldwide partners, EBI and Swiss Institute of Bioinformatics (SIB), were granted an award from the National Institutes of Health (NIH) to make UniProt, by merging the databases of PIR-PSD, SWISS-PROT, and TrEMBL (Fig. 1.9).

(B) UniProt

It comprises of two sections:

(a) SWISS-PROT
(b) Translated EMBL (TrEMBL)

Fig. 1.9 The home page of Protein Information Resource (http://pir.georgetown.edu/)

Fig. 1.10 The home page of UniProt (http://www.uniprot.org/)

(a) *SWISS-PROT*

SWISS-PROT (http://www.uniprot.org/) (Fig. 1.10), established in 1896, is the most widely used protein sequence database created by the University of Geneva and the EMBL, collaboratively. After 1994, the collaboration moved to EMBL's UK outstation, the EBI.

SWISS-PROT Format

Each line starts with a two-character line code, which specifies the kind of data contained in the line.

(b) *Translated EMBL*

TrEMBL benefits from the SWISS-PROT format and comprises translations of all coding sequences (CDS) in EMBL. It has two core divisions, designated SWISS-PROT-TrEMBL and REM-TrEMBL.

1.3.1.2 Structure Databases

- PDB (Protein Data Bank)
- MMDB (Molecular Modeling Database)
- VAST (Vector Alignment Search Tool)
- CDD (Conserved Domain Database)
- NDB (Nucleic acid Structure Database)

From the above databases, some of the database is shown below in detail.

(I) Protein Data Bank (PDB)

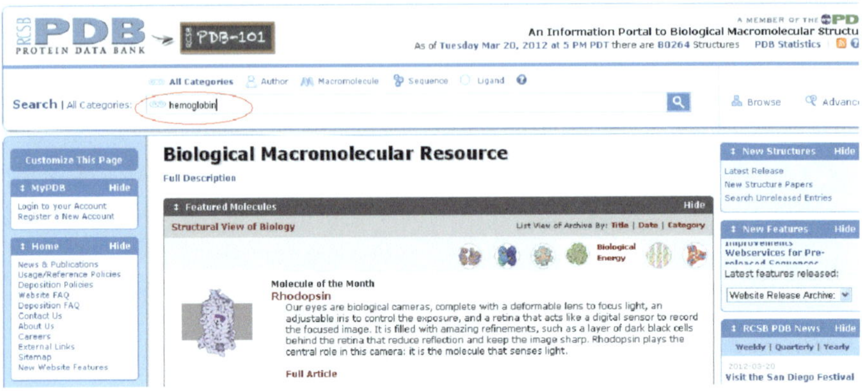

Fig. 1.11 The home page of PDB with the query Hemoglobin (http://www.rcsb.org/pdb/home/home.do)

The PDB (http://www.rcsb.org/pdb/home/home.do) in Fig. 1.11, a source for the three-dimensional structural data of huge biological molecules, includes proteins and nucleic acids. It was established in 1971 by the Research Collaborators for Structural Bioinformatics (RCSB). The data submitted by scientists from different parts of the world are easily without cost available through the Internet. The PDB is supervised by the Worldwide Protein Data Bank (wwPDB) (Berman 2008).

As on March 20, 2012 at 5 PM PDT, there were 80,264 structures. Each structure has been assigned a PDB ID, which contains four characters both alphabets and numerical. The first character is a numeral, while the last three characters can be either numerals or letters. Search results and structure for hemoglobin were showed in Figs. 1.11 and 1.12.

PDB File Format

This format was primarily practiced by the Protein data bank and previously was known as the PDB file format. The PDB also retains data on biological macromolecules, "macromolecular crystallographic information file format" (mmCIF), initiated to be phased in 1996. In the year 2005, an Extensible Markup Language (XML) version of PDBML was described (Westbrook et al. 2005).

Data Deposition Tool of PDB

Auto Dep Input Tool (ADIT) (http://deposit.rcsb.org/adit/) (Fig. 1.13) is developed by RCSB, and it is responsible for depositing structures to PDB in an efficient manner.

(II) *Nucleic Acid Structure Database (NDB)*

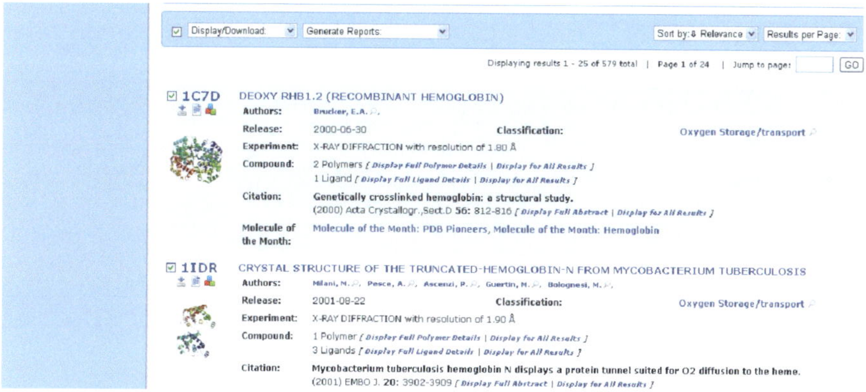

Fig. 1.12 Search result of Protein Data Bank (http://www.rcsb.org/pdb/results/results.do?qrid=57082E24&tabtoshow=Current)

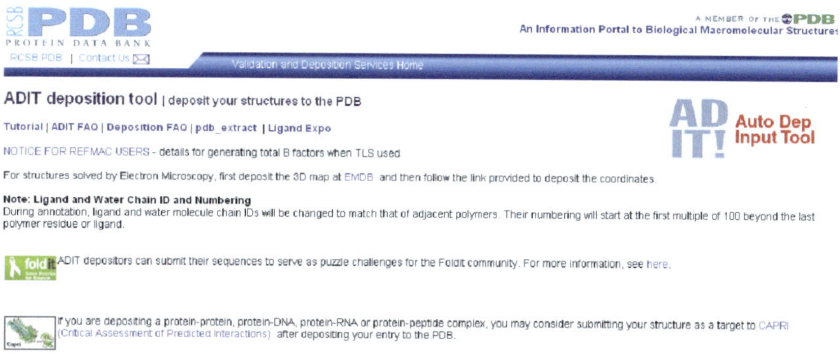

Fig. 1.13 The home page of Auto Dep Input Tool (http://deposit.rcsb.org/adit/)

This database (http://ndbserver.rutgers.edu/) (Fig. 1.14) provides us 3D structures of nucleic acids.

1.3.1.3 Literature Database

Literature databases provide us library of life science work done all over the world. Various literature databases available are the following:

- MEDLINE
- CiteXplore
- OMIM
- Patent abstracts
- FlyBase archives

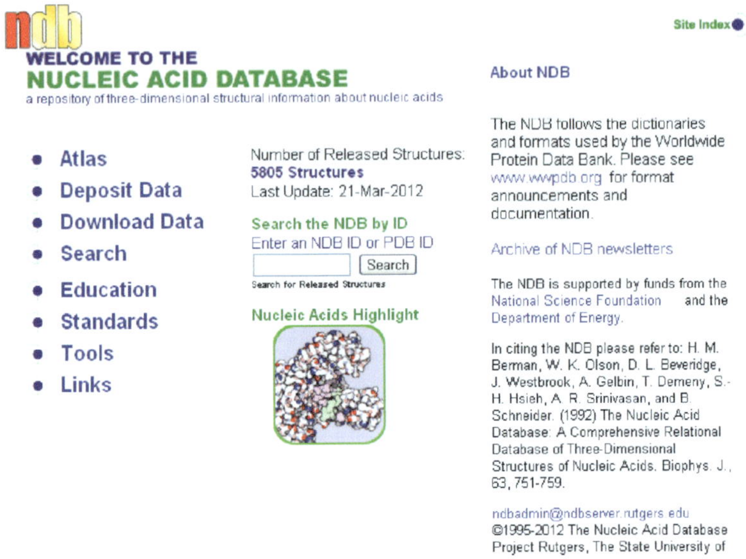

Fig. 1.14 The home page of nucleic acid structure database (http://ndbserver.rutgers.edu/)

1.3.1.4 Pathway Database

To comprehend molecular interactions and chemical reaction networks, the pathway database is used by pathway maps. Various pathway databases available are the following:

- BioCyc database collection comprising EcoCyc and MetaCyc
- KEGG PATHWAY Database (www.genome.jp/kegg/)
- MANET database
- Reactome (Laboratory of Cold Spring Harbor, EBI, Gene Ontology Consortium)

1.3.1.5 Chemical Database

A collection of the chemical information precisely planned is called chemical database. These are the few freely available chemical databases:

- Chemical Entities of Biological Interest (ChEBI)
- PubChem
- Zinc
- eMolecules
- DrugBank

1.3.1.6 Enzyme Database

Enzyme databases cover an extensive range of properties and functions, such as structure, occurrence, kinetics of enzyme-catalyzed reactions, and metabolic function. Various enzyme databases available are the following:

- ExPASy
- BRENDA
- REBASE
- EC enzyme database

1.3.1.7 Disease Database

The disease database provides all disease-related information; it is a cross-referenced index of diseases, symptoms, medications, signs, abnormal investigation findings, etc.

- OMIM
- OMIA

1.3.1.8 Domain Database

Domain database is a database for ancient domains and full-length proteins.

- CDD (Conserved Domain Database)

1.3.1.9 Structural Classification of Protein Database

It provides hierarchical classification of protein structure which defines the evolutionary association between proteins.

- The Structural Classification of Proteins (SCOP) (http://scop.mrclmb.cam.ac.uk/scop/).
- Class, architecture, topology, and homologous superfamily (CATH) is freely available to scientists (www.cathdb.info/).

1.3.1.10 Genome Database

Genome databases are a collection of genome sequences of many species; it interprets and examines them and provides free public access.

- Genome Databases at the National Center for Biotechnology Information (Index)
- Genome Databases at the National Center for Biotechnology Information (Entrez)
- Genome Databases at the National Center for Biotechnology Information (PMGif) Genome List in NIH

- Mitochondrial DNA Database (MitBASE)
- Mouse Genome Informatics
- Plant Genome Project maintained by the National Science Foundation
- Organelle Genome Sequences (PMGif)

1.3.2 Biological Databases Based on Database Source

This database is subdivided into two databases, primary and secondary.

1. *Primary*: databases comprising of data generated experimentally like nucleotide sequences and 3D structures are identified as primary databases.

Examples are GenBank, DDBJ, EMBL, PIR, PDB, NDB, UniProt, TrEMBL, SWISS-PROT, etc.

2. *Secondary*: it contains databases directly derived from the primary databases.

Examples are PROSITE, Pfam, Blocks, Prints, SCOP, CATH, OMIM, KEGG, etc.

1.3.3 Composite Databases

It combines various different primary database sources. This makes searching the query more efficient. So, composite database amalgamates various primary databases for easy access.

Examples are OWL, NRDB, MIPSX, SP, and TrEMBL.

1.3.4 Biological Databases Based on Database Design

This database is subdivided into two databases, object-oriented and relational databases.

1.3.4.1 Object Oriented

A database controlling system in which information is characterized in the form of objects. These databases are unlike table-oriented relational databases.

Objects mostly comprise of Attributes and Methods.

How Data Is Stored

There are two methods used for the storage of objects:

- Each object has an exclusive ID and is known as a subclass of a base class, by inheritance to explain attributes.
- For management and object storage, virtual memory mapping has been used.

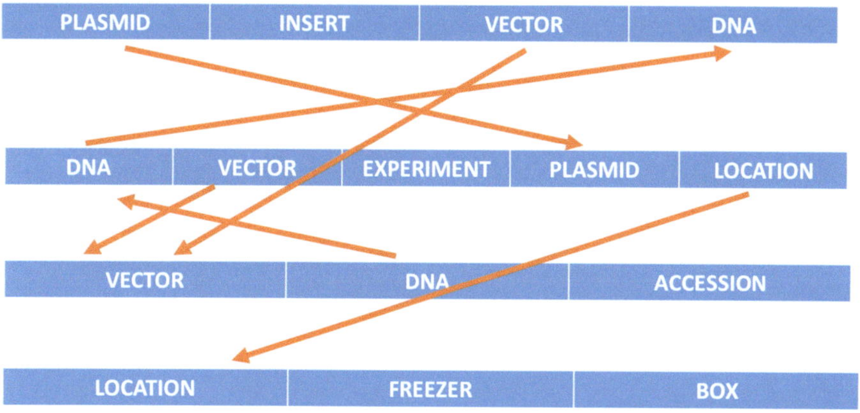

Fig. 1.15 Four tables are shown: plasmid, vector, DNA, and location. Arenas that reference other tables are mentioned to as links. Numerous factors have to be considered when designing a relational database (http://home.cc.umanitoba.ca/)

1.3.4.2 Relational Database

Relational databases can be assumed as comprehensive tables of data. Each record from a flat file could be applied as a row in a table. Although a relational database can be applied in a single large table or "relation," it is often helpful to split the database up into multiple tables (Fig. 1.15).

A benefit of relational databases is that by breaking up the database to various tables, in many circumstances, only one table needs to be rewritten when creating changes in fields. In other cases, addition of a record may need rewriting many or most tables.

References

Benton D (1990) Recent changes in the GenBank on-line service. Nucleic Acids Res 18 (6):1517–1520

Berman HM (2008) The protein data bank: a historical perspective. Acta Crystallogr A64:88–95

Dayhoff MO, N. B. R. Foundation (1973) Atlas of protein sequence and structure: supplement. National Biomedical Research Foundation

Dayhoff MO, N. B. R. Foundation (1976) Atlas of protein sequence and structure. National Biomedical Research Foundation

Foundation N. B. R. (1972) Atlas of protein sequence and structure. National Biomedical Research Foundation

George DG et al (1997) The protein information resource (PIR) and the PIR-International protein sequence database. Nucleic Acids Res 25(1):24–28

Liu L, Özsu MT (2009) Encyclopedia of database systems. Springer US

N. C. f. B. I (2013) The NCBI handbook. In: Mizrachi I (ed) NCBI handbook [Internet], 2nd edn. National Center for Biotechnology Information (US), Bethesda

Westbrook J et al (2005) PDBML: the representation of archival macromolecular structure data in XML. Bioinformatics 21(7):988–992

Bioinformatics in Next-Generation Genome Sequencing

2

Satendra Singh, Anjali Rao, Pallavi Mishra, Arvind Kumar Yadav, Ranjeet Maurya, Sukhdeep Kaur, and Gitanjali Tandon

Abstract

In the present era, genomics represent a crucial role in the field of life sciences. Advancements in genomics and the development of high-throughput techniques facilitate the characterization of a wide range of genes according to their functions such as regulation of genes, metabolic pathways, and their reconstruction. In the era of genomics we face the challenge of storage and analysis of a huge amount of important data. Even in this early stage of the era there are many commercial techniques and tools/software available to analyze next-generation sequencing (NGS) data. All of these programs can be used for many uses such as sequence alignments, polymorphism detection, and functional and structural comparative genomics. In this chapter, we focus on advances in bioinformatics and also computational biology in genome sequencing and NGS data analysis and on the potential applications for the efficient collection, storage, and analysis of the huge amount of genomic data generated by researchers and the information retrieved from different sources and web browsers in relation to NGS analysis.

Keywords

Bioinformatics · Genomics · Next-generation sequencing · Technologies

S. Singh (✉) · A. Rao · P. Mishra · A. K. Yadav · R. Maurya · S. Kaur · G. Tandon
Department of Computational Biology and Bioinformatics, Jacob Institute of Biotechnology and Bioengineering, Sam Higginbottom University of Agriculture, Technology and Sciences, Allahabad, Uttar Pradesh, India

G. Wadhwa et al. (eds.), *Current trends in Bioinformatics: An Insight*,
https://doi.org/10.1007/978-981-10-7483-7_2

2.1 Introduction

When Sanger invented the chain termination method in 1977, this led to scientists reproducing DNA in different ways. DNA was first explained by Oswald Theodore Avery in 1944, and in 1953 James D. Watson and Francis Crick determined that DNA had a double helical strand structure with each strand composed of four bases (A, T, G, C), leading to the basis of molecular biology. Most species have a unique form of DNA sequence which is why they are categorized into different groups and species (Church and Gilbert 1984). DNA sequencing technologies offer a broad range of applications in many fields such as plant breeding, cloning, finding pathogen-related genes, evolution, and comparative studies. These technologies are fast, precise, easy to control, and inexpensive and could help scientists in pharmacogenomics. In the last decade of genomics, DNA sequencing technologies and their applications have undergone tremendous development and act as the central engine of the genomic era, in which there is a plethora of genomic data, and thus are used in diverse areas of research and applications (Liu et al. 2012).

Before discussing next-generation sequencing, we first outline the evolutionary history of DNA sequencing in brief. In the 1970s, Maxam, Gilbert, Sanger, and colleagues (Maxam and Gilbert 1977; Sanger et al. 1977) developed a chain termination technique for DNA sequencing. This method deciphers the entire genome according to information related to the gene of interest (International Human Genome Sequencing Consortium 2004). The Human Genome Project, however, consumed enormous resources and time, inferring that faster, higher-throughput, and cheaper technologies were required (Schloss 2008). The new sequencing methods came with three major improvements. First, next-generation sequencing (NGS) cell-free libraries replaced the bacterial cloning of DNA fragments. Second, NGS produces several million sequencing reactions simultaneously. Third, the output is program-readable and can be processed directly without the need for electrophoresis (Van Dijk et al. 2014).

The very first NGS technology was the pyrosequencing method developed by 454 Life Sciences (now Roche) in 2005 (Margulies et al. 2005). The 454 Genome Sequencer can produce about 200,000 reads (20 Mb) of 110 base-pairs (bp). One year later, the Illumina sequencing platform was commercialized (Illumina acquired Solexa in 2007), followed by the release of Sequencing by Oligo Ligation Detection (SOLiD) by Applied Biosystems (now Life Technologies) in 2007 (Valouev et al. 2008). The SOLiD and Illumina sequencers generated a larger number of reads than 454 sequencing (30 million reads for SOLiD and 100 million reads for Ilumina). The Personal Genome Machine (PGM) was released by Ion Torrent (now Life Technologies) in 2010. PGM was developed by the founder of 454 Life Sciences, Jonathan Rothberg, and resembled the 454 system. PGM does not rely on fluorescence and camera scanning for optical detection of incorporated nucleotides, rather using semiconductor technology which is much higher in speed, lower in cost, and smaller in instrument size. The first PGM generated somewhat shorter reads than those generated by 454 sequencing: up to 270 Mb and 100 nt of sequence and length of reads, respectively.

Today, the need for sequencing is vast because of its application in different sectors, including applied medicine for diagnostics and therapeutics, forensic comparative genomics, and evolution and epidemiology (Metzker 2005). NGS can be utilized for more broad characterization of varied and intricate microbial communities such as microflora concomitant to animals (e.g., termite hindgut [Warnecke et al. 2007], cow rumen [Deng et al. 2008; Brulc et al. 2009], human saliva [Willner et al. 2011; Yang et al. 2012], and human intestinal tract [Turnbaugh et al. 2007]) making it useful for metagenomics to study complex pelagic microbial communities in soil.

NGS techniques are applicable in different biological studies; it aids metagenomics (Schuster 2007;Qin et al. 2010), variant detection of single nucleotide polymorphisms (SNPs) and genomic structural variants (Van Tassell et al. 2008;Alkan et al. 2009;Medvedev et al. 2009), personalized medicine (Auffray et al. 2009), messenger RNA (mRNA) expression analysis (Sultan et al. 2008), oncology (Guffanti et al. 2009), and DNA methylation studies (Taylor et al. 2007).

Each NGS run for all of the above-mentioned applications produces an enormous amount of genomic data and handling these data is a significant computational challenge (Horner et al. 2010;Miller et al. 2010). Thus, genomics studies using NGS require substantial computational resources, tools, core bioinformatics knowledge, and a bioinformatician who can infer important information from these data (Scholz et al. 2012).

2.2 Next-Generation Sequencing (NGS)

2.2.1 Basics of NGS

NGS works as a high-throughput sequencing technique, which is cheaper and faster than the previously available Sanger method. NGS involves the following four basic steps (Mardis 2008):

- Library preparation
- Cluster generation
- Sequencing
- Data analysis.

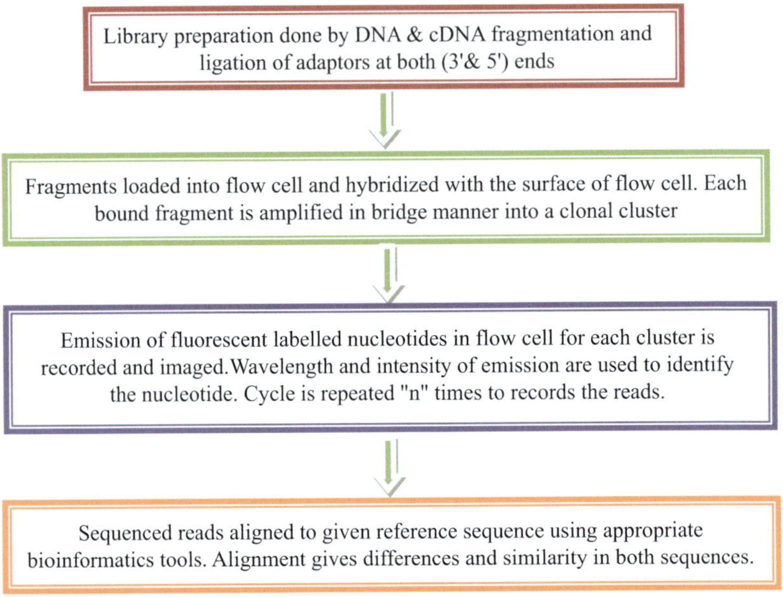

2.2.2 Advances in Sequencing Technology

2.2.2.1 Paired-End Sequencing (PE)

The development of paired-end sequencing (PE) is a major advancement in NGS. PE enables both ends of the desired DNA to be sequenced because the distance between each read is known. The alignment algorithm uses this information to create maps. Repetitions of regions are avoided.

2.2.2.2 Library Preparation

Library preparation in the NGS method is more accurate and rapid than in the Sanger method. Many improvements have been made in recent years in library preparation in the Illumina sequencing method. At present, preparation of the Nextera XT DNA library takes 90 min (King et al. 2014).

2.2.2.3 Multiplexing

In multiplexing, a large number of libraries are pooled and sequenced simultaneously during a single run.

2.3 Latest NGS Technologies

The following sections discuss the latest NGS technologies (Table 2.1).

Table2.1 Summary of four next-generation sequencing methods

	Roche 454	Illumina analyzer	SOLiD	Helicos heliscope
Sequencing method	Pyrosequencing	Reversible dye terminator	Ligation	Single molecule sequencing
Read length	400 bases	100 bases	50 bases	35 bases
Sequencing run time	10 h	10 days	11–12 days	30 days
Total bases per run	500 Mb	20 Gb	100 Gb	35 Gb
Error rate	0.1%	1.5%	4%	2–7%

SOLiD Sequencing by Oligo Ligation Detection

2.3.1 Illumina/Solexa Sequencing Method

In the Illumina sequencing method, the bp length should be 100–150 bp; longer fragments must be cut into small fragments and annealed with adaptors. Polymerase chain reaction (PCR) is then carried out to amplify each fragment (making copies). After this process, DNApolymerase terminator should be added in to the reaction to terminate the cycle. These nucleotides are fluorescently labeled and each base added one at a time.

2.3.2 Roche 454 Sequencing Method

The Roche 454 sequencing method is used for longer reads (1 kb) than the Illumina. Adaptors are added at both the ends and annealed to the beads. If one DNA fragments as one bead, PCR is then applied using adaptor-specific primers and DNA polymerase and sequencing buffers flooded with four NTP solutions are also added.

2.3.3 Ion Torrent

Ion torrent is another kind of NGS which contains fragments lengths of ~200 bp.

2.3.4 Sequencing by Oligo Ligation Detection (SOLiD)

Microbeads of emulsion PCR are used to enrich fragmented or mate-paired libraries. These microbeads are then adhered to a glass slide. Probes labeled with fluorophore are added to the flow cell. The first two positions comprise a fluorophore-specific di-base pair; these bases query the first and second positions and trailed the hybridized primer. Bases three to five are degenerate bases separated from bases at positions six to eight; these bases are attached by a phosphorothiolate linkage made up of universal inosine bases. A matching 1,2-probe is linked to the primer by DNA ligase. After the fluorescence imaging technique confirms the ligation process, silver ions cleave the phosphorothiolate link, thus regenerating the 5′ phosphate group for subsequent ligation (McKernan et al. 2011). The ligation, cleavage, and detection cycle is repeated *n* times, and the complementary strand is extended to a predefined length by the number of cycles. When a specific length is reached, the product is removed from the extension process and a new process is begun using the reset template. The new process starts with a primer complementary to the $n - 1$ position of the previous primers. The template is reset each time it is extended using ligation and reaches a specific length; this process takes place four times, resulting in assessment twice and balancing of the template base. This creates and aligns a series of color images analyzed through time and space and facilitates determination of the actual DNA sequences. SOLid techniques are used in resequencing (Ashelford et al. 2011), transcriptomics, and genomic sequencing (Silva Ascencio 2011).

2.3.5 Helicos Sequencing

The first commercially available single-molecule sequencing system (SMS) was provided by The Helicos Genetic Analysis System (http://www.helicosbio.com). Helicos technology uses fragmented, denatured DNA and works in two ways, either by hybridizing DNA with immobilizing oligonucleotide primers to a solid surface or direct immobilizing with the template and covalently bonding to a solid surface followed by universal primer attachment. Some nucleotides used as a virtual terminator were labeled using fluorescent dye and used with DNA polymerase in sequential washing over the DNA template. Washing takes place only one nucleotide at a time and the incorporated events are captured using fluorescence imaging. *n* numbers of cycles are carried out to reach specific read length. Helicos sequencing outputs a 35 bp average read length across 600–1000 million reads at a rate of more than 1 Gb/h, producing 21–35 Gb per run. Helicos sequencing also supports multiplexing and can use 96 samples per channel (4800 samples/run) (http://www.helicosbio.com).

2.4 NGSMethods (Table 2.2)

2.4.1 Whole-Genome Sequencing (WGS)

Disease-based studies mostly used genome-wide association studies that are based on microarray analysis. Whole-genome sequencing (WGS) is an efficient tool for researchers in many areas of biological study.

Sequencing of many species has been performed, and this is very important in the study of livestock, microbial studies, animal diseases, agriculture, animal breeding, population studies, etc. In 2011 in Europe, work relating to *Escherichia coli* promoted rapid research in the biological field. This solved many problems

Table 2.2 Next-generation sequencing-based DNA analysis tools and their availability

Tool	Web address	Available?
SomaticSniper	http://gmt.genome.wustl.edu/somatic-sniper	Yes
BWA	http://biobwa.sourceforge.net/	Yes
Novoaligns	http://www.novocraft.com	No
Bowtie 2	http://bowtiebio.sourceforge.net/bowtie2	Yes
CONTRA	http://sourceforge.net/projects/contra-cnv/	Yes
GATK (UnifiedGenotyper)	https://www.broadinstitute.org/gatk/download	No
SAM tools	http://samtools.sourceforge.net/	Yes
PINDEL	http://gmt.genome.wustl.edu/pindel/current/	Yes
Delly	https://github.com/tobiasrausch/delly	Yes

related to a disease which can be lethal for many organisms. NGS helps provide information about how transmissions and mutations occur in viruses and bacteria (Grad et al. 2012).

2.4.2 Exome Sequencing (ES)

The exome sequencing (ES) method is most widely used for protein coding sequencing. It can identify whether a protein sequence is involved in a disease-related protein, making it a very effective method in comparison with WGS.

2.4.3 De Novo Sequencing

De novo sequencing is used without aid of reference sequence. This method is known as a novel method as prior information or references are missing. Its main NGS application is to characterize the genome of a new species. NGS provides whole-genome characterization of model organisms such as yeast, mouse, *E. coli*, Arabidopsis, Drosophila, the human genome, and so on. BGI (Beijing Genomics Institute) has worked on some very large initiatives such as the 1000 Plant and Animal Reference Project and the Ten Thousand Microbial Genomes Project. There are many more initiatives using de novo sequencing, such as analysis of the genome of endangered and extinct species, e.g., pandas (Li et al. 2010), early humans, and mammoths (Rohland et al. 2010). All of these projects have produced a plethora of data that can be used in evolutionary studies and studied further (Reich et al. 2010).

2.4.4 Targeted Sequencing

In targeted sequencing a region of interest is sequenced from the genome. Targeted sequencing is helpful for researchers as it takes into account only genes of interest and is much cheaper in respect of both time and money than other conventional WGS methods. Kits relate to specific areas such as autism, cardiomyopathy, and cancer, which is based on specific probes. Illumina have two methods for this: target enrichment and amplicon generation (Lo and Chiu 2009; McEllistrem et al. 2009).

The amplicon method is applied on the bacterial 16S ribosomal RNA (rRNA) sequencing, which helps determine taxonomy from metagenomic samples (Ram et al. 2011). Both methods are shown in the flowchart.

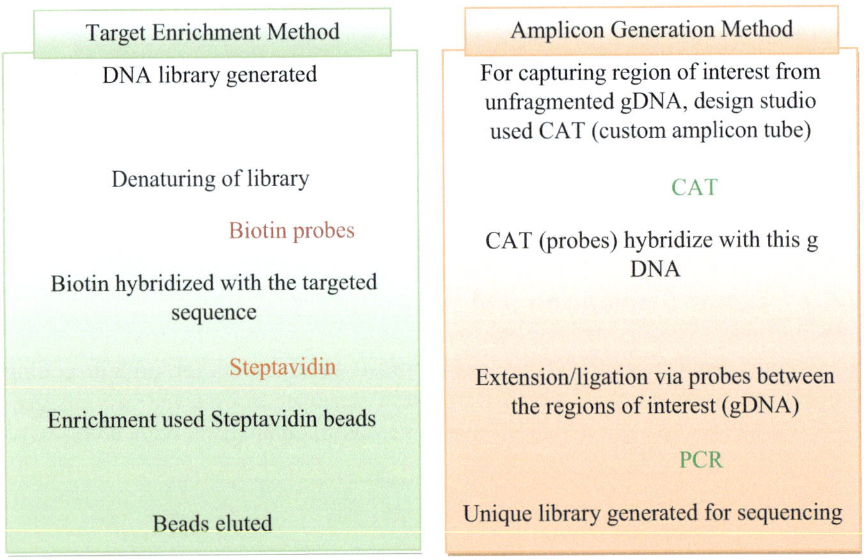

2.4.5 Transcriptomics

Transcriptomics is the study of the expression and sequencing of the transcriptome. In transcriptomics, complementary DNA (cDNA) are isolated, hybridized, and captured, followed by sequencing (Fig. 2.1). This method is also known as RNA sequencing (RNA-Seq) and provides a cost-effective and efficient method to study the genetic aspects of expression in a particular subset of the transcriptome, including mutant study, expression, and structural alteration (Gibbons et al. 2009). Study of the transcriptome can be reference-assisted and de novo in absence of a reference genome. Many assembly algorithms provide a platform for reconstruction of the transcriptome without a reference genome (Martin and Wang 2011). Large-scale tumor profiling studies have been carried out to identify novel fusion gene products and sequence variations with the help of targeted RNA-Seq analysis (Levin et al. 2009) (Fig. 2.1).

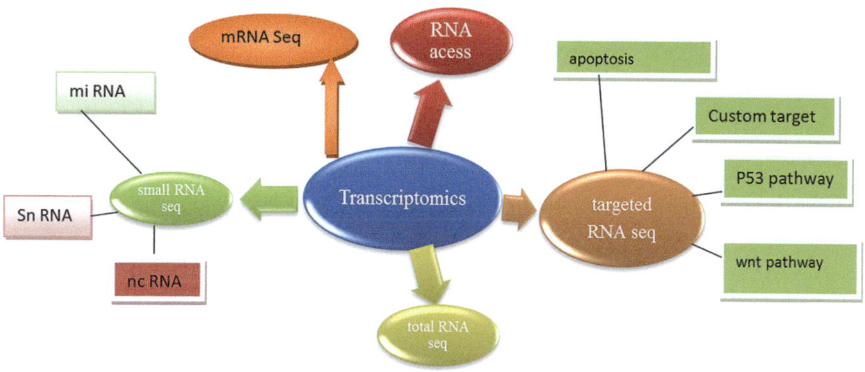

Fig. 2.1 Complete view of transcriptomics

2.4.6 Epigenomics

Epigenetics are very important phenomena when we talk about the genome. Epigenetics is the study of heritable changes that occur in the genome that do not involve changes in the DNA sequence. These changes have many causes such as DNA methylation, DNA–protein interactions, histone modifications, small RNA (sRNA)-mediated regulation, etc.

2.4.7 Methylation Sequencing

DNA methylation (5-methylcytosine [5mC]) acts at a specific area of regulation such as the promoter region and heterochromatin. Two methylation methods are very well-known: WGBS (whole genome bisulphite seq) and RRBS (reduced representation of bisulphite seq).

2.4.8 Chromatin Immunoprecipitation (ChIP) Sequencing

Another use of NGS technology is the study of protein binding sites or transcription factor binding sites in genomic DNA, which means an interaction study of protein and DNA based on the Chromatin Immunoprecipitation (ChIP) method followed by sequencing to understand the protein regulation in gene expression. ChIP sequencing is also known as ChIP-Seq (Blat and Kleckner 1999). In ChIP-Seq, ChIP-enriched sequences have been primarily studied using an array technique that is confined to the array content of sequences of interest, which is then replaced by selection of CpG islands and the promoter region. In ChIP-Seq the most crucial step is enrichment of the desired sequences and maintaining their purity. Even if high-quality enriched sequences are used for a ChIP assay, signal intensities often remain weak. To overcome this problem, sequencing of ChIP is used. Sequencing of enriched reads provides identification of sequences with signals and determination of sensitivity. Weak enriched sequences can also give good signals simply by increasing the number of reads in the sequencing step.

ENCODE (ENCyclopedia of DNA elements) (ENCODE Project Consortium 2004) and FANTOM5 (Functional Annotation of the Mammalian Genome) (Andersson et al. 2014) are two projects that are based on ChIP-seq and DNase I hypersensitive site (DHS) mapping to reveal genome-wide and binding sites.

2.4.9 Illumina NGS Workflow

Illumina is fully integrated for data analysis. It starts from library preparation of the whole data, which is based on user requirement. Data analysis is performed by the user using selected information from that available. There are many types of bioinformatics software available for data analysis; BaseSpace bioinformatics software is for the study of integrated NGS data that are obtained from different resources.

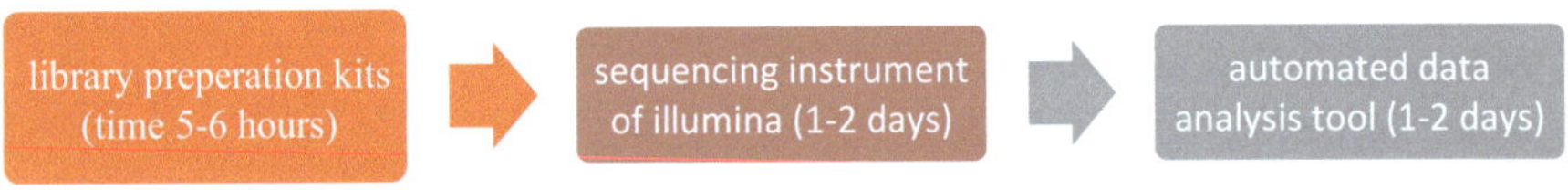

2.5 Conclusion

The advent of NGS brought about a revolution in biological sciences for researchers. NGS allows the handling of a huge amount of data in a very limited time and accurately. It also allows DNA sequencing to be performed in a very easy way. ES provides a clinical approach for biologists that aids research into cancer, autism, and lethal diseases. NGS also helps identify genetic diversity and aids taxonomy analysis. NGS provides a platform for the bioinformatician to solve those problems which currently don't have a solution.

Acknowledgements The authors are grateful to the Sam Higginbottom Institute of Agriculture, Technology and Sciences (Formerly Allahabad Agriculture Institute) (Deemed-to-be-University), Allahabad, India, for providing the facilities and support to complete the work.

References

Alkan C et al (2009) Personalized copy number and segmental duplication maps using next-generation sequencing. Nat Genet 41:1061–1067

Andersson R et al (2014) An atlas of active enhancers across human cell types and tissues. Nature 507:455–461

Ashelford K et al (2011) Full genome re-sequencing reveals a novel circadian clock mutation in Arabidopsis. Genome Biol 12:1

Auffray C et al (2009) Systems medicine: the future of medical genomics and healthcare. Genome Med 1:1

Blat Y, Kleckner N (1999) Cohesins bind to preferential sites along yeast chromosome III, with differential regulation along arms versus the centric region. Cell 98:249–259

Brulc JM et al (2009) Gene-centric metagenomics of the fiber-adherent bovine rumen microbiome reveals forage specific glycoside hydrolases. Proc Natl Acad Sci U S A 106:1948–1953

Church GM, Gilbert W (1984) Genomic sequencing. Proc Natl Acad Sci U S A 81:1991–1995

Deng W et al (2008) The use of molecular techniques based on ribosomal RNA and DNA for rumen microbial ecosystem studies: a review. Mol Biol Rep 35:265–274

ENCODE Project Consortium (2004) The ENCODE (ENCyclopedia of DNA elements) project. Science 306:636–640

Gibbons JG et al (2009) Benchmarking next-generation transcriptome sequencing for functional and evolutionary genomics. Mol Biol Evol 26:2731–2744

Grad YH et al (2012) Genomic epidemiology of the *Escherichia coli* O104: H4 outbreaks in Europe, 2011. Proc Natl Acad Sci U S A 109:3065–3070

Guffanti A et al (2009) A transcriptional sketch of a primary human breast cancer by 454 deep sequencing. BMC Genomics 10:1

Horner DS et al (2010) Bioinformatics approaches for genomics and post genomics applications of next-generation sequencing. Brief Bioinform 11:181–197

International Human Genome Sequencing Consortium (2004) Finishing the euchromatic sequence of the human genome. Nature 431:931–945

King JL et al (2014) High-quality and high-throughput massively parallel sequencing of the human mitochondrial genome using the Illumina MiSeq. Forensic Sci Int Genet 12:128–135

Levin JZ et al (2009) Targeted next-generation sequencing of a cancer transcriptome enhances detection of sequence variants and novel fusion transcripts. Genome Biol 10:1

Li R et al (2010) The sequence and de novo assembly of the giant panda genome. Nature 463:311–317

Liu L et al (2012) Comparison of next-generation sequencing systems. Biomed Res Int 2012:251364

Lo YD, Chiu RW (2009) Next-generation sequencing of plasma/serum DNA: an emerging research and molecular diagnostic tool. Clin Chem 55:607–608

Mardis ER (2008) Next-generation DNA sequencing methods. Annu Rev Genomics Hum Genet 9:387–402

Margulies M et al (2005) Genome sequencing in microfabricated high-density picolitre reactors. Nature 437:376–380

Martin JA, Wang Z (2011) Next-generation transcriptome assembly. Nat Rev Genet 12:671–682

Maxam AM, Gilbert W (1977) A new method for sequencing DNA. Proc Natl Acad Sci U S A 74:560–564

Mcellistrem MC et al (2009) Genetic diversity of the pneumococcal capsule: implications for molecular-based serotyping. Future Microbiol 4:857–865

Mckernan KW et al (2011) Reagents, methods, and libraries for bead-based sequencing. Patent no. US20110077169 A1

Medvedev P et al (2009) Computational methods for discovering structural variation with next-generation sequencing. Nat Methods 6:S13–S20

Metzker ML (2005) Emerging technologies in DNA sequencing. Genome Res 15:1767–1776

Miller JR et al (2010) Assembly algorithms for next-generation sequencing data. Genomics 95:315–327
Qin J et al (2010) A human gut microbial gene catalogue established by metagenomic sequencing. Nature 464:59–65
Ram JL et al (2011) Strategy for microbiome analysis using 16S rRNA gene sequence analysis on the Illumina sequencing platform. Syst Biol Reprod Med 57:162–170
Reich D et al (2010) Genetic history of an archaic hominin group from Denisova Cave in Siberia. Nature 468:1053–1060
Rohland N et al (2010) Genomic DNA sequences from mastodon and woolly mammoth reveal deep speciation of forest and savanna elephants. PLoS Biol 8:e1000564
Sanger F et al (1977) DNA sequencing with chain-terminating inhibitors. Proc Natl Acad Sci U S A 74:5463–5467
Schloss JA (2008) How to get genomes at one ten-thousandth the cost. Nat Biotechnol 26:1113
Scholz MB et al (2012) Next generation sequencing and bioinformatic bottlenecks: the current state of metagenomic data analysis. Curr Opin Biotechnol 23:9–15
Schuster SC (2007) Next-generation sequencing transforms today's biology. Nature 200:16–18
Silva Ascencio H (2011) The genome of woodland strawberry (Fragaria vesca). Nat Genet 43 (2):109–116
Sultan M et al (2008) A global view of gene activity and alternative splicing by deep sequencing of the human transcriptome. Science 321:956–960
Taylor KH et al (2007) Ultradeep bisulfite sequencing analysis of DNA methylation patterns in multiple gene promoters by 454 sequencing. Cancer Res 67:8511–8518
Turnbaugh PJ et al (2007) The human microbiome project: exploring the microbial part of ourselves in a changing world. Nature 449:804
Valouev A et al (2008) A high-resolution, nucleosome position map of C. elegans reveals a lack of universal sequence-dictated positioning. Genome Res 18:1051–1063
Van Dijk EL et al (2014) Ten years of next-generation sequencing technology. Trends Genet 30:418–426
Van Tassell CP et al (2008) SNP discovery and allele frequency estimation by deep sequencing of reduced representation libraries. Nat Methods 5:247–252
Warnecke F et al (2007) Metagenomic and functional analysis of hindgut microbiota of a wood-feeding higher termite. Nature 450:560–565
Willner D et al (2011) Metagenomic detection of phage-encoded platelet-binding factors in the human oral cavity. Proc Natl Acad Sci U S A 108:4547–4553
Yang F et al (2012) Saliva microbiomes distinguish caries-active from healthy human populations. ISME J 6:1–10

The Role of Bioinformatics in Epigenetics

3

Budhayash Gautam, Kavita Goswami, Neeti Sanan Mishra, Gulshan Wadhwa, and Satendra Singh

Abstract

Epigenetics is an upcoming field that studies the gene regulation of mitotically heritable genes which change the physiology of cells without altering the DNA sequence. Various epigenetic elements such as modification of histone proteins, methylation of DNA, chromatin modeling, and RNA-mediating silencing influence the regulation of genes at many levels, which leads to diseases such as cancer. All of these factors modulate gene expression in a tissue-specific manner. Bioinformatics is a successful approach in the field of molecular biology for studying epigenomics data. To generate these epigenomic data which can be analyzed using various bioinformatics tools and software, a variety of technologies are being used by researchers. Many biological databases which store a huge amount of information related to the modifications due to epigenetics are available online. With the help of these data, we can identify key target genes that can be manipulated to achieve some resistance against diseases caused by epigenetic factors.

B. Gautam (✉) · S. Singh
Department of Computational Biology and Bioinformatics, Jacob Institute of Biotechnology and Bioengineering, Sam Higginbottom University of Agriculture, Technology and Sciences, Allahabad, Uttar Pradesh, India
e-mail: budhayash.gautam@shiats.edu.in

K. Goswami · N. S. Mishra
Plant RNAi Biology Group, International Center for Genetic Engineering and Biotechnology, New Delhi, India

G. Wadhwa
Department of Biotechnology, Apex Bioinformatics Centre, Ministry of Science & Technology, New Delhi, India

G. Wadhwa et al. (eds.), *Current trends in Bioinformatics: An Insight*,
https://doi.org/10.1007/978-981-10-7483-7_3

Keywords
Epigenetics · Methylation · RNA silencing · Histone modification · Bioinformatics

3.1 Introduction

Epigenetics refers to the study of mitotically and meiotically inherited gene regulation that leads to phenotypic changes without altering the primary DNA sequence (computational epigenetics). Epigenetic processes entail genetic modification that reflects gene regulation by changing the DNA conformation. Conrad Waddington was the first person to introduce 'epigenetics'. He speculated that there is something more to DNA-based gene regulation which controls cell differentiation as well as enabling the integrity of genetic information during development. DNA methylation and histone acetylation play a crucial role in remolding of the DNA structure and regulation of gene expression (Bannister and Kouzarides 2011). Epigenetics can be either natural or induced by environmental factors and diseases. These epigenetic modifications may lead to a severe, harmful effect which results in diseases such as cancer (Jiang et al. 2004). The human body is made up of many different cells such as neurons, pancreatic cells, liver cells, inflammatory cells, as well as many others, each containing defined sets of genes that can be silenced or turned off by epigenetics (Brunet and Berger 2014).

Epigenetics refers to other factors that guide a gene regarding when and how it is to be expressed and that determine which proteins are to be synthesized, how the genetic message is encoded in the DNA, and how a single fertilized egg cell generates every cell of a multicellular organism with specialized function. Epigenetics works in a cell- and tissue-specific manner and varies over a time period. It plays a crucial role in aging or disease, because every cell has a specific gene expression pattern (Calvanese et al. 2009). Epigenetics has become one of the fastest growing areas in the field of life sciences for studying the development and biological factors of disease. Epigenetic mechanisms work via any of the four systems in the cell—covalent post-translational protein histone modification, DNA methylation, chromatin modeling, and RNA-associated silencing—which can interact with each other to silent the gene and regulate gene function during growth, development, and differentiation (Fig. 3.1) (Bowman and Poirier 2014). Such epigenetic factors are able to answer why cells, despite the same genetic material being in each, differentiate into different cell types with variable fates. These processes can change the expression pattern of any gene by influencing the expression of others (Bird 2002).

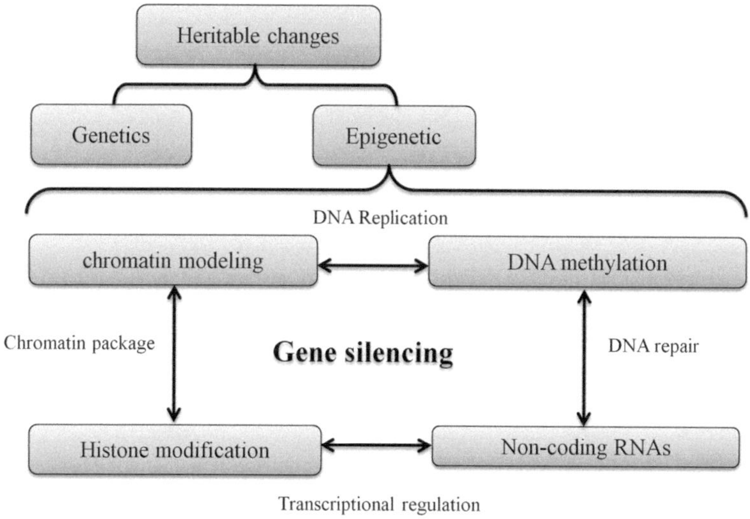

Fig. 3.1 Interaction of different processes due to the epigenetics

3.2 Histone Modification and Chromatin Modeling

Histone proteins are the key component of chromatin found in the nuclei of eukaryotic cells. This is important because they help to compact DNA by forming the nucleosome and regulate chromatin, affecting gene transcription and other chromatin-based processes. The histone octamer of the nucleosome molecule looks like beads on a string that are connected by linker DNA (Berger 2002). Histones control transcriptional machinery by modifying the DNA packing, which allows for additional gene regulatory control. Gene expression and DNA replication are both controlled by the structural state of chromatin. In eukaryotes, regulation of the transcriptional machinery relies on the histone proteins in various essential ways. Histone N-terminal tails and fold domains of different lengths are the feature of core histones, facilitated by post-translational modifications (PTMs). PTMs include changes to DNA and its connected proteins, which are a key component of epigenetics (Dupont et al. 2009). Histone acetylation and methylation are one of the best examples of post-translational histone modification. It is a reversible modification of specific residues in histone tails. A wide variety of modifications are involved in silencing of genes, transcriptional activation, chromatin assembly, and DNA replication, which are caused by enzymes such as

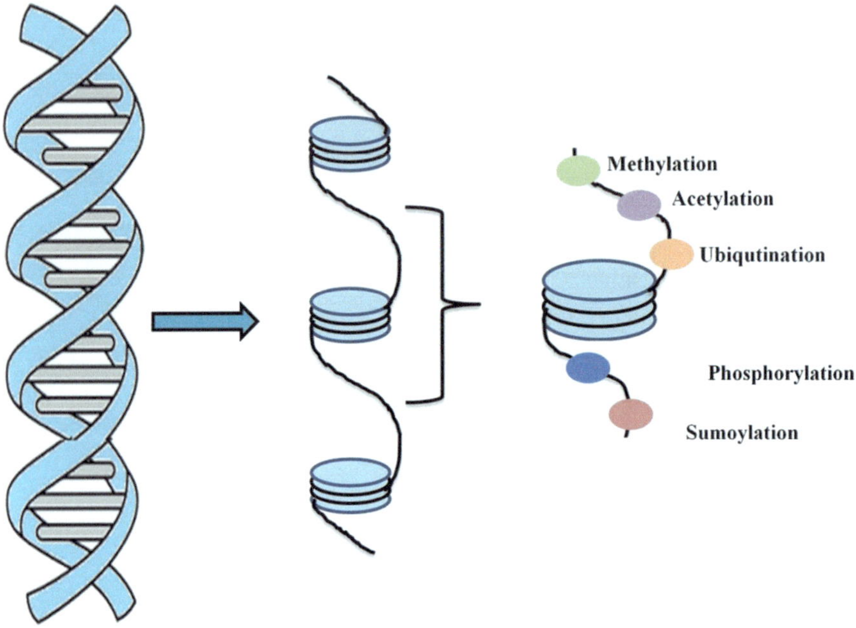

Fig. 3.2 Histone modification: modification of the histone protein and modulation of chromatin

histone acetyltransferases (HATs), histone methyltransferases (HMTs), and histone deacetylases (HDACs), etc., which control modification of histones and also control covalent modifications such as acetylation (Ac), ubiquitination (Ub), methylation (Me), SUMOylation (Su), and phosphorylation (P) (Fig. 3.2) (Deaton and Bird 2011).

3.2.1 Acetylation

Histone acetylation and deacetylation play a crucial role in gene regulation (Jaenisch and Bird 2003); these modifications are normally catalyzed by a particular enzyme, namely HAT and HDAC, respectively. Acetylation is the approach that helps to unwrap the DNA from histone by activating the HAT enzyme; it transfers the acetyl group to the N-terminal and results in unwrapped DNA. In contrast, deacetylation prompts the opposite reaction, removing the acetyl group from a molecule to wrap the DNA in a compact form with histone (Eberharter and Becker 2002).

3.2.2 Ubiquitination

Ubiquitination is a process in which the ubiquitin protein is attached via an isopeptide bond to lysine residues on a protein and leads to PTM on histones H2A (K119) and H2B (Zou and Mallampalli 2014). This is a very large modification process consisting of three steps:

1. Activation of ubiquitin protein by ubiquitin-activating enzymes (E1 enzymes).
2. Binding by ubiquitin-conjugating enzymes (E2s).
3. Ubiquitin ligation—attachment to the substrate protein by ubiquitin (E3s).

Additional factors are recruited to chromatin to keep the chromatin open physically via a 'wedging' process, given its huge size (Egger et al. 2004).

3.2.3 SUMOylation

SUMOylation is a post-translational modification which plays an important role in many cellular processes; it is a major regulator of protein function. In simulation, members of the SUMO (small ubiquitin-like modifiers) family are covalently attached to lysine residues in specific target proteins through an enzymatic cascade which is analogous to ubiquitin but different from the ubiquitination pathway. SUMOylation modification leads to structural damage and subcellular localization of the protein. All H4, H2A, and H2B histone proteins are sites for this epigenetic modification. SUMOylation antagonizes both acetylation and ubiquitylation, which occur on the same lysine residue, and consequently this modification is a repressive one for transcription (Wilkinson and Henley 2010).

3.2.4 Phosphorylation

The addition of a phosphate group (PO_4^{3-}) to a molecule is known as phosphorylation. Many amino acids such as serine, threonine, and tyrosine are favorable sites for histone phosphorylation. Histone phosphorylation constitutes an essential part of the 'histone code', or combinatory function of post-transcription modifications on chromatin. Little is known about histone phosphorylation and gene expression. Phosphorylation of H2A(X) is an important histone modification that plays a major role in diverse DNA damage response pathways such as homologous recombination and non-homologous end joining, and is also involved in replication-coupled DNA repair. Phosphorylation occurs in all phases of the cell cycle (Kadonaga 2004).

Histone modifications can also act as transcriptional co-activators and co-repressors. Histone modifications play an important role in recruiting other proteins that induce the space among histones and allocate transcription factors or polymerase to associate with open DNA and alter the chromatin compaction and sequester promoter element. Alteration in DNA compactness, which controls the level of the accessible coding region for further transcription by allowing the binding of RNA polymerase enzyme and recruitment of other important factors of transcription machinery, is also caused by histone modification (Jones 2012).

Deregulation of the histone protein can lead to aberrant gene expression and can cause tumorogenesis. Due to these modifications, the nucleosome assembly is influenced by modulating the DNA-binding affinity and chromatin remodeling, which leads to silencing of significant portions of chromatin. The transcriptional machinery during specific periods of time and at precise locations needs accessible DNA in a tissue-specific manner which is regulated and maintained by chromatin (Holoch and Moazed 2015).

These modifications are the epigenetic changes, which do not alter the DNA sequence but modify gene expression. Besides histone proteins, non-histone proteins can sometimes also modulate chromatin structure in various manners such as interacting with histones and DNA. Chromatin remolding complex acts as an adenosine triphosphate (ATP)-dependent molecular machinery that can displace the nucleosome along with the DNA (Gangaraju and Bartholomew 2007).

3.3 DNA Methylation or Modification and Its Effect on Gene Expression

DNA methylation, also known as chemical modification, occurs with the addition of a methyl group (–CH3) to the DNA molecule such as the addition of a cytosine or guanine nitrogenous base at the fifth carbon atom 5-methylcytosine (5mC) promoter region. The promoter region contains ~70% CpG content in eukaryotes. In prokaryotes, this addition may be at the sixth carbon atom of adenine nitrogenous base 6-methyladesine (6mA). These CpG residues are DNA methylation sites that are also known as CpG islands and act as the specific sites for methylation (Arand et al. 2012). CpG is the linking of cytosine with guanine by a phosphate bond in a nucleotide sequence. CpG islands are usually >2000 base pairs (bp) in length and include a high quantity of CpG sites found in repetition. In eukaryotes, DNA methyltransferase (DNMT) enzymes perform this methylamine process, which transfers a methyl group to cytosine base at fifth position, resulting in a 5mC (Fig. 3.3). These methylases can be sub-classified into two: those that usually methylate cytosine are known as DNA cytosine methylase (DCM) and those that methylate adenine are known as DNA adenine methylase (DAM) in prokaryotes (Hackett and Surani 2013).

DNA methylation is a vital epigenetic tool used by cells to block the gene. It plays a crucial role in the silencing of tissue-specific genes, averting them from being expressed in the wrong tissue. DNA methylation is essential for silencing retroviral elements and an imperative factor in various cellular processes, including genomic imprinting, embryonic development, chromosome stability preservation, and X-chromosome inactivation. During cell division, originated cells from embryonic stem cells are tissue specified and the modification in DNA can alter expression of these cells (Wolffe and Matzke 1999). This alteration of gene expression is

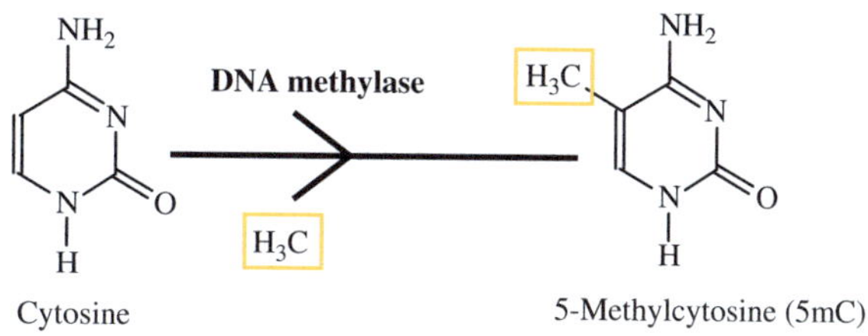

Fig. 3.3 DNA methylation process: adding a methyl group to a nitrogenous base

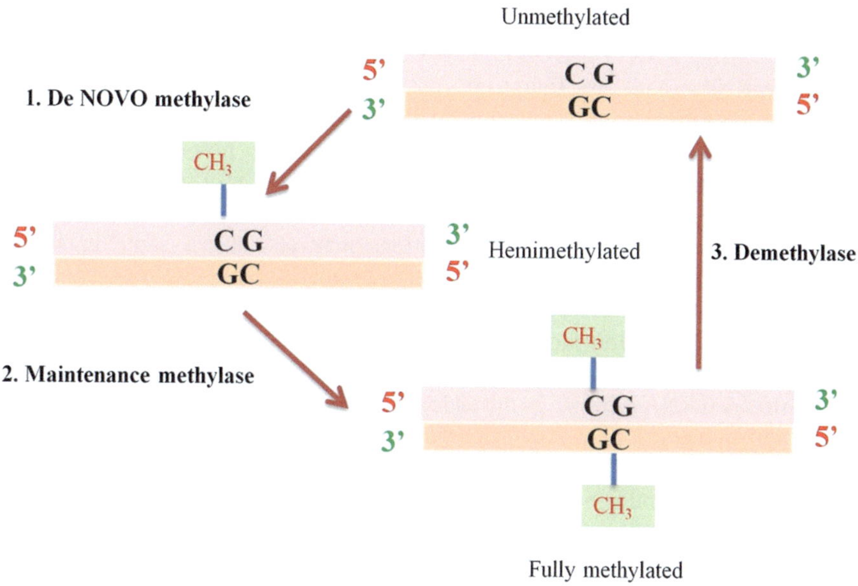

Fig. 3.4 DNA methylation cycle

permanent unless DNA methylation is stopped during zygote formation as restoration is possible during development. Methylation provides time for selection of which gene replica is inherited from which parent, also called imprinting (Wolffe 1998).

The process of methylation occurs in a cyclic structure where three enzymes work together, carrying forward this process: de novo methylase, maintenance methylase, and demethylase. In the first place the methyl group attachment to the unmethylated DNA takes place with the help of a enzyme de novo methylase, which attaches a methyl group to the old strand of DNA; the new strand of DNA remains unmethylated right until the replication. This is the first type of methylase and the DNA strand is known as hemi-methylated DNA. This kind of methylation can be carried out by less effective methylase simply known as maintenance methylase. This is the second type of methylation, which results in both the strands being methylated (Fig. 3.4). However, in embryonic stem cells less CG methylation is observed. CpG islands are the remaining unmethylated CG dinucleotides, which are usually found near gene promoters in dense clusters. CG-methylation is performed to identify whether or not to express a gene in a particular tissue. When expression of a tissue-specific gene is required, it needs to be demethylated. This function is performed by another type of enzyme known as demethylase which removes the methyl group and the gene returns to its original state.

3.4 RNA-Associated Silencing

A class of non-coding RNAs, such as such as microRNAs (miRNAs), small RNAs (piwi-interacting RNAs [piRNAs], guide RNAs, small-interfering RNA [siRNA]), and large RNAs (antisense transcript), can turn off gene expression through the silencing (RNA interference) pathway by associating with Argonaut protein (Yang and Wu 2007). It can silence or modulate the genes at a post-transcriptional or transcriptional level and can also affect gene expression by formation of heterochromatin, or by altering histone protein, X-chromosome inactivation, DNA methylation, and paramutation. These regulations may also affect some important genomic functions, such as the structure of chromatin, chromosome assortment, RNA editing, RNA stability, and protein synthesis. Small RNA independent recruitment of chromatin-modifying complexes is facilitated by certain signals present on several long non-coding RNAs. miRNAs are 18–24 nucleotides long, single-stranded RNAs, known as the endogenous small RNA gene, that reduce gene expression by binding to the complementary sequence at the 3′UTR (3′ untranslated region) of messenger RNA (mRNA) at the post-transcriptional level. Several epigenetic factors and genomic mutations are involved in miRNA biogenesis that can alter the expression of miRNAs at various stages by changing the expression and function of the nuclear microprocessor complex, formed by Drosha and DGCR8 (known as Pasha) and the RNA III enzyme Dicer (Fig. 3.5). It has been reported that alteration in the miRNA biogenesis/processing, such as mutation of the primary transcript, can alter the miRNA expression level and single nucleotide polymorphisms (SNPs) can avert the interaction between miRNA and mRNA because their presence plays an important role in miRNA functioning, which interrupts the homeostasis of cells. Additionally, miRNAs can change the gene expression by modifying the histone protein and DNA (DNA methylation) of the promoter region with the help of an RNA-induced Initiation of Transcriptional Silencing (RITS) complex. Binding of RITS protein complex with miRNA executes post-translational modifications in histone tails, as reported by Villeneuve and Natarajan (2010) who noted that heterochromatin formed histone H3 lysine 9 by methylation (H3K9me) which leads to transcriptional repression. Transcriptional RNA silencing is a RNA silencing process, but it is not very well-understood. What is understood, however, it is that it involves the RITS complex, which

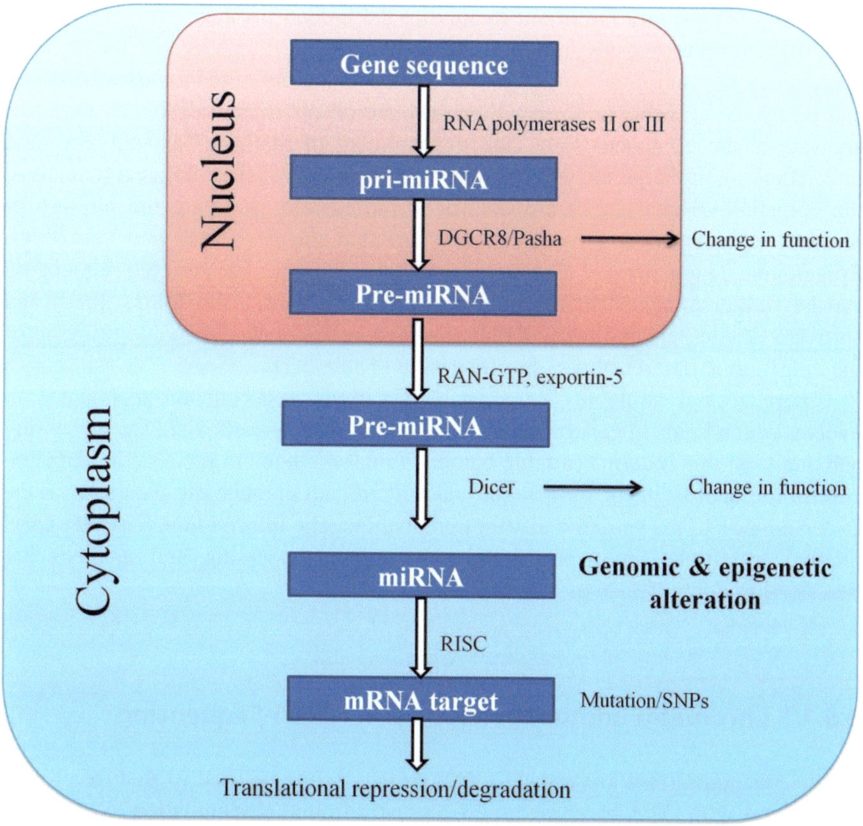

Fig. 3.5 microRNA biogenesis

functions to down-regulate genes pre-transcriptionally via modification of histones and leads to heterochromatin formation. It contains chromodomain protein Chp1, Argonaute homolog Ago1, and an uncharacterized protein (Tas3). RITS use siRNAs and Ago1 to cleave RNA. It binds to chromatin and is thought to destroy newly produced RNAs in *cis*.

3.5 Bioinformatics and Epigenetics

In the field of molecular biology, epigenetics has recently emerged as a key approach for studying the heritable gene and environmental factors that can influence the function of an organism without changing the DNA. Bioinformatics or computational analysis have become useful approaches to understand epigenetic mechanisms and analyze the epigenomic data (Yang and Lee 2004). Many tools and databases are available now that help to analyze epigenetic data, and these are becoming increasingly important in solving the hidden mysteries of gene regulation. Bioinformatics tools and techniques help in various ways in the field of epigenetics, the major uses of which include understanding the biology of diseases caused by the epigenome so that therapies can be optimized, design of tools for analysis of the epigenetic data, and development of methods for analyzing and understanding the large epigenome datasets. Many approaches are used to analyze the epigenomic data which are helpful in aiding understanding the biology of diseases such as cancer as well as many heritable changes (Yang and Lee 2004). Epigenomic data generated using experimental techniques such as chip sequencing can be further analyzed using bioinformatics tools (Hirst and Marra 2010) and software (Table 3.1) and many databases have been created to enable retrieval of the epigenomic dataset of any specific tissue (Table 3.2).

Computational strategies have enabled various experimental techniques to evolve, which helps in generating a large amount of epigenetic data for processing and analysis; this requires suitable computational methods to process the data and control the quality of the epigenomic data. Predicted epigenomic data help in the understanding of the genomic distribution of epigenetic information. Several experimental techniques involved in epigenetic data processing and analysis are described in Sects. 5.1–5.4.

3.5.1 Chromatin Immunoprecipitation (ChIP)-Sequencing

ChIP-sequencing (also known as ChIP-Seq) is a technique used to analyze protein interactions with DNA by combining Chromatin Immunoprecipitation (ChIP) with massively parallel DNA sequencing to identify the binding sites of DNA-associated proteins (Jothi et al. 2008). It is used to determine the gene expression of a protein–DNA interaction and to discern biological processes and disease states (Furey 2012). ChIP has become a valuable and extensively used approach for determination of the location of the protein-binding genomic region, which can be transcription factors, modified histone protein, and DNA-binding enzymes in living cells (Visel et al. 2009).

3.5.2 ChIP-on-Chip

Also known as ChIP-chip, ChIP-on-chip analyzes interactions between proteins and DNA by combining ChIP with the DNA microarray ('chip'). Protein–DNA interactions play an essential role in many biological processes such as transcription, replication, DNA repair, and splicing. This is well-suited approach for inclusive analysis of histone modification patterns, transcription factor-binding sites, and nucleosome tenancy. In particular, it allows the detection of the cistrome for DNA-binding proteins on a genome-wide basis (Carey et al. 2009).

Table 3.1 List of tools, software, and browsers for analysis of epigenomic datasets

Tool/software/browsers	Description	Link
Aclust	Algorithm to detect sets of neighboring CpG sites	https://github.com/brentp/aclust/
BLAST	Tool for searching homology	http://blast.ncbi.nlm.nih.gov/Blast.cgi
CoSBI	Used to check chromatin modification patterns in the human genome	http://www.healthcare.uiowa.edu/labs/tan/CoSBIWebpage.html
Epidaurus	Integrative analyses of epigenetic data promise a deeper understanding of the epigenome	http://bioinformaticstools.mayo.edu:8080/Epidaurus/
EpiExplorer	Web tool that allows use of large reference epigenome datasets for analysis	http://epiexplorer.mpi-inf.mpg.de/
EpiGRAPH	Software available for genome and epigenome analysis	http://epigraph.mpi-inf.mpg.de/WebGRAPH/
VizHub	Visualization hub displaying sequencing data from the *Roadmap Epigenomics* project	http://vizhub.wustl.edu/
WashU Epigenome Browser	New-generation genome browser for integrative visualization of genomic information	http://epigenomegateway.wustl.edu/browser/
MethBLAST/MethPrimerDB	Web tools for PCR-based methylation analysis	http://medgen.ugent.be/methblast
MethPipe	Computational pipeline for analyzing bisulfite sequencing data (WGBS and RRBS)	http://smithlabresearch.org/software/methpipe/
MOABS	MOABS (MOdel-based Analysis of Bisulfite Sequencing) data	https://code.google.com/p/moabs/
BiQ analyzer	Tool for visualization and quality control of DNA methylation data from bisulfite sequencing	http://biq-analyzer.bioinf.mpi-inf.mpg.de/
NCBI Epigenomics sample browser	Unique interface for intuitive browsing and searching of biological datasets	www.ncbi.nlm.nih.gov/epigenomics
PANTHER (Protein ANalysis THrough Evolutionary Relationships)	Gene analysis tools	http://pantherdb.org/

NCBI National Center for Biotechnology Information, *PCR* polymerase chain reaction, *RRBS* reduced representation bisulfite sequencing, *WGBS* whole-genome bisulfite sequencing

Table 3.2 List of databases and repositories of epigenetic data

Database/repository	Description	Link
COXPRESdb	Co-expressed gene database identifying new gene functions or functional modules in metabolic pathways and signaling pathways	http://coxpresdb.jp/
CREMOFAC (Chromatin Remodeling Factors)	Chromatin-remodeling factor database	http://www.jncasr.ac.in/cremofac/
DBCAT (DataBase of CpG islands and Analytical Tool)	A database to recognize comprehensive methylation profiles of DNA alteration in human cancer	http://dbcat.cgm.ntu.edu.tw/
DDMGD (Dragon Database for Methylated Genes and Diseases)	Provides a comprehensive repository of information related to genes methylated in diseases	http://www.cbrc.kaust.edu.sa/ddmgd/
DiseaseMeth	Web-based resource focused on the aberrant methylomes of human diseases	http://202.97.205.78/diseasemeth/intro.html
EpiFactors	Manually curated database providing information about epigenetic regulators, their complexes, targets, and products	http://epifactors.autosome.ru/
Epigenome Browser	This site contains the reference sequence and working draft assemblies for a large collection of genomes	http://www.epigenomebrowser.org/
EpimiR	Repository of mutual regulation between epigenetic modifications and miRNAs	http://202.97.205.78:8080/EpimiR/
*HI*stome	Post-translational modifications and modifying enzymes database of human histones	http://www.actrec.gov.in/histome/index.php
MeInfoText	Web server for gene methylation and cancer associations based on text-mining and maximum entropy	http://bws.iis.sinica.edu.tw:8081/MeInfoText2/
MethBank	DNA methylome programming database	http://www.dnamethylome.org/
Methbase	A central reference methylome database	http://smithlabresearch.org/software/methbase/
MethDB	Database for DNA methylation data	http://www.methdb.de
MethHC	The human pan-cancer methylation database	http://methhc.mbc.nctu.edu.tw/php/index.php
MethyCancer database	Database of human DNA methylation and cancer	http://methycancer.psych.ac.cn/
MethylomeDB	Brain Methylome Database	http://www.neuroepigenomics.org/methylomedb/

(continued)

Table 3.2 (continued)

Database/repository	Description	Link
miRBase	Searchable database of published miRNA sequences and annotation	http://www.mirbase.org/
PCMdb (Pancreatic Cancer Methylation Database)	Database of methylated genes found in pancreatic cancer cell lines and tissues	http://crdd.osdd.net/raghava/pcmdb/
PEpiD (Prostate Epigenetic Database)	Repository for epigenetic data relating to prostate cancer of human, mouse, and rat	http://wukong.tongji.edu.cn/pepid
PubMeth	Cancer methylation database combining text-mining and expert annotation	http://www.pubmeth.org
The Histone Database	Searchable collection of full-length sequences and structures of histones and non-histone proteins	http://genome.nhgri.nih.gov/histones

miRNA microRNA

3.5.3 Bisulfite Sequencing

Bisulfite genomic sequencing is a primary method of DNA methylation which gives single nucleotide resolution for identification and quantification of methylation in different research and clinical settings. After DNA modification by bisulfite conversion, cytosine residues are converted to uracil but 5-methylacytosine remains unaffected (Patterson et al. 2011). This conversion protocol becomes the basis for most of the methods used for analysis of DNA methylation, accounting for the majority of recent data on DNA methylation (Ehrich et al. 2007).

3.5.4 Methylated DNA Immunoprecipitation (MeDIP-Seq)

Methylated DNA immunoprecipitation (MeDIP) is a technology used to analyze methylated DNA such as 5mC and 5-hydroxymethylcytosine (5hmC) and targets the methylome data in bulk using an antibody. MeDIP-seq is capable of generating approximately 80% of the 28 million CpGs in the human genome (Lee et al. 2006).

To analyze the experimental data, many bioinformatics approaches such as tools, software, data repositories, and databases are facilitated. Epigenetics databases are available for histone modification, DNA methylation, chromatin modeling, and non-coding RNAs (Lim et al. 2010).

3.6 Conclusions and Perspectives

Epigenomics is a relatively new approach to understanding the hidden layers of gene regulatory machinery in both prokaryotes and eukaryotes and for various aspects of cell physiology among cells of a normal or diseased state which is gaining immense attention among researchers worldwide. The post-genomics era has seen a boost in high-throughput sequencing data, and the capturing of epigenetic information has been revolutionized with more accurate prediction capabilities. Much has been done to develop the algorithms and software related to methylation studies so that a global picture of inherent epigenomic information can be decoded for a particular organism. More computational strategies need to be developed to gain useful insights into the epigenetic code of an organism in order to generate better tools to fight deadly diseases such as cancer. For this, greater research effort needs to be focused in this direction to encourage appropriate progress in the field of epigenetics.

Acknowledgments The authors are grateful to the Sam Higginbottom University of Agriculture, Technology & Sciences, Allahabad, India, for providing the facilities and support to complete the present research work.

References

Arand J et al (2012) In vivo control of CpG and non-CpG DNA methylation by DNA methyltransferases. PLoS Genet 8:e1002750

Bannister AJ, Kouzarides T (2011) Regulation of chromatin by histone modifications. Cell Res 21:381–395

Berger SL (2002) Histone modifications in transcriptional regulation. Curr Opin Genet Dev 12:142–148

Bird A (2002) DNA methylation patterns and epigenetic memory. Genes Dev 16:6–21

Bowman GD, Poirier MG (2014) Post-translational modifications of histones that influence nucleosome dynamics. Chem Rev 115:2274–2295

Brunet AS, Berger L (2014) Epigenetics of aging and aging-related disease. J Gerontol A Biol Sci Med Sci 69:S17–S20

Calvanese V et al (2009) The role of epigenetics in aging and age-related diseases. Ageing Res Rev 8:268–276

Carey MF et al (2009) Chromatin immunoprecipitation (ChIP). Cold Spring Harb Protoc,pdb prot5279

Deaton AM, Bird A (2011) CpG islands and the regulation of transcription. Genes Dev 25:1010–1022

Dupont C et al (2009) Epigenetics: definition, mechanisms and clinical perspective. Semin Reprod Med 27:351–357

Eberharter A, Becker PB (2002) Histone acetylation: a switch between repressive and permissive chromatin. EMBO Rep 3:224–229

Egger G et al (2004) Epigenetics in human disease and prospects for epigenetic therapy. Nature 429:457–463

Ehrich M et al (2007) A new method for accurate assessment of DNA quality after bisulfite treatment. Nucleic Acids Res 35:e29

Furey TS (2012) ChIP-seq and beyond: new and improved methodologies to detect and characterize protein-DNA interactions. Nat Rev Genet 13:840–852

Gangaraju VK, Bartholomew B (2007) Mechanisms of ATP dependent chromatin remodeling. Mutat Res/Fundam Mol Mech Mutagen 618:3–17

Hackett JA, Surani MA (2013) DNA methylation dynamics during the mammalian life cycle. Philos Trans R Soc Lond Ser B Biol Sci 368:20110328

Hirst M, Marra MA (2010) Next generation sequencing based approaches to epigenomics. Brief Funct Genomics 9:455–465
Holoch D, Moazed D (2015) RNA-mediated epigenetic regulation of gene expression. Nat Rev Genet 16:71–84
Jaenisch R, Bird A (2003) Epigenetic regulation of gene expression: how the genome integrates intrinsic and environmental signals. Nat Genet 33:245–254
Jiang YH et al (2004) Epigenetics and human disease. Annu Rev Genomics Hum Genet 5:479–510
Jones PA (2012) Functions of DNA methylation: islands, start sites, gene bodies and beyond. Nat Rev Genet 13:484–492
Jothi R et al (2008) Genome-wide identification of in vivo protein-DNA binding sites from ChIP-Seq data. Nucleic Acids Res 36:5221–5231
Kadonaga JT (2004) Regulation of RNA polymerase II transcription by sequence-specific DNA binding factors. Cell 116:247–257
Lee TI et al (2006) Chromatin immunoprecipitation and microarray-based analysis of protein location. Nat Protoc 1:729–748
Lim SJ et al (2010) Computational epigenetics: the new scientific paradigm. Bioinformation 4:331–337
Patterson K et al (2011) DNA methylation: bisulphite modification and analysis. J Vis Exp 56:3170
Villeneuve LM, Natarajan R (2010) The role of epigenetics in the pathology of diabetic complications. Am J Physiol Renal Physiol 299:F14–F25
Visel A et al (2009) ChIP-seq accurately predicts tissue-specific activity of enhancers. Nature 457:854–858
Wilkinson KA, Henley JM (2010) Mechanisms, regulation and consequences of protein SUMOylation. Biochem J 428:133–145
Wolffe AP (1998) Packaging principle: how DNA methylation and histone acetylation control the transcriptional activity of chromatin. J Exp Zool 282:239–244
Wolffe AP, Matzke MA (1999) Epigenetics: regulation through repression. Science 286:481–486
Yang HH, Lee MP (2004) Application of bioinformatics in cancer epigenetics. Ann N Y Acad Sci 1020:67–76
Yang Z, Wu J (2007) MicroRNAs and regenerative medicine. DNA Cell Biol 26:257–264
Zou C, Mallampalli RK (2014) Regulation of histone modifying enzymes by the ubiquitin-proteasome system. Biochim Biophys Acta 1843:694–702

Three Dimensional Structures of Carbohydrates and Glycoinformatics: An Overview

4

K. Veluraja, J. Fermin Angelo Selvin, A. Jasmine, and T. Hema Thanka Christlet

Abstract

Carbohydrates are regarded as the interesting molecules of nature because of their structural diversity and functional characteristics. The nature of existence of carbohydrates in varied forms and conformations is crucial in understanding their functional features in living systems. The dynamical behavior of carbohydrates in free or bound state with other biological molecules influences their functional role in biological systems. In N- and O-glycosylation, sequence, structure, and conformation of carbohydrates play a vital role. Hence, necessity arises for the complete understanding of the three-dimensional structures of carbohydrates. One of the theoretical ways of studying the structural and conformational aspect of carbohydrates is by molecular dynamics simulation. Not only the structure and conformation but also the interaction of carbohydrates with its conjugated forms can be investigated. The resources for carbohydrates in the form of databases available are discussed. Sialic acid-containing oligosaccharides which have an important role in molecular recognition phenomena are attributed to their sequence, structure, and conformational diversity. A three-dimensional structural database for sialic acid-containing carbohydrates (3DSDSCAR) developed based on molecular dynamics simulation results is

K. Veluraja (✉)
Department of Physics, School of Advanced Sciences, VIT University, Vellore, Tamil Nadu, India
e-mail: veluraja.k@vit.ac.in

J. Fermin Angelo Selvin
Department of Physics, Nadar Mahajana Sangam S. Vellaichamy Nadar College, Madurai, Tamil Nadu, India

A. Jasmine
Department of Physics, Manonmaniam Sundaranar University, Tirunelveli, Tamil Nadu, India

T. Hema Thanka Christlet
Department of Physics, Dr. Ambedkar Government Arts College, Chennai, Tamil Nadu, India

G. Wadhwa et al. (eds.), *Current trends in Bioinformatics: An Insight*,
https://doi.org/10.1007/978-981-10-7483-7_4

discussed in detail. Glycoinformatics, knowledge about carbohydrates or glycans, is still a field of informatics to be explored more.

Keywords
Glycoinformatics · Carbohydrate database · Molecular dynamics simulation

4.1 Introduction

Carbohydrates are the richest and widely diverse groups of biomolecules in nature. As their name suggests, they are made up of carbon, oxygen, and hydrogen. Carbohydrates constitute a major share of our diet serving mainly as energy reserves as starch in plants and glycogen in animals. Carbohydrates are the metabolic precursors of virtually all other biomolecules. They also provide structural stability to plants in the form of cellulose, to arthropods and fungi as chitin, and to bacteria as peptidoglycans (Paulson 1989; Karlsson 1995; Crocker and Feizi 1996; Varki 1998; Olofsson and Bergström 2005). Industrial use of carbohydrates covers a wide spectrum that includes food processing, agrochemicals, pharmaceuticals, and paints and paper industries (Lonngren 1989; Wilson and Itzstein 2003; Werz and Seeberger 2005). A large number of carbohydrates and carbohydrate derivatives are being used in therapeutics and diagnosis over centuries. Even though carbohydrates are exploited extensively for their versatile applications, their role in the biological functions and their structural elegance has not been understood and studied until recently. In this chapter, three-dimensional structures of carbohydrates, their functional roles in context with their structures, and glycoinformatics will be discussed.

4.1.1 Functional Aspects of Carbohydrates

Carbohydrates engage themselves in innumerable biological functions of the living organisms, and their roles are briefly listed but not limited to the following four categories.

1. Energy supply and storage
2. Proteins paring
3. Lipid metabolism and ketosis prevention
4. Biological recognition including protein-carbohydrate interaction and carbohydrate-mediated processes

Dietary carbohydrates provide glucose as a major source of energy which can be utilized by the cells, and this energy is needed for the body to perform functions including physical activities. Also, carbohydrates spare proteins and help the fatty acids breakdown to avoid ketosis. Lactose is one of the essential components of milk. Blood group oligosaccharides are commonly made up of monosaccharides such as D-galactose, D-galactosamine, L-fucose, D-glucosamine, and N-acetylneuraminic acid. These sugars are found to exist in saliva, mucous lining of various tracts and systems, cystic fluid, blood serum, and other body secretions. Carbohydrate when conjugated to proteins in erythrocytes forms the A, B, O, Rh, and other antigens and thereby helps in the differentiation of blood groups. Also, when they occur as proteoglycans, they act as receptors, carrier of macromolecules, and regulators of cell growth, essential component for the absorption by fibroblast and influences protein synthesis as well. Chondroitin sulfates A and C constitute the major structural components of cartilage, tendons, and bones. They are related with the structure of collagen. Carbohydrates like dextran, heparin, and hyaluronic acid have major impact in human medicine over the decades (Biswas and Rao 1980, 1982; Rao and Biswas 1981; Cagas and Bush 1990; Bush 1992; Schauer and Kamerling 1995; Poveda and Jiménez-Barbero 1998; Muramatsu 2000; Olofsson and Bergström 2005; Varki and Freeze 2009; Moskalewski and Jankowska-Steifer 2011). In addition to the above-stated roles, carbohydrates also act as allergenic agents and produce immune responses to parasitic infections (Afferni et al. 1999). Carbohydrate is the only molecule in the biological system that can occur on its own like polysaccharides and can occur with other biological macromolecules such as with lipids as glycolipids, with proteins as glycoproteins, and with DNA and RNA molecules as deoxyribose and ribose sugars.

Typically every cell surface is coated with complex carbohydrates, and these carbohydrates are potentially important for cell recognition processes. Due to their extensive roles in biological recognition phenomena, cell surface carbohydrates are regarded as *molecules of molecular recognition*. Often the cell surface carbohydrates act as either inhibitors or receptors for the invading pathogens. The property of being utilized for recognition makes the functional role of carbohydrates manifold, and this functional elegance complies with the structural versatility of these sugar molecules.

4.2 Monosaccharides

Monosaccharides are the basic building blocks of polysaccharides and long-chain carbohydrates. Monosaccharides are regarded as the fundamental sugar units since they cannot be hydrolyzed into simpler units further. They are nothing but the aldehydes or ketones having two or more hydroxyl groups. Depending upon the number of carbon atoms present in the backbone, monosaccharides are classified into trioses, tetrose, pentose, hexose, heptose, octose, nonose, and decose. Typically, monosaccharides are found in acyclic forms and cyclic forms. Open-chain form of aldoses and ketoses is not the predominant form in solution. The acyclic forms cyclize in solution due to the formation of hemiacetals. The formation of hemiacetals creates an additional asymmetric carbon known as anomeric carbon. Pentoses and higher sugars exist in water as tautomeric ring forms in dynamic equilibrium with small amounts of acyclic forms.

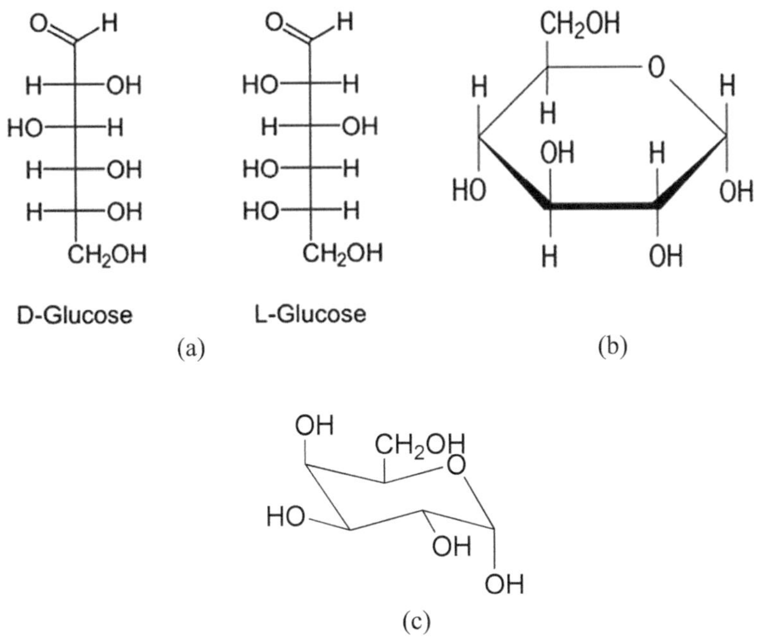

Fig. 4.1 Linear and cyclic forms of D-glucose. (**a**) Linear form of D-glucose. (**b**) Cyclic form of α-D-glucose (Haworth projection). (**c**) Cyclic form of α-D-glucose in $^4C_1$ chair form

The open-chain form and the cyclic forms of D-glucose are given in Fig. 4.1. The cyclic forms are given in both Haworth projection and chair conformation.

Apart from serving as energy sources, monosaccharides perform specialized functions in the recognition and regulation processes performed in living systems when they form part and parcel of the oligosaccharide structures. A few biologically significant monosaccharides are listed in Table 4.1.

4.3 Disaccharides

Disaccharides are formed by the conjugation of two monosaccharide units linked covalently by an O-glycosidic bond. The formation of O-glycosidic bond is due to the reaction of hydroxyl group of one (mono) sugar with the anomeric carbon of the other. If two monosaccharides are linked through their anomeric centers, the disaccharide formed is a nonreducing disaccharide. Sucrose and trehalose are examples of nonreducing disaccharides. If one monosaccharide is linked by one of its other hydroxyl groups, then the anomeric center is unsubstituted, and a reducing disaccharide is formed. Cellobiose, gentiobiose, maltose, isomaltose, and lactose are a few examples of reducing disaccharides.

The structure and conformation of a disaccharide is dictated by the glycosidic torsion around the glycosidic linkage. The variation in glycosidic torsional angle

Table 4.1 Significant monosaccharides that exist in biological systems

S. No	Name of the monosaccharide	Structure
1	α-D-Xylose	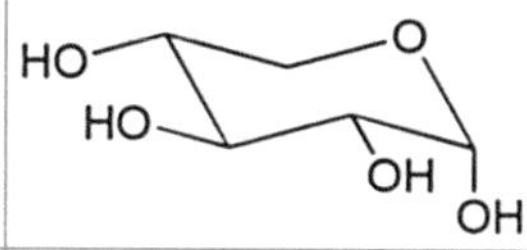
2	α-D-Mannose	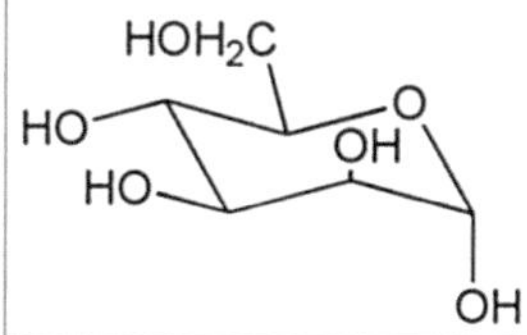
3	α-D-Galactose	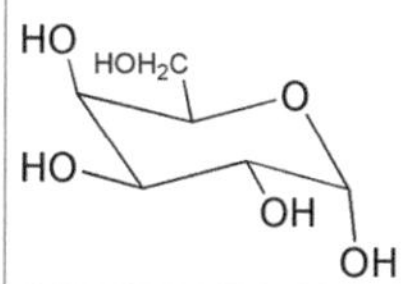
4	α-D-Glucose	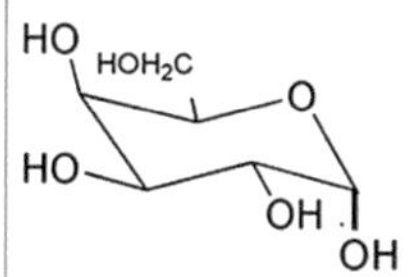
5	α-L-Fucose	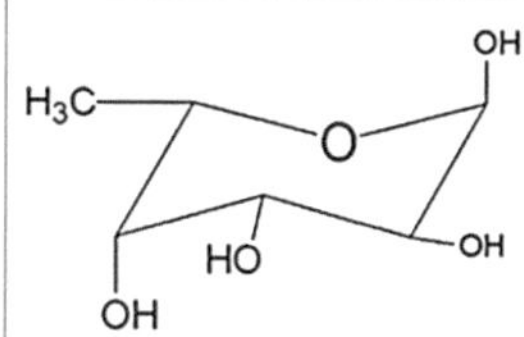
6	α-D-Neu5Ac N-Acetylneuraminic acid	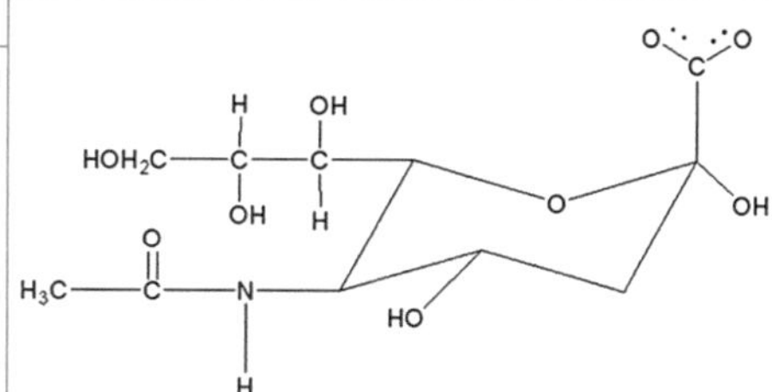
7	β-D-Xylose	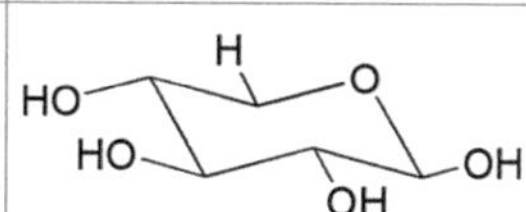
8	β-D-Mannose	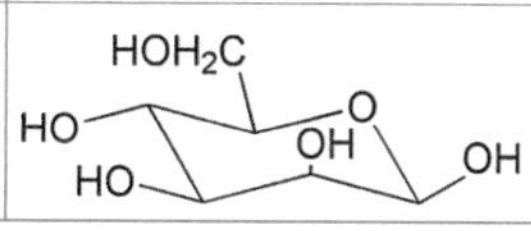

(continued)

Table 4.1 (continued)

S. No	Name of the monosaccharide	Structure
9	β-D-Galactose	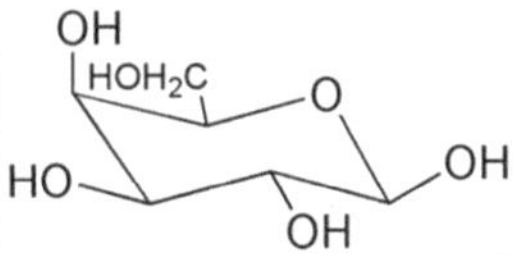
10	β-D-Glucose	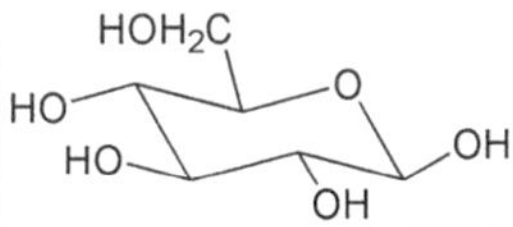

leads to change in conformation of the disaccharide which in turn influences the specificity of protein-carbohydrate interactions and carbohydrate-carbohydrate interactions. For example, α-mannobiose is a disaccharide formed by two mannose residues joined together through α(1→4) glycosidic linkage. The spread of glycosidic torsion angles based on molecular dynamics simulation of 10 ns time duration is shown in the glycosidic torsional space map (Fig. 4.2).

From the distribution of glycosidic torsional angles, it is observed that α-mannobiose prefers a single rigid conformation (Φ, Ψ) around (−60°, −50°) region. The conformation is stabilized by the formation of a hydrophobic cluster between the hydrogen atoms attached to the ring carbon atoms C-3 and C-5 (first residue) and C-4 and C-6 (second residue) and is given in Fig. 4.3.

On the contrary to its α-anomer, for β-mannobiose which is a disaccharide formed by two mannose residues joined together through β(1→4) glycosidic linkage, the glycosidic torsional space map, obtained from 10 ns MD simulation in aqueous solution, shows two conformational regions as shown in Fig. 4.4.

From the conformational map, it can be observed that β-mannobiose prefers to exist in two distinct conformations at (−50°, −50°) and (70°, −20°). The conformation at (−50°, −50°) is stabilized by water-mediated hydrogen bonding between O-2 of first residue and O-6 of second mannose. The other conformation is stabilized by a direct hydrogen bonding between O-5 of first mannose and O-3 of second mannose. The interaction diagrams of the two conformers are given in Fig. 4.5.

The conformational study on Mannobiose disaccharides indicate that the disaccharide conformations are greatly influenced by the anomeric configuration of the constituent monosaccharide units [M. Phil. Dissertation, M. S. University, Tirunelveli]. Another typical example of disaccharide conformational flexibility is the case of sialyldisaccharide Neu5Acα(2→3)Gal which is a significant segment of many sialic acid-containing oligosaccharides playing an important role in molecular recognition processes. Molecular dynamics simulations of 10 ns duration are carried out in order to explore the flexibility at glycosidic torsion, and the conformational space map (Fig. 4.6) indicates that the glycosidic torsional angle

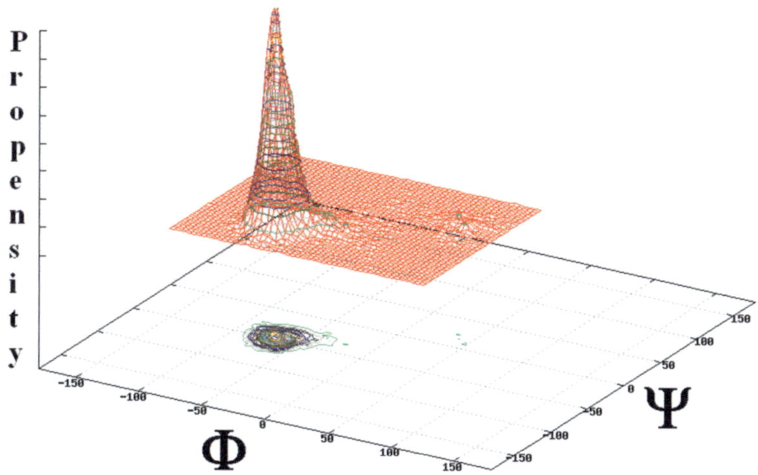

Fig. 4.2 Conformational space map of α-mannobiose

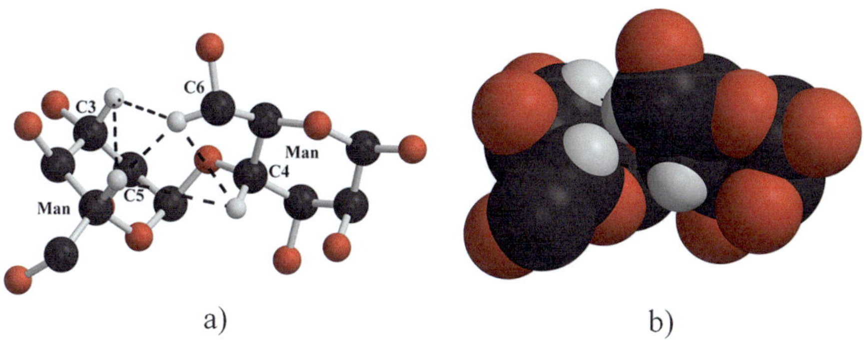

Fig. 4.3 A single and rigid conformer of α-mannobiose. (**a**) Ball and stick view. (**b**) Space-filling view

values (Φ, Ψ) prefer three unique conformational regions around (−100°, −50°), (−70°, 0°), and (−150°, −30°). The conformation at (−100°, −50°) region is stabilized by direct hydrogen bond between the carboxylic acid group of sialic acid and O-4 of galactose. The conformation at (−70°, 0°) region is stabilized by direct hydrogen bond between O-6 of sialic acid and O-2 of galactose. A water-mediated hydrogen bond between the carboxylic acid group of sialic acid and O-4 of galactose stabilizes the conformation at (−150°, −30°). The conformational models are given in Fig. 4.7.

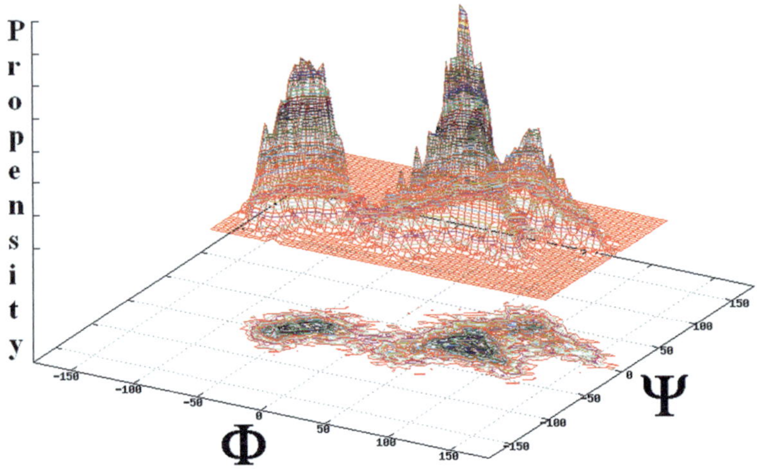

Fig. 4.4 Conformational map of β-mannobiose

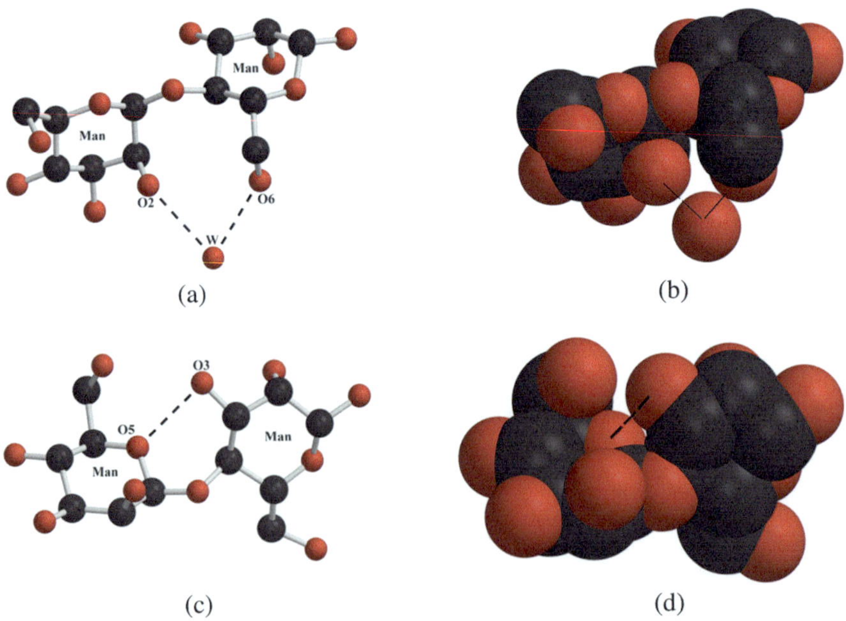

Fig. 4.5 Two conformers of β-mannobiose. (**a**) Ball and stick model of conformer 1. (**b**) Space-filling model of conformer 1. (**c**) Ball and stick model of conformer 2. (**d**) Space-filling model of conformer 2

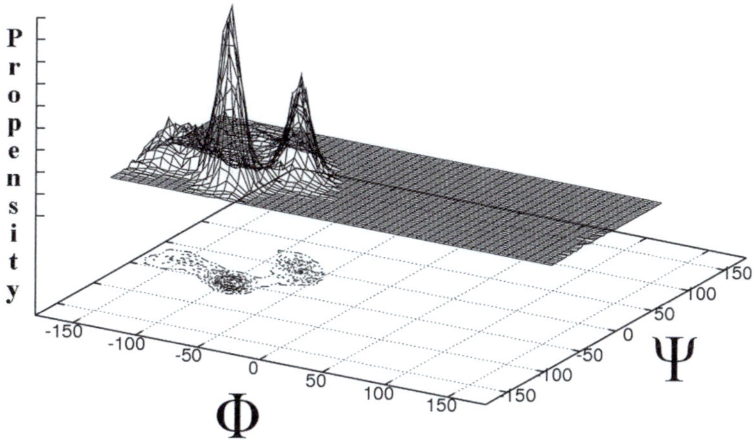

Fig. 4.6 The glycosidic conformational space map of Neu5Acα(2→3)Gal

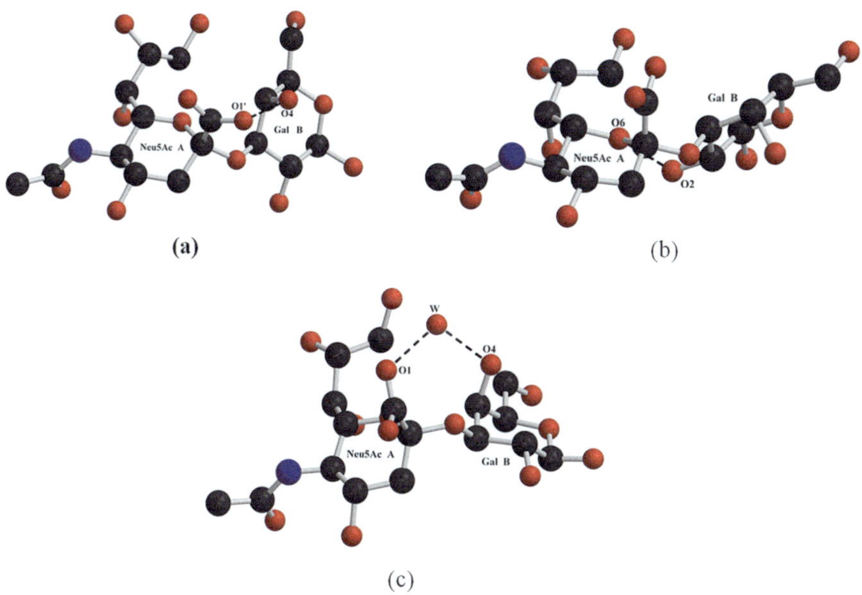

Fig. 4.7 Three conformational models of Neu5Acα(2→3)Gal based on10ns molecular dynamics simulation. (**a**) Conformer1of Neu5Acα(2→3)Gal. (**b**) Conformer2 of Neu5Acα(2→3)Gal. (**c**) Conformer3 of Neu5Acα(2→3)Gal

4.4 Oligosaccharides

Oligosaccharides are short-chain carbohydrates which exhibit manifold variations in the monosaccharide structures and linkage types. Often they are found covalently linked with other classes of biomolecules to make glycoconjugates like glycoproteins and glycolipids (Rao 1998). Oligosaccharide chains attached to proteins can give substantial mass, and hence they play pivotal roles in governing the overall structure and function of the glycoproteins. The glycans of glycoproteins vary widely in their structure, and hence their functional roles in biological phenomena are indispensable. Oligosaccharides tend to exist in multiple conformational states in biological environment. Human milk contains a high concentration of diversely soluble oligosaccharides which are similar to protein concentration and exceeds the lipids content (Bode 2006).

The building blocks of human milk oligosaccharides (HMO) are the five monosaccharides, namely, D-glucose (Glc), D-galactose (Gal), N-acetylglucosamine (GlcNAc), L-fucose (Fuc), and N-acetylneuraminic acid (Neu5Ac). Sialyl-Lewis X tetrasaccharide plays crucial role in the adhesion of leukocytes and neutrophils to vascular endothelial cells which is associated with normal and pathogenic inflammatory processes (Revelle et al. 1996). The recognition of oligosaccharides by proteins and other biomolecules is a crucial phenomenon since it is involved in many biological processes including cell signaling, cell-cell interaction, and many more. The conformation of oligosaccharides may vary in both their bound and free states (Rao 1998). The overall flexibility of the oligosaccharide structure is dictated by the flexibility around the glycosidic linkages between the constituent monosaccharides. Because of the presence of many interacting groups, oligosaccharides can adopt branched structure. Figure 4.8 shows the structure of two unbranched oligosaccharides.

4.5 Polysaccharides

Polysaccharides are polymers of saccharides which contain repeating units of monosaccharides or more than one type of saccharides. When the constituting monosaccharides are of the same species, the polymer is called homopolysaccharide, and when the monosaccharides differ, the polymer is called heteropolysaccharide (Rao 1998). Cellulose is one of the polymer of D-glucose units which are $\beta(1 \rightarrow 4)$ linked. It is the major constituent of plant cell walls (Fig. 4.9). The polymeric shapes of cellulose are like a twisted ribbon and are stabilized by intra- and interchain hydrogen-bonding interactions between the sugar residues. A few other polysaccharides to mention are starch, chitin, glycogen, xylan, amylose, guar gum and xanthan gum. Polysaccharides show interesting physical properties, and they are extensively used in industrial applications such as paints, mineral suspensions, fertilizers, and pigments.

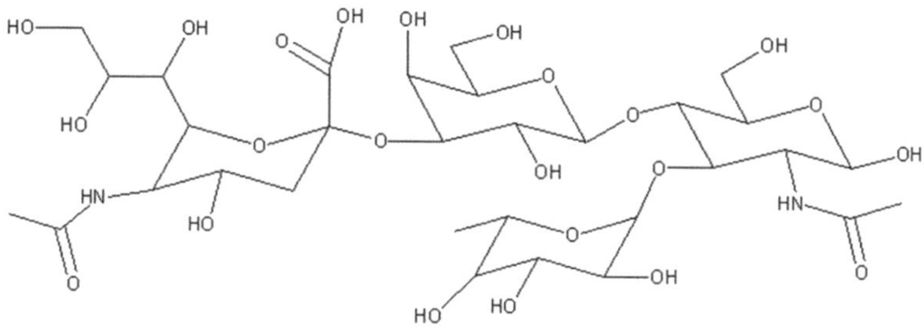

(a)

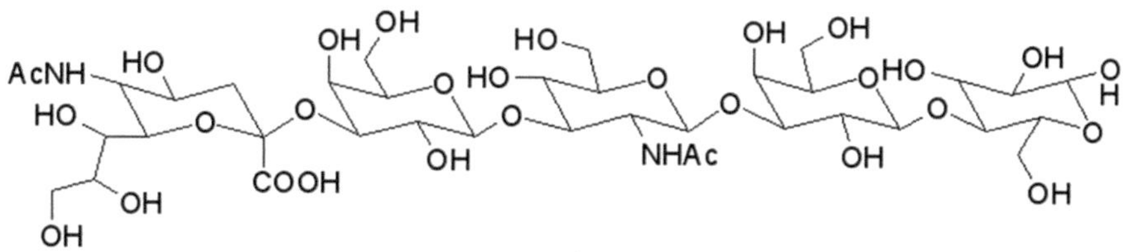

(b)

Fig. 4.8 Structure of oligosaccharides. (**a**) Sialyl-Lewis X, a blood group antigen tetrasaccharide. (**b**) Sialyl N-tetrose, a human milk pentasaccharide

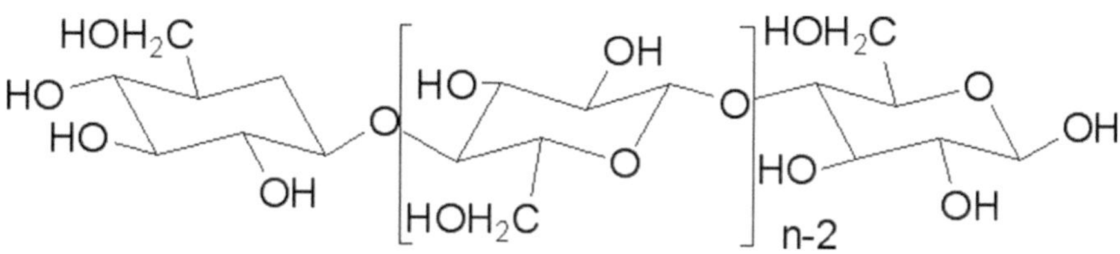

Fig. 4.9 Structure of cellulose, a polysaccharide with β(1→4) linked D-glucose units

4.6 Glycoconjugates

Apart from their existence as glycans, carbohydrates can coexist with other classes of biomolecules to provide structural stability and contribute to their biological functions. Glycans or oligosaccharide chains are attached to proteins and lipids to form glycoconjugates. Glycoconjugates can be categorized into categories such as glycosaminoglycans, glycoproteins, glycopeptides, peptidoglycans, glycolipids, and lipopolysaccharides. Due to their exceptionally multitude structural and functional features, glycoconjugates are regarded as crucial molecules in a vast range of biological processes that include cell-cell recognition and cell-matrix interactions.

4.6.1 Glycosaminoglycans

Glycosaminoglycans (GAG) are long unbranched polymers which form the major portion of the extracellular matrices. GAGs are made up of repeating disaccharide units primarily of either *N*-acetylglucosamine (GlcNAc) or *N*- acetylgalactosamine (GalNAc) and an uronic acid or a hexose. GAGs are usually negatively charged molecules adopting an extended conformation and are linked with the core proteins in extracellular matrix. Hyaluronic acid is the only GAG which is not linked to core proteins, and it exists as pure polysaccharide, and its structure has been reported (Atkins et al. 1974; Cael et al. 1976). Hyaluronic acid, chondroitin sulfate, dermatan sulfate, heparin, heparin sulfate, and keratin sulfate are some GAGs of physiological significance (Lonngren 1989). The three-dimensional structure of the GAGs are crucial in recognition events, and the three-dimensional structure and conformation of GAGs, especially the heparin and heparin sulfate, had been studied using spectroscopic methods and molecular mechanics calculations (Mulloy and Forster 2000).

4.6.2 Glycoproteins

Oligosaccharide units are covalently bonded to the polypeptide side chains of proteins to form glycoproteins. The carbohydrate moiety of the glycoprotein can contribute to the mass of the glycoprotein from less than 1% and up to 80%. Glycoproteins are regarded for their extremely diverse glycans structures and functions. Glycoproteins exhibit structural functions such as constituents of cell walls and connective tissues. They act as protective barrier and lubricant. They are also present in large amount in the blood plasma and play in numerable roles in regulation of biochemical processes. *Glycosylation* is the most common and essential co- and posttranslational modification found in most of the eukaryotic proteins, to which a carbohydrate moiety is conjugated to the side chains of particular amino acid residues. In the glycosylation process, the carbohydrates are linked N-glycosidically to asparagine or O-glycosidically to serine/threonine of the constituent protein (Lis and Sharon 1993).

4.7 Glycosylation and Protein-Carbohydrate Interactions

Understanding the phenomenon of protein-carbohydrate interactions at the molecular level is of key interest to glycobiologists, bioinformatists, biotechnologists, and biophysicists. Nature uses carbohydrate-protein interactions in order to mediate the multiplex functions of the biological system (Bundle and Young 1992; Toone 1994; Crocker and Feizi 1996; Varki 1998). The array of such functions includes cell adhesion, signal transduction, cell recognition, immune responses, growth control, and apoptosis (Roy 1996; Lis and Sharon 1998; Simanek et al. 1998). In carbohydrate-protein interactions, the hydrogen bonds play a dominant role in stabilizing the structures (Vyas et al. 1991; Christlet et al. 1999). In addition to this, aromatic side chain and saccharide stacking as well as hydrophobic interactions tend to stabilize the complexes (Quiocho 1989). In several carbohydrate-protein complexes, water molecule plays a crucial role (Bourne and Cambillau 1993).

Carbohydrate-protein interactions play significant role in the formation of carbohydrate-protein complexes and also influence the functional properties of glycoproteins. Both the structure and dynamics of glycoproteins can be analyzed by means of NMR spectroscopy (Imberty and Pérez 1995; Wyss et al. 1995). The carbohydrate moiety linked to the polypeptide induces local conformational changes in the polypeptide (Live et al. 1999; Wu et al. 1999). Studies on glycoproteins will provide useful information and understanding on targeting drugs and enzyme replacement therapy for genetic disorders.

In glycoproteins, the attached oligosaccharide chains to a single protein vary from 1 to 30 or more, and the sugar chains possess one or two residues to much larger structures. Most of the mammalian plasma proteins are glycosylated except serum albumin. Most of the membrane-bound proteins and cell surface receptors of the eukaryotes are glycosylated. The covalent bonding of carbohydrates to proteins affects the protein structure, function, and local conformation (O'Connor and Imperiali 1996). The precise sequence of the glycans is more important in determining its function and the site of attachment of glycan to the proteins (Gagneux and Varki 1999).

4.8 Bioinformatics of Glycosylation

4.8.1 N-linked Glycosylation

N-glycosylation can be called as a co-translational event where the carbohydrate moiety N-acetylglucosamine (GlcNAc) gets conjugated to asparagine residue (Fig. 4.10) as soon as the polypeptide chain is synthesized. For N-glycosylation, there exists a consensus sequence motif Asn-X-Ser/Thr, where X can be any amino acid except proline (Hunt and Dayhoff 1970; Bause and Legler 1981).

There exist only a small percentage of sites with Asn-X-Ser/Thr triplet sequences as glycosylated ones in glycoproteins (Hunt and Dayhoff 1970), and all the consensus sequence is not glycosylated. Hence, it is not requisite condition for N-glycosylation. The schematic representation of the Asn-X-Ser/Thr peptide fragment along with the dihedral angles Ψ^{N}, Φ^{X}, Ψ^{X}, and $\Phi^{S/T}$ which fix the mutual orientation of the side chains of Asn and Ser/Thr is given in Fig. 4.11. Irrespective of the peptide conformational preference, N-glycosylation might be expected to alter the structure of the glycoprotein (Bause and Legler 1981; Imperiali 1997). Upon surveying the stereochemistry of 44 sites, Imberty and Perez concluded that 25% of N-glycosylation sites are found to occur on β-turns (Imberty and Pérez 1995).

The frequency of occurrence of amino acid residues at position *X* of the consensus Asn-*X*-Ser/Thr motif is carried out on 696 consensus glycosylating

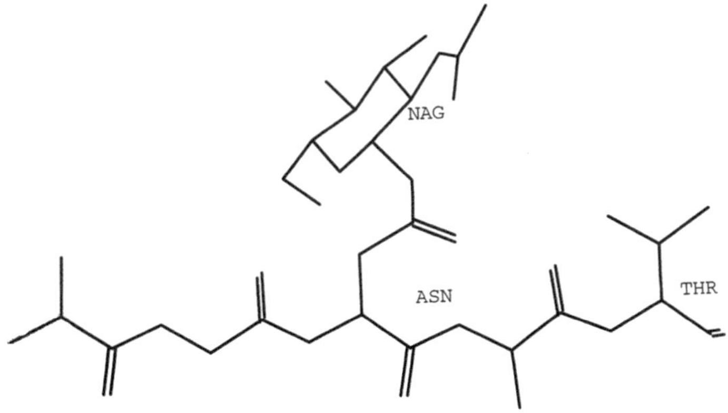

Fig. 4.10 Schematic representation of N-glycosylation

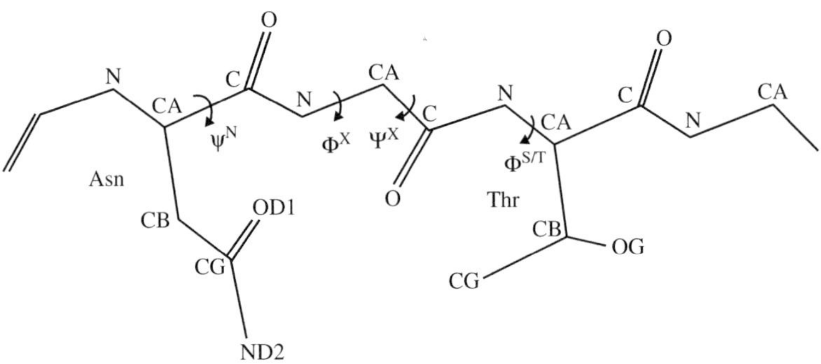

Fig. 4.11 The schematic representation of the Asn-X-Ser/Thr peptide

sequence from 488 nonhomologous proteins from Protein Data Bank (Christlet et al. 1999). It is observed that nearly 65% of Asn residues are found to lie on the surface of the protein, and it indicates their potential to get glycosylated. A deviation parameter (DP) is defined as a measure of preference (positive) or non-preference (negative).

$$\mathrm{DP}(A) = 100[P_{\mathrm{observed}}(A) - P_{\mathrm{expected}}(A)]/P_{\mathrm{expected}}(A)$$

$$P_{\mathrm{expected}}(A) = \sum N_i(A)/\sum T_i$$

$$P_{\mathrm{observed}}(A) = N_X(A)/m$$

The amino acid residues glycine, asparagine, and phenylalanine have statistically significant positive DP values at the *X* position of the consensus motif. The high preference value for asparagine at the *X* position reveals the occurrence of homodoublets, while the high preference of phenylalanine might be due to the

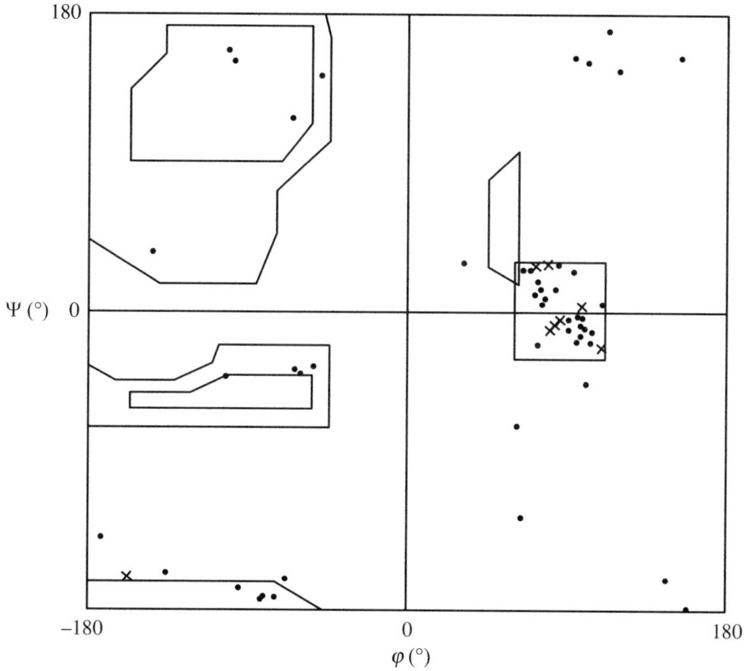

Fig. 4.12 Ramachandran plot for glycine in the consensus sequence (clustered region)

stacking interaction of its aromatic ring with the glycan. Also, the amino acid glycine has high preference at the *X* position of the consensus glycosylating sequence, and it may be functionally significant because of its preference and percentage of occurrence in proteins.

The conformational analysis of the Ramachandran angles (Φ^G, Ψ^G) are calculated for the available 52 glycine-containing consensus sequences. These glycine residues are found to occur in the conformational space which is disallowed for non-glycyl residues (Fig. 4.12). There occurs a clustered region around $\Phi^G \approx +60°$ to $+110°$ and $\Psi^G \approx -30°$ to $+30°$. Of which, nearly 50% of the glycine residues of the consensus sequences are found on the surface of the protein. A conformational analysis on the eight confirmed Gly-containing glycosylated sequences (on the surface) revealed that seven of the (Φ^G, Ψ^G) angles lie in this region. Similarly, analysis on the 12 available Gly-containing sequences from the interior of the protein indicates that only three (25%) of the (Φ^G, Ψ^G) angles to be in this region.

Of the 52 available Asn-Gly-Ser/Thr sequences, there is a possibility of hydrogen bond between OD1 of Asn with OG (Ser/Thr) or OD1 (Asn) with NH (Ser/Thr) (Fig. 4.13) or ND2 (Asn) and OG (Ser/Thr). It is clear that the backbone conformation of glycine tends to occur in the specified clustered region whenever the hydrogen-bonding interactions prevail. In few of the consensus sequences, the angles (Φ^G, Ψ^G) fall in the clustered region, and there involves no hydrogen bond formation with the backbone conformation Ψ^N around $25 \pm 30°$.

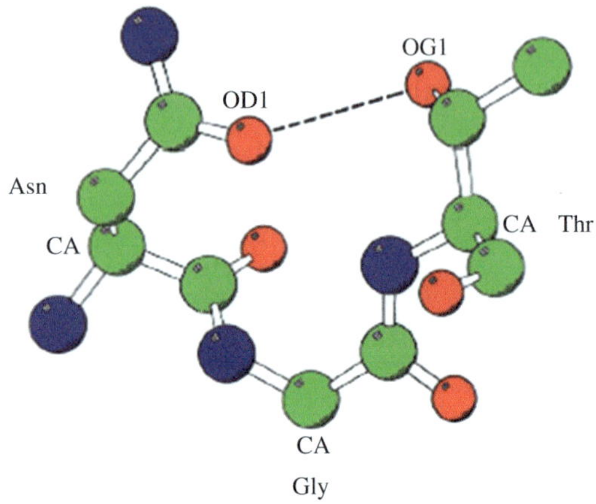

Fig. 4.13 Illustration of the direct hydrogen bond between OD1 and OG

In almost all sequences, irrespective of what amino acid is at position X, this study shows the existence of direct or water-mediated hydrogen bond (Fig. 4.14) between the side chain of Asn and Ser/Thr plays a dominant role in the process of glycosylation in addition to the backbone conformation. This study also shed interests in direct or water-mediated hydrogen bonds of glycoproteins, which can be exploited by structural biologists, glycobiologists, and protein engineers.

4.8.2 O-Glycosylation

O-Glycosylation is a posttranslational modification of protein, where a carbohydrate moiety is covalently linked to the hydroxyl group of the hydroxy amino acids, serine, or threonine (Fig. 4.15) after protein folding. Unlike N- glycosylation where there exists a (Asn-X-Ser/Thr) consensus sequence, there is no consensus motif identified for O-glycosylation (Hansen et al. 1995; Christlet and Veluraja 2001).

It is found that the amino acid proline is found around the sites of glycosylation with considerable frequency (O'Connell et al. 1991; Hansen et al. 1995; Yoshida et al. 1997; Christlet and Veluraja 2001). A statistical analysis proposed that the number of proline, serine, and threonine residues is apparently high around the glycosylation site (Wilson et al. 1991). There are several examples of O-glycosylated sites which lack proline within the surrounding flanking region (Rall et al. 1982; Gooley et al. 1991). Some reports have stated that peptides with serine residues were not glycosylated in vitro (Hughes et al. 1988; Yoshida et al. 1997).

In the work by Christlet and Veluraja (Bause and Legler 1981; Christlet and Veluraja 2001), statistical analysis is carried out to explore the sequential aspects of O-glycosylation around the sites of Ser/Thr amino acids. From the 992 sequences of O-GLYCBASE database, the frequency of occurrence of the amino acids around the O-glycosylated Ser/Thr is calculated. It resulted in the frequency of occurrence of proline, serine, threonine, alanine, glycine, and valine residues to be high around

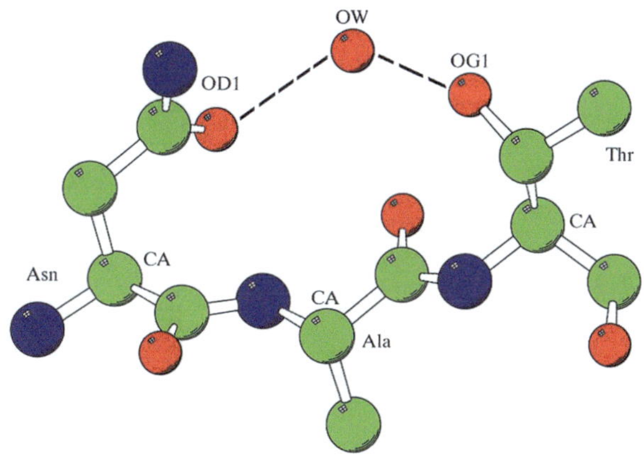

Fig. 4.14 Water-mediated hydrogen bond between side chain of Asn and Ser/Thr

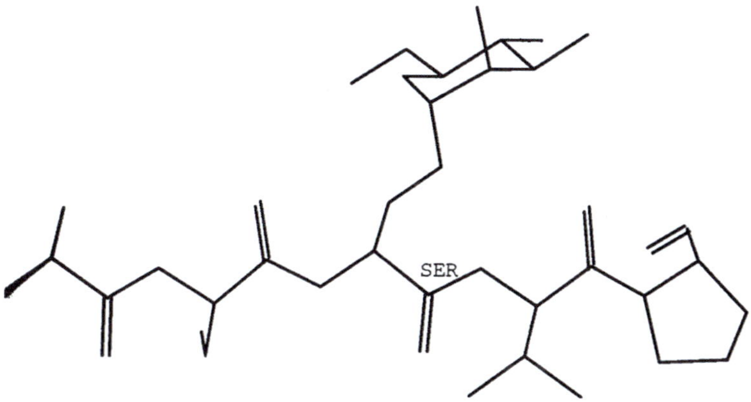

Fig. 4.15 Schematic representation of O-glycosylation

the sites of O-glycosylation (Fig. 4.16). Additionally the frequency of occurrence of proline, serine, and threonine residues are found moderate in the dataset containing non-glycosylated sequences. Also the computed values of deviation parameter (DP) show that the amino acid proline has the maximum preference around the *O*-glycosylation site. Irrespective of the number of glycosylation sites, the presence of amino acid proline at the +3 position and/or −1 position strongly favors glycosylation. In addition, serine and threonine are preferred around the multiple glycosylation sites due to the effect of clusters of closely spaced glycosylated Ser/Thr. Aromatic amino acids, cysteine, and amino acids with bulky side chains inhibit *O*-glycosylation.

In order to emphasize the profound preference of proline around the *O*-glycosylation sites, the DP values are calculated for various window sizes ranging from 7 to 21, and the results are plotted in Fig. 4.17a–d. From the figures, it is clear that the presence of proline enhances threonine glycosylation rather than serine glycosylation. Also the presence of proline at −2 and +2 positions favors glycosylation

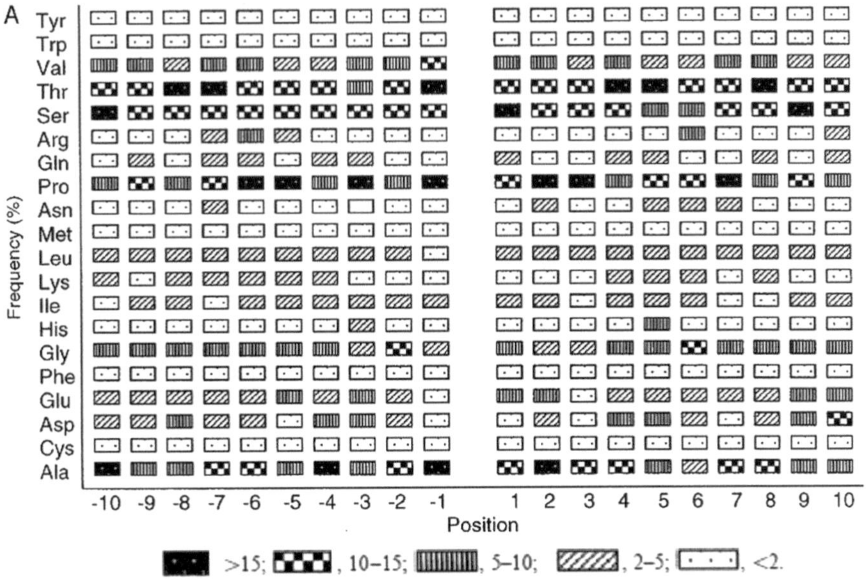

Fig. 4.16 Frequency of occurrence of amino acids around glycosylated Ser/Thr. The frequency of each amino acid is expressed as a percentage in this plot for positions −10 to 10. The shading at each location indicates the percentage frequency of a residue >15; , 10–15; , 5–10; , 2–5; , <2.

of serine. There is an abundance of proline residues at −1 and +3 positions, which is not altered even though the window size is changed. Significantly, proline has smaller DP values when positioned farther away from the glycosylation site, i.e., the preference is reduced beyond ±3 positions. Also, proline is not preferred around the non-glycosylated Ser/Thr, and thus it provides further evidence for the high significance of proline near glycosylated Ser/Thr.

Sequence motifs tend to possess proline at the +3 position repeatedly. During the analysis it is observed that there are 287 sequences which carry proline at the +3 position. Of these, about 16% of the sequences contain *Thr-Ala-Pro-Pro* fragment, and it is found to be glycosylated. Similarly, the fragment of sequences Ser-Ala-Pro-Pro is identified, but they are found unglycosylated. No defined sequence fragments are identified in case of non-glycosylated proteins. Also, few motifs are found to exist for the proline containing (at +3) sequence motifs, and they are as follows. Thr-Val-X-Pro, Ser/Thr-Pro-X-Pro, and Thr-Ser-Ala-Pro are preferred favorably (8%, 15%, and 15%), where X can be any amino acid.

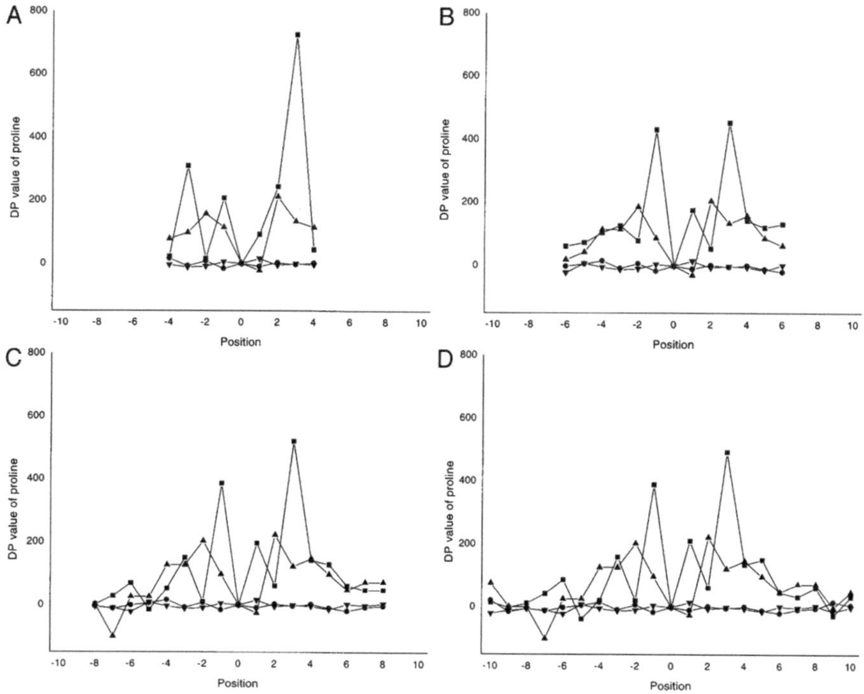

Fig. 4.17 (**a–d**) Positional preference of proline around the *O*-glycosylation site with different window sizes. ■ DP value of proline around glycosylated threonine; • DP value of proline around non-glycosylated threonine; ▲ DP value of proline around glycosylated serine; ▼DP value of proline around non-glycosylated serine

4.9 Glycolipids

Glycolipids and glycoconjugates are family of lipid molecules that contains oligosaccharide chains attached to ceramides. The ceramide molecule is bound tightly with the lipid bilayer of the cell, and the oligosaccharide portion is protruding out of the cell surface. The main function of glycolipids is to serve as markers for cell recognition processes and as a source of energy. The structure and conformation of the oligosaccharide chains present in glycolipids are extensively used in the specific recognition of sugars by pathogens (Rao 1998). Glycolipids can broadly be categorized into three types, galactolipids, glycosphingolipids, and glycosylphosphatidylinositols. Galactolipids contain galactose as their sugar moiety and are mostly found in membrane lipids in plants. They are distinguished from the glycosphingolipids by the absence of nitrogen. Glycosphingolipids are sphingosine containing glycolipids which can be divided into cerebrosides, globosides, and gangliosides. Gangliosides are sphingolipids which feature the presence of sialic acids, a versatile sugar molecule which is involved in molecular recognition processes. Cerebrosides have a single sugar molecule attached to ceramide, and globosides contain more than one neutral sugar molecule attached to the ceramide. The three-dimensional structure of glycolipids had been studied using theoretical methods (Pascher and Sundell 1977; Jarrell et al. 1987; Poppe et al. 1992).

4.10 Peptidoglycans

Peptidoglycans are linear chains of oligosaccharides interconnected by peptides. The oligosaccharide chain consists of β(1, 4) linked *N*-acetylglucosamine and *N*-acetylmuramic acid. The amino acid chains of peptidoglycans can interact with each other, and this leads to the mesh-like structure of peptidoglycans. Presence of peptidoglycans in cell wall gives structural strength and also supports the cell wall to withstand the osmotic pressure of the cytoplasm (Tipper 1970; Moir and Smith 1990). Hence essentially they are the important constituents of the cell walls in bacteria. The structure and conformation of peptidoglycans in biological environments had been investigated using experimental and theoretical methods (Kelemen and Rogers 1971; Klaić and Domenick 1990).

4.11 Experimental Structure Determination of Carbohydrates

In a way to understand the molecular mechanisms of carbohydrate-mediated recognition processes, it is an essence to investigate and understand the structural flexibility and conformational dynamics of oligosaccharides in solution. In general, nuclear magnetic resonance (NMR) spectroscopy and X-ray crystallography are the foremost experimental techniques that are used in the structural determination of the biomolecules. Structure determination of carbohydrates using these methods is challenged by the inherent nature of carbohydrates that they are highly flexible at the glycosidic linkages. Crystallization of carbohydrates is too difficult, and many of the carbohydrate chains in glycoproteins are removed from the proteins before crystallization of protein because of the fact that the presence of carbohydrates may hinder the growth of the protein crystal itself. Since the glycosidic linkages are highly flexible, oligosaccharide chains present themselves in more than one conformational state in solution. With the limited freedom of experimental carbohydrate structure determination, efforts were taken to obtain three-dimensional structural data of carbohydrates, and these data are insufficient to define and explain the varied conformations of glycans in aqueous environment. Carbohydrate structure is greatly influenced by the type of the glycosidic linkages, the size and functional groups of monosaccharides involved, and the conformational freedom around the glycosidic linkages. Experimental methods like NMR and X-ray crystallography cannot provide the complete details about the oligosaccharide chains, and hence there is always scarcity of three-dimensional structural data of carbohydrates through experimental methods (Lemieux and Koto 1974; Pérez and Marchessault 1978; Yan and Bush 1990; Carver 1991; Rice et al. 1993; Van

Halbeek 1994; Woods 1995, 1998; Peters and Pinto 1996; Imberty 1997; Imberty and Pérez 2000; Karplus and McCammon 2002).

4.12 Role of Theoretical Methods in Carbohydrate Structure Prediction

The structure of oligosaccharides cannot readily be determined by experimental methods because of the reasons given earlier. Hence one has to depend on theoretical methods in the structural determination of carbohydrates. In most of the time, the experimental structural data frequently needs to be supported by theoretical calculations. The foremost computational technique that is used in the biomolecular structure determination is molecular dynamics simulations. Other computational methods include molecular mechanics calculations, quantum mechanical calculations, hard sphere energy calculations, and Monte Carlo simulations. Molecular dynamics simulations are computational methods which predict the evolution and behavior of a given system over a time. The dynamics and structural preferences of a biomolecule can also be studied using MD simulations performed over nanosecond scales. Even though the use of MD simulations dates back to1957 (Alder and Wainwright 1957), the first biomolecular simulation is reported in 1977 on bovine pancreatic trypsin inhibitor (BPTI) (McCammon et al. 1977). The subsequent years witnessed evolution of an era in MD simulations and the development of force fields for biomolecular simulations. However the carbohydrate theoretical study started with the report from Brady in1986 on the MD simulation of alpha-D-glucose in gas phase (Brady 1986). From then, the use of MD simulations in the conformational analysis of carbohydrates is constantly growing. Force fields for simulating carbohydrates are available now, and the quality of the force fields is continuously improved with new parameterizations using advanced methods. Molecular dynamics simulations are used in finding the conformational preferences of mono-, di-, oligo-, and polysaccharides in solution as the aqueous solutions can mimic the biological environment (Veluraja and Rao 1983, 1984; Cumming and Carver 1987; Brady 1991; Brocca et al. 2000; von der Lieth et al. 2002; von der Lieth et al. 2009).

4.12.1 Conformational Analysis of Sialic Acid and Sialylglycans: An Example of Theoretical Study of Three-Dimensional Structures of Carbohydrates

Sialic acids are the class of acid sugars with nine-carbon atom which consists of a family of more than 40 naturally occurring derivatives of N-acetylneuraminic acid (Neu5Ac), N-glycolylneuraminic acid (Neu5Gc), and keto deamino neuraminic acid (KDN). All the cell surface carbohydrates invariably contain Neu5Ac in order to mediate recognition processes. To gain insights into the molecular recognition processes, knowledge about the structure and conformation of Neu5Ac is essential. The conformational analysis of sialic acid and sialylglycans is an example of conformational analysis of carbohydrates through molecular dynamics simulations. Even though crystal structure of sialic acid ester is available through X-ray diffraction and geometry through NMR technique, the accurate structure and geometry have not been attempted through theoretical methods. The initial study of the conformation of Neu5Ac was carried out by Veluraja and Rao using hard sphere calculations in 1980 (Veluraja and Rao 1980). The results were consistent with the earlier X-ray and NMR results (Brown et al. 1975; Czarniecki and Thornton 1977). Despite its importance, almost no attempt had been made on the structure and conformation of Neu5Ac except the one by Sawada in 2006 in which sialic acid is treated as a neutral one irrespective of its negative charge (Sawada et al. 2006). Very recently, the structure and conformation of Neu5Ac has been studied by Priyadarzini et al. using molecular dynamics simulation combined with quantum mechanical calculations (Priyadarzini et al. 2012). Molecular dynamics simulations showed the atomic level interactions and the various hydrogen-bonding schemes which stabilize the structure in solution. A further geometry refinement using quantum mechanical calculations resulted in exploring the preference of conformation by considering the amphiprotic nature of water.

Sialylglycans constitute an important class of glycans and has been of greater interest in many conformational studies. The initial study on the conformation of sialyldisaccharides was carried out by Veluraja and Rao using molecular mechanics method (Veluraja and Rao 1984a, b). Later Xavier and Veluraja carried out molecular dynamics simulations of 1 ns duration on four important sialyldisaccharides and found out the conformational preferences (Suresh and Veluraja 2003). The conformational preferences of sialyldisaccharides are reprised by the 10 ns molecular dynamics simulations carried out by the same group (Veluraja et al. 2010) revealed the conformational preferences and water-mediated hydrogen-bonding interaction schemes which stabilize the distinct conformational states. A recent molecular dynamics study on the sialyldisaccharides for the extended duration of 20 ns has given insights into the conformational preferences of the sialyldisaccharides in terms of the water mediation, occupancy of water, and relative energies of conformational states [results not published yet].

Sialyloligosaccharides are also studied using molecular dynamics simulations for their solution conformations (Veluraja et al. 2001; Veluraja and Margulis 2005; Veluraja and Seethalakshmi 2008). The oligosaccharide portions of gangliosides are studied using 10 ns MD simulations in order to understand their structure-function relationship. Almost all the simulation results complement the NMR of X-ray crystallography results of conformations of oligosaccharides. The conformational study on monosialo-, disialo-, and higher gangliosides had been carried out by using MD simulations (Vasudevan and Balaji 2001, 2002; Sharmila and Veluraja 2004a, b, 2006). The interaction patterns obtained from the MD simulations can throw light on the specific recognition processes which exhibit preferences over conformations, and hence knowing the three-dimensional structure of gangliosides would be of immense use in the design of inhibitors.

4.13 Glycoinformatics: Database Analysis and Development of Databases

On comparison with the other classes of biomolecules, the data mining work executed on carbohydrates is very less. A statistical analysis has been reported on the available X-ray diffraction data for the conformations of oligosaccharide around its glycosidic linkages (Petrescu et al. 1999; Wormald et al. 2002). The available oligosaccharides in the complex carbohydrate structural database (CCSD) were analyzed to find out their preference of occurrence (Berteau and Stenutz 2004).

The number of databases and the amount of structural data available for carbohydrates lag far behind the data available for proteins. Since there are too many limitations for the experimental methods and the theoretical methods consume more computational power, the outcome of three-dimensional structural data for carbohydrates is also very less. Few initiatives are taken to explain the nature and diversity of sugars, their sequential, and spatial structures (Berteau and Stenutz 2004). Complex carbohydrate structural database (CCSD), known as CarbBank, was the first data base in glycobiology. It was a sequential database for carbohydrates which existed until 1997 (Doubet et al. 1989). SugaBase is a database for carbohydrates which contains NMR-derived data with proton and carbon chemical shift values and contains information combined with CCSD (van Kuik et al. 1992). GlycomeDB is an integrated open-access database for carbohydrate (www.glycome-db.org) (Ranzinger et al. 2008), and it is a meta-database. W3-SWEET is an online tool to which chemical sequence information can be given to get a reliable three-dimensional model (Bohne et al. 1998) for the desired molecule. However, W3-SWEET generates only one conformation out of a manifold. GLYCAN is a collection of experimentally determined glycan structures available from the literature (Hashimoto et al. 2006). Database existing for sugar-binding proteins is CAZy-carbohydrate-active enzymes. A number of web-based tools for analyzing the glycan data are developed in the recent past and are available online (Marchal et al. 2003; Lütteke et al. 2004, 2006).

But quite a few databases are available for the three-dimensional structure of glycans. The center for study of vegetal macromolecules has remarkable information on the study of biomolecular structure. Three-dimensional structural information on monosaccharides, disaccharides, oligosaccharides, polysaccharides, lectins, glycosyltransferases, and GAG-binding proteins are collected and organized into a database called Glyco-3D and is available at http://glyco3d.cermav.cnrs.fr/glyco3d/ (Pérez and Mulloy 2005). GLYCAM is a carbohydrate database which can generate a single conformation for a given input carbohydrate sequence (Woods Group 2005–2010).

However not even a single database gives the information about the solution conformations of a given sugar molecule. Since the complexities associated with deriving and understanding conformations of carbohydrates is high, no attempt has been made to integrate the three-dimensional structural information into a single database.

4.14 Development of 3DSDSCAR: Carbohydrate Three-Dimensional Structural Database

Recently a three-dimensional structure database for sialic acid-containing carbohydrates (3DSDSCAR) has been developed and is available at the URL http://www.3dsdscar.org (Veluraja et al. 2010). The database of 3DSDSCAR contains all possible conformations of a given oligosaccharide which are deduced using molecular dynamics simulations. To the best of our knowledge, 3DSDSCAR is the foremost three-dimensional structural database for carbohydrates which contain all possible conformers for a given glycan generated through analysis of simulation trajectories. Currently 3DSDSCAR database contains 89 conformational models belonging to 24 different biologically important glycans. The Cartesian coordinates are given for all the conformational models, and these structural models can be used in the study of carbohydrate modeling, protein-carbohydrate interactions, and carbohydrate-based therapeutics. The number of times this database accessed is 46,900 as on August 2016.

4.14.1 Typical Data File in 3DSDSCAR

The carbohydrate three-dimensional data file has been formatted to resemble the data file widely used for proteins (PDB file). The file contains annotation section and coordinates section. The information about sequential structure, methodology adopted, and results of the theoretical calculations are given in annotation section. The coordinate section contains the Cartesian coordinates (x,y,z) of all the atoms of the given molecule.

```
HEADER   GANGLIOSIDES
TITLE    GM3 - MONOSIALO GANGLIOSIDE
MOLEID   GM3
SEQSTR
SEQSTR       3          2        1
SEQSTR    Neu5Ac(2-3)Galβ(1-4)Glc-Cer
SEQSTR
SEQSTR              (OR)
SEQSTR         3                2              1
SEQSTR   α-D-Neu5Acp-(2→3)-β-D-Galp-(1→4)-β-D-Glcp-(1→1)-Cer
SEQSTR
SEQSTR   RESIDUE 1 - Glc         Glc         -   Glucose
SEQSTR   RESIDUE 2 - Gal         Gal         -   Galactose
SEQSTR   RESIDUE 3 - Neu5Ac      Neu5Ac      -   N-Acetylneuraminic Acid
SEQSTR                           Cer         -   Ceramide
GLYTOR
GLYTOR       3          2        1
GLYTOR         (φ2,ψ2)   -  (φ1,ψ1)
GLYTOR
AUTHOR   K. VELURAJA AND T. R. K. PRIYADARZINI
AUTHOR   DEPARTMENT OF PHYSICS,MANONMANIAM SUNDARANAR UNIVERSITY
AUTHOR   TIRUNELVELI - 627012, TAMILNADU, INDIA.
AUTHOR   EMAIL: kvraja@sancharnet.in
METHOD   MOLECULAR DYNAMICS SIMULATION
METHOD   SIMULATION DETAILS
METHOD   FORCE FIELDS USED
METHOD   GAFF AND FF99 - AMBER9
METHOD   MOLECULAR DYMANICS SOFTWARE USED
METHOD   NAMD (NANOSCALE MOLECULAR DYNAMICS)
METHOD   VISUALISATION SOFTWARE USED
METHOD   VMD (VISUAL MOLECULAR DYNAMICS)
METHOD   DURATION OF THE SIMULATION : 10 NANO SECONDS
METHOD   TIME STEP : 1 FEMTO SECOND
METHOD   TRAJECTORIES COLLECTED OVER : 1 PICO SECOND
METHOD   SIMULATION TEMPERATURE : 300 K
METHOD   NUMBER OF SUGAR RESIDUES : 03
METHOD   TOTAL NUMBER OF ATOMS : 80
METHOD   TOTAL NUMBER OF WATER MOLECULES INVOLVED IN THE SIMULATION:920
MDCONF   RESULTS OF SIMULATION
MDCONF   NUMBER OF CONFORMATIONS OBTAINED = 3
MDCONF   DIHEDRAL VALUES OF CONFORMERS AROUND THE GLYCOSIDIC LINKAGES
MDCONF   FOR EACH CONFORMERS
MDCONF   CONFORMER       (φ1,ψ1)  (φ2,ψ2)   RELATIVE ENERGY (kcal/mol)
MDCONF   CONFORMER I     (80,0)  (-100,-50)        0.00
MDCONF   CONFORMER II    (80,0)  (-70,0)           3.50
MDCONF   CONFORMER III (80,0)  (-150,-30)          3.80
MDCONF   THE  GIVEN COORDINATE FILE IS = CONFORMER I
REMARK   THE MOLECULE CAN BE VISUALISED BY USING ANY AVAILABLE MOLECULAR
REMARK   VIEWER WHICH CAN USE PDB FORMAT.
REMARK   THE COORDINATES ARE GENERATED IN AN ORBITRARY FRAME OF REFERENCE
REMARK   FORMAT(A4,3X,I4,2X,A4,A3,2X,I4,4X,3F8.3)
ATOM       1  C1  GLC      1        9.326  -2.557   5.795
ATOM       2  C2  GLC      1        9.320  -2.666   4.276
ATOM       3  C3  GLC      1        8.257  -1.757   3.679
.....................................................
.....................................................
END
```

4.14.2 Screenshots of 3DSDSCAR

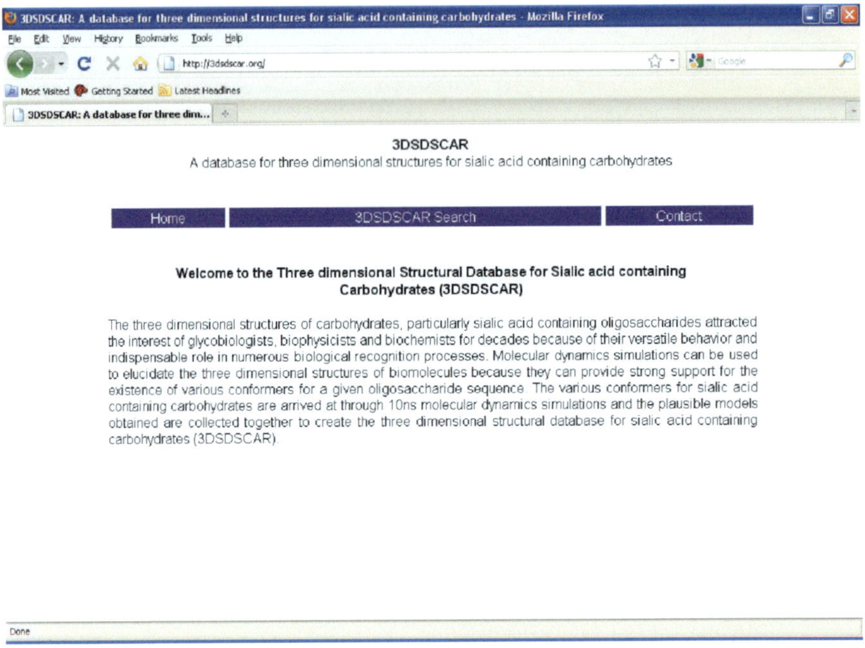

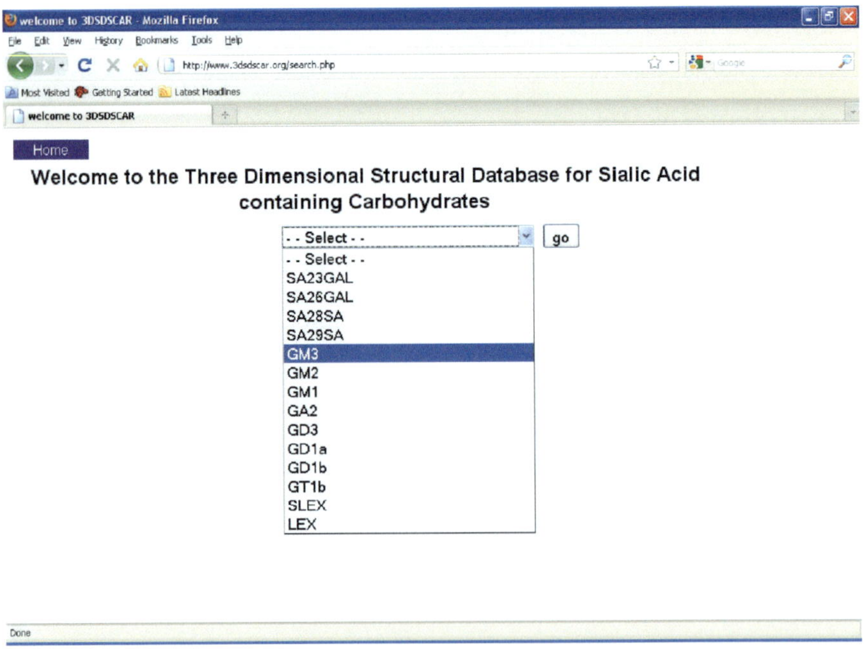

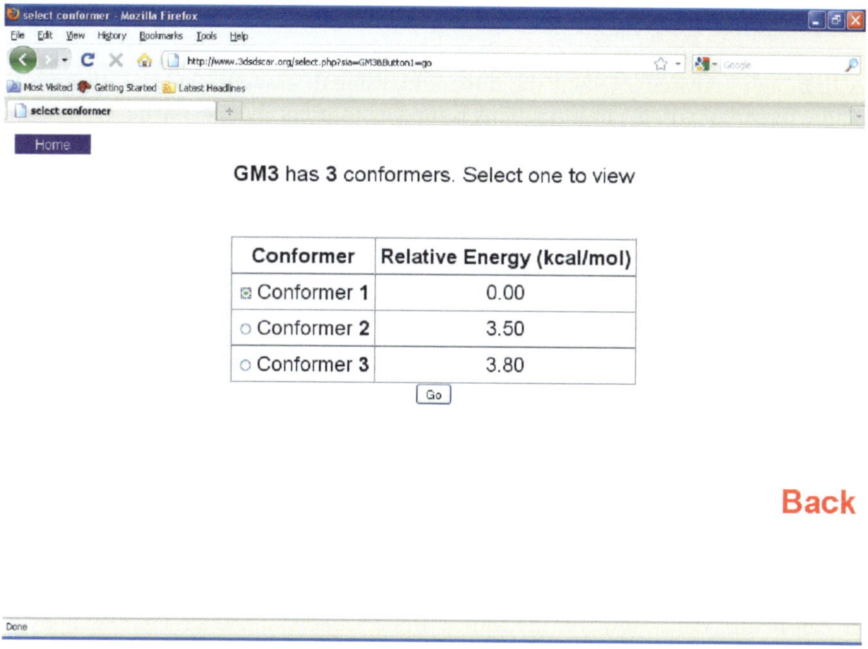
select conformer - Mozilla Firefox
File Edit View History Bookmarks Tools Help
http://www.3dsdscar.org/select.php?sia=GM3&Button1=go
Most Visited Getting Started Latest Headlines
select conformer
Home
GM3 has 3 conformers. Select one to view
Conformer
Relative Energy (kcal/mol)
Conformer 1
0.00
Conformer 2
3.50
Conformer 3
3.80
Go
Back
Done

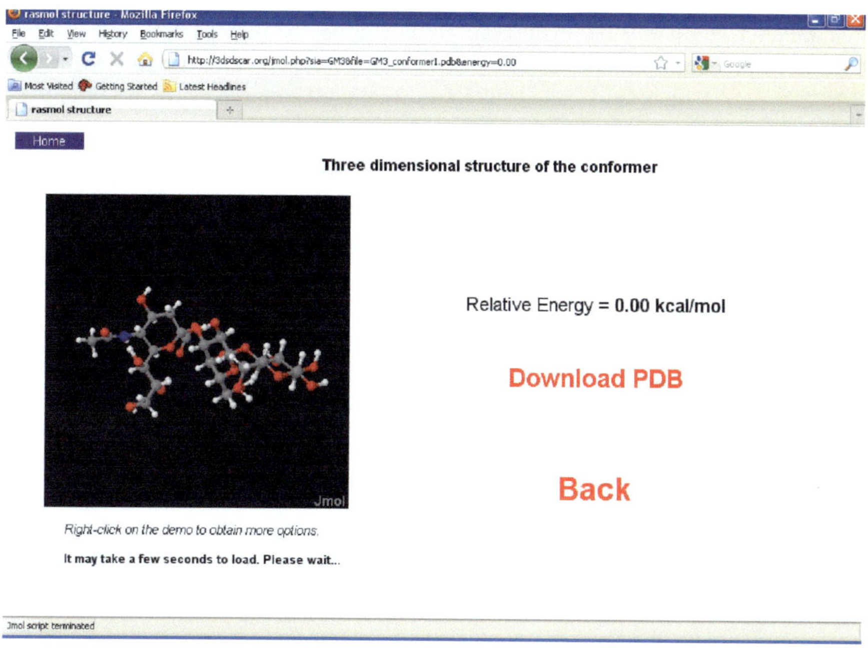
rasmol structure - Mozilla Firefox
File Edit View History Bookmarks Tools Help
http://3dsdscar.org/jmol.php?sia=GM3&file=GM3_conformer1.pdb&energy=0.00
Most Visited Getting Started Latest Headlines
rasmol structure
Home
Three dimensional structure of the conformer
Jmol
Relative Energy = 0.00 kcal/mol
Download PDB
Back
Right-click on the demo to obtain more options.
It may take a few seconds to load. Please wait...
Jmol script terminated

4.15 Conclusion

A thorough knowledge about the three-dimensional structure and the dynamical nature of carbohydrate conformations is the key to properly understand the numerous biological phenomena in which carbohydrates are involved. Carbohydrate three-dimensional structural data obtained from experimental methods such as X-ray crystallography and NMR spectroscopy are often complemented by theoretical calculations like molecular dynamics simulations, molecular mechanics, and quantum mechanical calculations in order to validate the structural details. The foremost computational method that has been used in the biomolecular structure determination is molecular dynamics simulations, and they are commonly used to study the conformational analysis of carbohydrates. Three-dimensional structural information of the carbohydrates can be extracted from the molecular dynamics simulation trajectories, and conformational models can be built based on that information. Protein Data Bank currently possess a significant collection of experimentally determined carbohydrate structures from which some primary information about the preference of torsional angles around glycosidic linkages in the bound state can be understood and used for further modeling. The advent of glycan microarray technology has triggered a dramatic increase in the generation of glycan-related data. However the carbohydrate sequential and structural data exist as islands, and there is no common nomenclature for a universal representation. The carbohydrate databases are not interconnected well with themselves as well as with other biomolecular databases. All the sequential and structural data must be manipulated efficiently for exploiting carbohydrates in therapeutics and antigen development. Bioinformatics approaches employed in glycobiology have led to the emergence of a new field called glycoinformatics. Comprehensive resources on carbohydrate data must be developed, and research should be initiated in organizing and exploiting the carbohydrate data. Glycomics is expected to dominate the therapeutics over the coming decade. Hence there is a deliberate need to develop new carbohydrate sequential and structural data and to unify the existing carbohydrate resources. A more common platform to integrate glycomics with proteomics and genomics should be developed with in a foreseeable time.

Acknowledgments Veluraja acknowledges the agencies DST, DBT, and DBT-India-AIST-Japan for funding his various research projects. JFA Selvin acknowledges DST for JRF, DBT for Research Assistantship, and CSIR for SRF. Jasmine acknowledges UGC-BSR for JRF and SRF. All the authors acknowledge the use of Bioinformatics Infrastructure Facility housed at Manonmaniam Sundaranar University, Tirunelveli, Tamilnadu.

References

Afferni C et al (1999) Role of carbohydrate moieties in IgE binding to allergenic components of Cupressus arizonica pollen extract. Clin Exp Allergy 29(8):1087–1094

Alder B, Wainwright T (1957) Phase transition for a hard sphere system. J Chem Phys 27(5):1208

Atkins E et al (1974) X-ray fibre diffraction of cartilage proteoglycan aggregates containing hyaluronic acid. Biochem J 141(3):919–921

Bause E, Legler G (1981) The role of the hydroxy amino acid in the triplet sequence Asn-Xaa-Thr (Ser) for the N-glycosylation step during glycoprotein biosynthesis. Biochem J 195 (3):639–644

Berteau O, Stenutz R (2004) Web resources for the carbohydrate chemist. Carbohydr Res 339 (5):929–936

Biswas M, Rao V (1980) Conformational analysis of the milk oligosaccharides. Biopolymers 19 (8):1555–1566

Biswas M, Rao V (1982) Conformational studies on the ABH and Lewis blood group oligosaccharides. Carbohydr Polym 2(3):205–222

Bode L (2006) Recent advances on structure, metabolism, and function of human milk oligosaccharides. J Nutr 136(8):2127–2130

Bohne A et al (1998) W3-SWEET: carbohydrate modeling by internet. Mol Model Ann 4 (1):33–43

Bourne Y, Cambillau C (1993) The role of structural water molecules in protein-saccharide complexes. Water and biological macromolecules. Springer, pp 321–337

Brady J (1986) Molecular dynamics simulations of. alpha.-D-glucose. J Am Chem Soc 108 (26):8153–8160

Brady JW (1991) Theoretical studies of oligosaccharide structure and conformational dynamics. Curr Opin Struct Biol 1(5):711–715

Brocca P et al (2000) Modeling ganglioside headgroups by conformational analysis and molecular dynamics. Glycoconj J 17(5):283–299

Brown EB et al (1975) Cell-surface carbohydrates and their interactions: I. NMR of N-acetyl neuraminic acid. Biochimic Biophys Acta (BBA)-Gen Subj 399(1):124–130

Bundle DR, Young NM (1992) Carbohydrate-protein interactions in antibodies and lectins. Curr Opin Struct Biol 2(5):666–673

Bush CA (1992) Experimental determination of the three-dimensional structure of oligosaccharides. Curr Opin Struct Biol 2(5):655–660

Cael JJ et al (1976) Polarized infrared spectra of crystalline glycosaminoglycans. Carbohydr Res 50(2):169–179

Cagas P, Bush CA (1990) Determination of the conformation of Lewis blood group oligosaccharides by simulation of two-dimensional nuclear overhauser data. Biopolymers 30 (11–12):1123–1138

Carver JP (1991) Experimental structure determination of oligosaccharides. Curr Opin Struct Biol 1(5):716–720

Christlet THT, Veluraja K (2001) Database analysis of O-glycosylation sites in proteins. Biophys J 80(2):952–960

Christlet THT et al (1999) A database analysis of potential glycosylating Asn-X-Ser/Thr consensus sequences. Acta Crystallogr D Biol Crystallogr 55(8):1414–1420

Crocker PR, Feizi T (1996) Carbohydrate recognition systems: functional triads in cell—cell interactions. Curr Opin Struct Biol 6(5):679–691

Cumming DA, Carver JP (1987) Virtual and solution conformations of oligosaccharides. Biochemistry 26(21):6664–6676

Czarniecki MF, Thornton ER (1977) Carbon-13 nuclear magnetic resonance spin-lattice relaxation in the N-acylneuraminic acids. Probes for internal dynamics and conformational analysis. J Am Chem Soc 99(25):8273–8279

Doubet S et al (1989) The complex carbohydrate structure database. Trends Biochem Sci 14 (12):475–477

Gagneux P, Varki A (1999) Evolutionary considerations in relating oligosaccharide diversity to biological function. Glycobiology 9(8):747–755

Gooley AA et al (1991) Glycosylation sites identified by detection of glycosylated amino acids released from Edman degradation: the identification of Xaa-Pro-Xaa-Xaa as a motif for Thr-O-glycosylation. Biochem Biophys Res Commun 178(3):1194–1201

Hansen JE et al (1995) Prediction of O-glycosylation of mammalian proteins: specificity patterns of UDP-GalNAc: polypeptide N-acetylgalactosaminyltransferase. Biochem J 308(3):801–813

Hashimoto K et al (2006) KEGG as a glycome informatics resource. Glycobiology 16(5):63R–70R

Hughes RC et al (1988) Substrate recognition by UDP-N-acetyl-α-d-galactosamine: polypeptide N-acetyl-α-d-galactosaminyltransferase. Effects of chain length and disulphide bonding of synthetic peptide substrates. Carbohydr Res 178(1):259–269

Hunt LT, Dayhoff MO (1970) The occurrence in proteins of the tripeptides Asn-X-Ser and Asn-X-Thr and of bound carbohydrate. Biochem Biophys Res Commun 39(4):757–765

Imberty A (1997) Oligosaccharide structures: theory versus experiment. Curr Opin Struct Biol 7 (5):617–623

Imberty A, Pérez S (1995) Stereochemistry of the N-glycosylation sites in glycoproteins. Protein Eng 8(7):699–709

Imberty A, Pérez S (2000) Structure, conformation, and dynamics of bioactive oligosaccharides: theoretical approaches and experimental validations. Chem Rev 100(12):4567–4588

Imperiali B (1997) Protein glycosylation: the clash of the titans. Acc Chem Res 30(11):452–459

Jarrell HC et al (1987) Determination of conformational properties of glycolipid head groups by deuterium NMR of oriented multibilayers. Biochemistry 26(7):1805–1811

Karlsson KA (1995) Microbial recognition of target-cell glycoconjugates. Curr Opin Struct Biol 5 (5):622–635

Karplus M, McCammon JA (2002) Molecular dynamics simulations of biomolecules. Nat Struct Mol Biol 9(9):646–652

Kelemen M, Rogers H (1971) Three-dimensional molecular models of bacterial cell wall mucopeptides (peptidoglycans). Proc Natl Acad Sci 68(5):992–996

Klaić B, Domenick RL (1990) 1 Hn. mr studies of a natural immunoadjuvant peptidoglycan monomer: proposed structure in solution in methyl sulfoxide. Carbohydr Res 196:19–27

Lemieux R, Koto S (1974) The conformational properties of glycosidic linkages. Tetrahedron 30 (13):1933–1944

Lis H, Sharon N (1993) Protein glycosylation. Eur J Biochem 218(1):1–27

Lis H, Sharon N (1998) Lectins: carbohydrate-specific proteins that mediate cellular recognition. Chem Rev 98(2):637–674

Live DH et al (1999) Probing cell-surface architecture through synthesis: an NMR-determined structural motif for tumor-associated mucins. Proc Natl Acad Sci 96(7):3489–3493

Lonngren J (1989) Carbohydrates and the pharmaceutical industry. Pure Appl Chem 61 (7):1313–1314

Lütteke T et al (2004) Data mining the protein data bank: automatic detection and assignment of carbohydrate structures. Carbohydr Res 339(5):1015–1020

Lütteke T et al (2006) GLYCOSCIENCES. de: an internet portal to support glycomics and glycobiology research. Glycobiology 16(5):71R–81R

Marchal I et al (2003) Bioinformatics in glycobiology. Biochimie 85(1):75–81

McCammon JA et al (1977) Dynamics of folded proteins. Nature 267(5612):585–590

Moir A, Smith DA (1990) The genetics of bacterial spore germination. Ann Rev Microbiol 44 (1):531–553

Moskalewski S, Jankowska-Steifer E (2011) Hydrostatic and boundary lubrication of joints – nature of boundary lubricant. Ortop Traumatol Rehabil 14(1):13–21

Mulloy B, Forster MJ (2000) Conformation and dynamics of heparin and heparan sulfate. Glycobiology 10(11):1147–1156

Muramatsu T (2000) Protein-bound carbohydrates on cell-surface as targets of recognition: an odyssey in understanding them. Glycoconj J 17(7–9):577–595

O'Connor SE, Imperiali B (1996) Modulation of protein structure and function by asparagine-linked glycosylation. Chem Biol 3(10):803–812
O'Connell B et al (1991) The influence of flanking sequences on O-glycosylation. Biochem Biophys Res Commun 180(2):1024–1030
Olofsson S, Bergström T (2005) Glycoconjugate glycans as viral receptors. Ann Med 37 (3):154–172
Pascher I, Sundell S (1977) Molecular arrangements in sphingolipids. The crystal structure of cerebroside. Chem Phys Lipids 20(3):175–191
Paulson JC (1989) Glycoproteins: what are the sugar chains for? Trends Biochem Sci 14 (7):272–276
Pérez S, Marchessault RH (1978) The exo-anomeric effect: experimental evidence from crystal structures. Carbohydr Res 65(1):114–120
Pérez S, Mulloy B (2005) Prospects for glycoinformatics. Curr Opin Struct Biol 15(5):517–524
Peters T, Pinto BM (1996) Structure and dynamics of oligosaccharides: NMR and modeling studies. Curr Opin Struct Biol 6(5):710–720
Petrescu AJ et al (1999) A statistical analysis of N-and O-glycan linkage conformations from crystallographic data. Glycobiology 9(4):343–352
Poppe L et al (1992) The solution conformation of sialyl-α (2→ 6)-lactose studied by modern NMR techniques and Monte Carlo simulations. J Biomol NMR 2(2):109–136
Poveda A, Jiménez-Barbero J (1998) NMR studies of carbohydrate–protein interactions in solution. Chem Soc Rev 27(2):133–144
Priyadarzini TR et al (2012) Molecular dynamics simulation and quantum mechanical calculations on α-D-N-acetylneuraminic acid. Carbohydr Res 351:93–97
Quiocho FA (1989) Protein-carbohydrate interactions: basic molecular features. Pure Appl Chem 61(7):1293–1306
Rall S et al (1982) Human apolipoprotein E. The complete amino acid sequence. J Biol Chem 257 (8):4171–4178
Ranzinger R et al (2008) GlycomeDB–integration of open-access carbohydrate structure databases. BMC Bioinforma 9(1):1
Rao VR (1998) Conformation of carbohydrates. CRC Press, Boca Raton
Rao V, Biswas M (1981) Conformations and interactions of blood-group oligosaccharides. Biochem Soc Trans 9(6):508–510
Revelle BM et al (1996) Structure-function analysis of P-selectin-Sialyl Lewis binding interactions mutagenic alteration of ligand binding specificity. J Biol Chem 271(8):4289–4297
Rice KG et al (1993) Experimental determination of oligosaccharide three-dimensional structure. Curr Opin Struct Biol 3(5):669–674
Roy R (1996) Syntheses and some applications of chemically defined multivalent glycoconjugates. Curr Opin Struct Biol 6(5):692–702
Sawada T et al (2006) Conformational study of α-N-acetyl-D-neuraminic acid by density functional theory. J Carbohydr Chem 25(5):387–405
Schauer R, Kamerling JP (1995) Chemistry, biochemistry and biology of sialic acids. New Compr Biochem 29:243–402
Sharmila DJS, Veluraja K (2004a) Disialogangliosides and their interaction with cholera toxin—investigation by molecular modeling, molecular mechanics and molecular dynamics. J Biomol Struct Dyn 22(3):299–313
Sharmila DJS, Veluraja K (2004b) Monosialogangliosides and their interaction with cholera toxin—investigation by molecular modeling and molecular mechanics. J Biomol Struct Dyn 21(4):591–613
Sharmila DJS, Veluraja K (2006) Conformations of higher gangliosides and their binding with cholera toxin-investigation by molecular modeling, molecular mechanics, and molecular dynamics. J Biomol Struct Dyn 23(6):641–656
Simanek EE et al (1998) Selectin-carbohydrate interactions: from natural ligands to designed mimics. Chem Rev 98(2):833–862

Suresh MX, Veluraja K (2003) Conformations of terminal sialyloligosaccharide fragments—a molecular dynamics study. J Theor Biol 222(3):389–402
Tipper DJ (1970) Structure and function of peptidoglycans. Int J Syst Evol Microbiol 20 (4):361–377
Toone EJ (1994) Structure and energetics of protein-carbohydrate complexes. Curr Opin Struct Biol 4(5):719–728
Van Halbeek H (1994) NMR developments in structural studies of carbohydrates and their complexes. Curr Opin Struct Biol 4(5):697–709
van Kuik JA et al (1992) A 1H NMR database computer program for the analysis of the primary structure of complex carbohydrates. Carbohydr Res 235:53–68
Varki A (1998) Factors controlling the glycosylation potential of the Golgi apparatus. Trends Cell Biol 8(1):34–40
Varki A, Freeze HH (2009) Glycans in acquired human diseases (Chapter 43), Essentials of Glycobiology, 2nd edn. Cold Spring Harbor Laboratory Press, Cold Spring Harbor
Vasudevan SV, Balaji PV (2001) Dynamics of ganglioside headgroup in lipid environment: molecular dynamics simulations of GM1 embedded in dodecylphosphocholine micelle. J Phys Chem B 105(29):7033–7041
Vasudevan SV, Balaji PV (2002) Comparative analysis of ganglioside conformations by MD simulations: implications for specific recognition by proteins. J Mol Struct THEOCHEM 583 (1):215–232
Veluraja K, Margulis CJ (2005) Conformational dynamics of sialyl lewisx in aqueous solution and its interaction with selectine. A study by molecular dynamics. J Biomol Struct Dyn 23 (1):101–111
Veluraja K, Rao V (1980) Theoretical studies on the conformation of β-DN-acetyl neuraminic acid (sialic acid). Biochim Biophys Acta (BBA)-Gen Subj 630(3):442–446
Veluraja K, Rao V (1983) Theoretical studies on the conformation of monosialogangliosides and disialogangliosides. Carbohydr Polym 3(3):175–192
Veluraja K, Rao V (1984a) Studies on the conformations of sialyloligosaccharides and implications. J Biosci 6(5):625–634
Veluraja K, Rao V (1984b) Theoretical studies on the conformations of higher gangliosides. Carbohydr Polym 4(5):357–375
Veluraja K, Seethalakshmi AN (2008) Dynamics of sialyl Lewis a in aqueous solution and prediction of the structure of the sialyl Lewis a–selectinE complex. J Theor Biol 252(1):15–23
Veluraja K et al (2001) Molecular modeling of sialyloligosaccharide fragments into the active site of influenza virus N9 neuraminidase. J Biomol Struct Dyn 19(1):33–45
Veluraja K et al (2010) 3DSDSCAR-a three dimensional structural database for sialic acid-containing carbohydrates through molecular dynamics simulation. Carbohydr Res 345 (14):2030–2037
von der Lieth CW et al (2002) Molecular dynamics simulations of glycoclusters and glycodendrimers. Rev Mol Biotechnol 90(3):311–337
von der Lieth CW et al. (2009) Bioinformatics for glycobiology and glycomics: an introduction. Wiley Online Library
Vyas N et al (1991) Comparison of the periplasmic receptors for L-arabinose, D-glucose/D-galactose, and D-ribose. Structural and functional similarity. J Biol Chem 266(8):5226–5237
Werz DB, Seeberger PH (2005) Carbohydrates as the next frontier in pharmaceutical research. Chem Eur J 11(11):3194–3206
Wilson J, Itzstein M (2003) Recent strategies in the search for new anti-influenza therapies. Curr Drug Targets 4(5):389–408
Wilson I et al (1991) Amino acid distributions around O-linked glycosylation sites. Biochem J 275 (2):529–534
Woods RJ (1995) Three-dimensional structures of oligosaccharides. Curr Opin Struct Biol 5 (5):591–598

Woods RJ (1998) Computational carbohydrate chemistry: what theoretical methods can tell us. Glycoconj J 15(3):209–216
WoodsGroup (2005–2010) GLYCAM Web, from http://www.glycam.com
Wormald MR et al (2002) Conformational studies of oligosaccharides and glycopeptides: complementarity of NMR, X-ray crystallography, and molecular modelling. Chem Rev 102 (2):371–386
Wu WG et al (1999) Structural study on O-glycopeptides: glycosylation-induced conformational changes of O-GlcNAc, O-LacNAc, O-sialyl-LacNAc, and O-sialyl-lewis-X peptides of the mucin domain of MAdCAM-1. J Am Chem Soc 121(11):2409–2417
Wyss DF et al (1995) Conformation and function of the N-linked glycan in the adhesion domain of human CD2. Science 269(5228):1273
Yan ZY, Bush CA (1990) Molecular dynamics simulations and the conformational mobility of blood group oligosaccharides. Biopolymers 29(4–5):799–811
Yoshida A et al (1997) Discovery of the shortest sequence motif for high level mucin-type O-glycosylation. J Biol Chem 272(27):16884–16888

Epigenome: The Guide to Genomic Expression

5

Ajit Kumar and Gulshan Wadhwa

Abstract

Epigenetic research endeavours to empathize traditional gene regulation not under direct encoding of DNA sequence. Histone modifications, DNA methylation and binding of nonhistone proteins are well-identified mechanisms of epigenetic control of cellular phenotype by gene expression regulations. Environmental factors cause, wholly or partly, different human diseases. Environmental chemicals have long been accepted to cause many diseases through alterations in the genome or genetic effects. Epigenomics (i.e. beyond genomics) encompasses amalgamation of customary genomics with other branches of science like mathematics, computer science, biochemistry, chemistry, proteomics and molecular biology. It looks for the comprehensive analysis of heritable phenotypic changes, alterations in gene function/expression that are not independent of gene sequence. The epigenomic science offers and beckons novel opportunities to help and elevate our understanding of nuclear organization, regulation of transcription, developmental phenomena and diseases at molecular level. This article presents a comprehensive report about the existing computational strategies and approaches for studying the different factors of epigenetics, with special focus on important computational tools and biological databases. In addition, a brief introduction into epigenetics have also been outlined.

A. Kumar (✉)
Centre for Bioinformatics, Maharshi Dayanand University, Rohtak, India

G. Wadhwa
Department of Biotechnology, Apex Bioinformatics Centre, Ministry of Science & Technology, New Delhi, India

G. Wadhwa et al. (eds.), *Current trends in Bioinformatics: An Insight*,
https://doi.org/10.1007/978-981-10-7483-7_5

Keywords
Computational epigenomics · DNA methylation · Epigenetics · Histone modification

5.1 Introduction

The term 'epigenetic landscape' was coined by Conrad Waddington and defined epigenetics as the synergistic interaction of genes with their products that bring out the phenotype into being. Epigenetics is the study of mitotically and meiotically heritable changes in gene function that does not bring out change in DNA sequence.

Genetically, it has been viewed for long that DNA expresses through genes that help us to survive, reproduce and develop, and it is all set in stone, i.e. this behaviour is settled in stone and the environment. Traditional genetics fails to explain many of our questions related to differential gene expressions in multicellular organisms.

The vertical flow of information, stored in DNA, is in general answered by classical genetics and genomics, but they fail to explain the control of horizontal flow of information within a cell, which has given rise to division of labour at multicellularity level, throughout the evolution of life. Epigenetics, as a name suggests, goes beyond the classical genetics and deals with the study about which genetic element is to be expressed along with where and when it is to be expressed. These questions are answered by the structural features of the genetic element and not by the sequence of the DNA as the later deals with what is to be expressed. To cut whole story short, we can say that it is epigenome – the real boss of DNA deciding its functionality. To detail the concept of epigenetics, we can take the example of identical twins. The identical twins have the characteristics like eye colour, hair colour, etc. depicted by the DNA sequences within their genome, and this is what we called genetics, but it is not like that way. Genetically identical twins can be very different; one might be normal while other may be suffering from any disease. This can't be explained on the basis of genetics because the DNA is identical in both the twins. To look beyond and understand the regulation of gene expression that is not directly encoded in the DNA sequences is epigenetics (Bock and Lengauer 2007). In other words, it is the chemical modifications in a gene that is heritable and influences the activity of DNA but lies outside of genomic sequences and does not alter the gene expression. It actually muddles the nature and the nurture.

5.2 Epigenetic Machinery

Epigenetic mechanism involves a number of processes that are critical for diversity of cell type during embryogenesis and gametogenesis and also crucial for tissue-specific gene expression and global gene silencing. Crucial developmental process from fertilization till the end of gastrulation phase can be altered and may result in structural abnormalities (Piplani et al. 2016). Epigenetic mechanisms encompass the mechanism that includes DNA and the nucleosomes and are known to be the carrier of this process (Espada and Esteller 2007; Rivera and Ren 2013). The three most common chemical and biological modifications that alter the genetic expressions are chromatin structure and modification, DNA methylation and miRNA.

5.2.1 DNA Methylation

DNA methylation is a crucial mechanism for cell differentiation and organism development. It alters the gene expression through methylation of DNA strand itself. This biochemical process involves the cytosine bases of eukaryotic DNA being converted into 5-methylcytosine and thus represses the transcription. This typically occurs at CpG island (Jones 2012; Virani et al. 2012).

This process thus passes on from generation to generation by the process of cell division. During zygote formation, DNA methylation is removed and re-established during cell divisions. The DNA methyltransferase (DNMT) enzymes are responsible for maintaining and establishing this unique pattern on CpG island (Dodge et al. 2002). Till date five DNMTs have been reported that are involved in catalysing this process (Roberts et al. 2003; Serman et al. 2006; Putiri and Robertson 2011) as summarized in Table 5.1.

5.2.2 Histone Modifications

These are another most common and complex epigenetic phenomenon that includes acetylation, phosphorylation, sumoylation, methylation and ubiquitination (Bannister and Kouzarides 2011). Histones are conserved proteins and can be modified at amino acid residues on N- and C- terminals. The C-terminus forms globular domains packed in core, while the N-terminal is comparatively flexible

Table 5.1 DNA methylating enzymes

Enzyme	Omim no.	Function
Dnmt1	126375	Maintain methyltransferase (Pradhan and Esteve 2003; Eastvaran et al. 2004)
Dnmt2	602478	Play role in DNA methylation but doesn't have DNA methyltransferase activity (Okano et al. 1998)
Dnmt3a	602769	Participate in de novo methylation (Gonzalgo Jones 1997; Okano et al. 1999)
Dnmt3b	602900	Participate in de novo methylation
Dnmt3L	606588	For methylation of most imprinted loci in germ cells (Kaneda et al. 2004)

Table 5.2 Histone modification enzymes

Enzyme	Function
Histone acetyltransferases (HATs)	Acetylase conserved lysine on histone protein by transfer acetyl group from acetyl Co-A to form ε-*N*-acetyllysine (Marmorstein 2001)
Histone deacetylase (HDAC)	Removes acetyl group from ε-N-acetyllysine on histone, allow histone to wrap DNA tightly (Song et al. 2005; Halkidou et al. 2004)
Histone methyltransferases (HMTs)	Catalyse the transfer of methyl groups to lysine and arginine of histone proteins(Campagna-Slater et al. 2011)
Histone demethylases (HDMs)	Removes methyl group from histone (Kawamura et al. 2010)

and interacts directly with DNA and other proteins within nucleus and is important to maintain chromatin stability. Unlike the rest of the modifications, the methylation and acetylation on arginine and lysine residues repress the transcription. These modifications are achieved by different histone-modifying enzymes as summarized in Table 5.2. The histone modifications alter the chromosome structure by altering the electrostatic properties or by producing binding affinities for protein recognition module (Iizuka and Smith 2003; Brooks and Shi 2014).

5.2.3 miRNA

miRNAs are ~22 nucleotide long non-coding RNAs that negatively control gene expression post transcriptionally. miRNAs are highly conserved in plant and animal species and play critical roles in a variety of biological processes including pattern formation and developmental timing, cell signalling, carcinogenesis, etc. (Sato et al. 2011). About 50% of miRNA genes are housed in the fragile genome regions and are very sensitive to deletion, translocation or duplication. miRNAs have already been looked upon as new therapeutic agents and/or targets for different diseases (Chuang and Jones 2007).

5.2.4 Binding of Nonhistone Proteins

The proteins in chromatin which remain even after the removal of histones are commonly known as nonhistone proteins. These proteins interact with the chromosome structure and remodel the chromatin structure, thus regulating the silencing of genes. The two nonhistone proteins, polycomb and trithorax, are epigenetic regulators and affect gene expressions. Polycomb protein induces gene silencing, while the trithorax protein induces gene inactivation (Pullirsch et al. 2010).

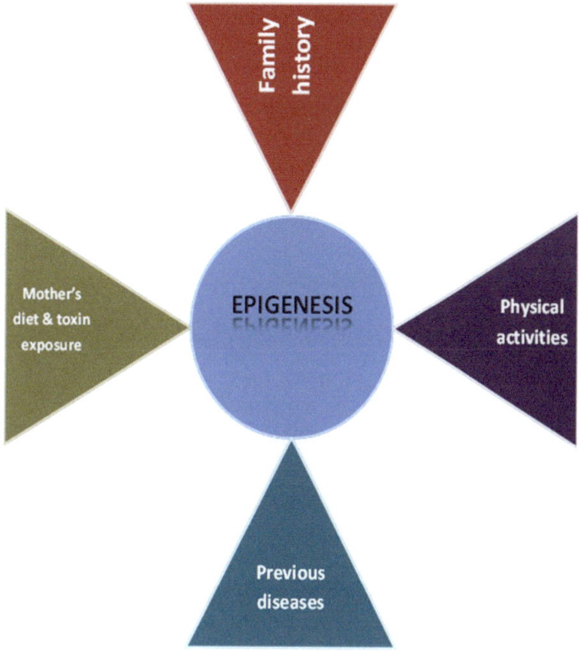

Fig. 5.1 External factors influencing epigenetic traits

5.2.5 Environmental Factors

Various dietary factors for mothers and environmental factors have been linked to epigenetic modifications. The conditions like stress, nutrition and environment deal with how our DNA behaves in present and next generation(s). Epigenetic mechanism regulates throughout the life. The nutrition of the mother affects the foetus; stress hormone also travels from mother to foetus. Social interactions, physical activities, exposure to toxins and diet are also the major factors shaping the epigenome (Kubota et al. 2012). The living place, consumption of alcohols and exposure to various drugs are potentially able to alter the changes in epigenetic status (Aguilera et al. 2010; Choi and Friso 2010). The external factors influencing epigenetic traits are depicted in Fig. 5.1.

5.3 Analysis of Epigenetic Modifications

5.3.1 In Vitro Methods

Various techniques are in extensive use for the interpretation of epigenetic data. Bisulphite sequencing method is used to determine the DNA methylation. This technique uses methylation restrictive enzymes that convert DNA with bisulphite to 5-methylcytosine (Jones and Takai 2001). Another commonly used method is 'chromatin immunoprecipitation' that determines the DNA binding sites on the genome for a particular protein (Collas 2010). It also predicts the DNA protein interactions, either inside the nucleus or within the living cell. The last decade witnessed rapid development of microarray-based technology. It analyses the bisulphite-treated DNA methylation sites (Schumacher et al. 2006) and utilizes the pairs of oligonucleotide hybridization probes for targeting the CpG sites.

5.3.2 In Silico Methods

With the developing technologies and advancement in the field of computational biology, several techniques have now developed to focus on identification of epigenetic modifications, such as Support Vector Machine (SVM), artificial neural networks (ANNs), hidden Markov model (HMM) and clustering analysis. SVM is used to diagnose the cytosine methylation in CpG nucleotides (Bhasin et al. 2005; Robinson et al. 2014). It is a successful machine learning technique to evaluate the pattern, but the problem is the lack of experimentally available public data. ANN performs the prediction algorithm for the human-specific methylation sites (de Pretis and Pelizzola 2014). This method is designed to mimic the architecture of the brain. The data that is needed to be evaluated is differentiated into processing units called neurons. These neurons process the data using a variety of mathematical evaluations (Marchevsky et al. 2004). Another widely used computational technique is HMM which is used to detect the CpG islands (Robinson and Pelizzola 2015). This method is deliberately used for the sequence analysis (Wu et al. 2010). Clustering analysis helps in reflecting the true distribution of gene space (Fazzari and Greally 2004). Many reported attempts have been successfully made to analyse epigenetic modifications using in silico methods (Fig. 5.2) (Gitan et al. 2002; Collins et al. 2003;

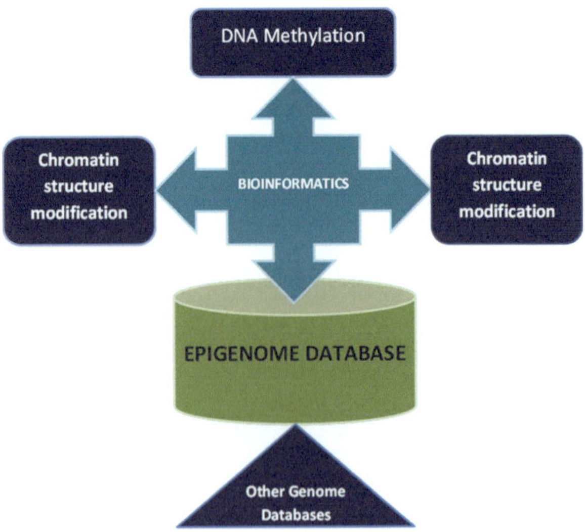

Fig. 5.2 Integration of technology to generate epigenome database

Laird 2003; Yan et al. 2004; Meissner et al. 2005; Pfister et al. 2007; Petrossian and Clarke 2009).

5.4 Epigenomics: The Computational View of Epigenetics

Post-'Human Genome Project' era has witnessed a tsunami of epigenetic data due to rapid development in technological applications in biological sciences. Bioinformatics has evolved at a parallel rate to support, store, manage, manipulate and exploit the biological data flood. As a result, a new jargon has been added to bioinformatics vocabulary, namely, 'epigenomics'. It deals in the computational analysis of experimentally determined epigenetic data and is emerging as a separate but prominent frontier of biological sciences. It involves in silico collection of data, maintaining databases and analysing the stored data in a computationally intensive manner, related to alterations of genetic expressions and gene activity independent of gene sequence. Epigenetics deals in stable alterations in genetic expressions and gene activity independent of gene sequence, both in the same generation (horizontal flow) and long-term changes in transcriptional potential of a cell that are generally not heritable. Epigenomics differs from epigenetics in emphasizing on global analyses of sequence-independent genetic changes throughout the genome, while the latter looks for studying the same for a gene or a gene set (Tammen et al. 2013). There have been many epigenomic efforts being carried out worldwide of medium/large scale; a major few are as listed in Table 5.3.

Various bioinformatics methods help in identifying gene expression by studying the contribution of DNA methylation, identifying the CpG islands and studying the in silico modelling and dynamics of epigenetic processes. Above all, the most important is the epigenomic data collection for solving the mystery. Algorithms like BLAST (Altschul et al. 1990), BLAT (Kent 2002) and Clustal W (Thompson et al. 1994) allow for the sequence analysis for inference of functional, structural and evolutionary relationships. The scientific literature database like PubMed and molecular databases like DDBJ, EMBL and GenBANK serves as repositories for nucleotide and protein sequence of different species. The databases, like GEO, GENSAT, StemBase, etc., allow for the identification of dynamic changes in gene expression in different cell types. The epigenomic data is getting enriched every day; a few important resources are summarized in Table 5.4.

5.5 Recent Insights into Disease Epigenomics

The epigenomics supersedes epigenetic studies in the manner that the former has comprehensive approach as compared to single-gene association to a disease in epigenetics (Feinberg 2010). Candidate gene-disease association studies have now been replaced by whole-genome disease association studies and are accepted with more scientific appreciations (Lieberman-Aiden et al. 2009). Methylation of cytosine residues of DNA has been attributed as the central stage player in epigenomic

Table 5.3 Large- and medium-scale epigenomic efforts

Projects	Aim	Web address
European epigenomic project	Epigenome Network of Excellence fosters the epigenetic research community in Europe, on technical aspects of epigenomics and development of tools	http://www.epigenome-noe.net/WWW/researchtools/protocols.php http://www.epigenome-noe.net/WWW/index.php
Canadian and Australian projects	The Canadian Institutes of Health and Research (Ottawa) and Australian Alliance for Epigenetics have come forward to elevate the understanding on 'epigenetics, environment and health'	http://www.epialliance.org.au/
Asian projects	Yonsei University (Seoul), the Japanese National Cancer Center, the Shanghai Cancer Institute and the Genome Institute of Singapore developed various technological platforms to generate large sequencing data and to promote interactions and collaborations in epigenomic research	
The ENCODE project	The Encyclopedia of DNA Elements (ENCODE) and modENCODE projects use a wide array of different assays to identify functional elements, and epigenomic profiling is thus an important component of the programmes	http://www.genome.gov/10005107
ICGC projects	International Cancer Genome Consortium (ICGC) investigating genomic changes with the goal of obtaining a comprehensive description of genomic, transcriptomic and	http://www.icgc.org/

	epigenomic changes in 50 different tumour types and/or subtypes	
NIH roadmap epigenomic programme	The consortium is developing standards and best practices for individual and integrative analyses of the different data types to provide a reference for the larger epigenomic community. Another role is to include development of various new techniques of epigenomic analysis and imaging, identification of new epigenetic modifications and investigations	http://www.roadmapepigenomics.org
The International Human Epigenome Consortium	IHEC builds on the NIH effort in epigenomics and created a truly global epigenomic project. IHEC develops best practices and standards for epigenomic data generation and analysis	http://ihec-epigenomes.org

Table 5.4 Information resources for epigenomics

Database	Information	URL
MethDB	Contains information on 19,905 DNA methylation content data and 5382 methylation patterns for 48 species, 1511 individuals, 198 tissues and cell lines and 79 phenotypes. Provides valuable search for patterns and profiles, 5mC content, phenotypes	http://www.methdb.de/
PubMeth	Contains over 5000 records on methylated genes in various cancer types	http://www.pubmeth.org
REBASE	Contains over 22,000 DNA methyltransferases genes derived from GenBank	http://rebase.neb.com/rebase/rebase.html
MeInfoText	Contains gene methylation information across 205 human cancer types	http://mit.lifescience.ntu.edu.tw/
MethPrimerDB	Contains 259 primer sets from human, mouse and rat for DNA methylation analysis	http://medgen.ugent.be/methprimerdb
The HistoneDB	Contains 254 sequences from histone H1, 383 from histone H2, 311 from histone H2B, 1043 from histone H3 and 198 from histone H4, altogether representing at least 857 species	http://research.nhgri.nih.gov/histones/
ChromDB	Contains 9341 chromatin-associated proteins, including RNAi-associated proteins, for a broad range of organisms	http://www.chromdb.org/
CREMOFAC	Contains 1725 redundant and 720 nonredundant chromatin-remodelling factor sequences in eukaryotes	http://www.jncasr.ac.in/cremofac/
CAMH: The Krembil Family Epigenetics Laboratory	Contains DNA methylation data of human chromosomes 21, 22, male germ cells and DNA methylation profiles in monozygotic and dizygotic twins	http://www.camh.ca/en/
MethylomeDB	Experimental information from over 14,000 entries and 175 high-throughput data sets from a wide number of sources also includes gene-centric methylation data of 72 human diseases	http://bioinfo.hrbmu.edu.cn/diseasemeth

modification and plays a pivotal role in cellular processes including regulation of genes, organism development and disease (Lister et al. 2009). The recent studies on different cancer-associated genes and its epigenetic machinery have revealed the requisites of epigenomic correlation search for unexplained diseases (Frigola et al. 2006; Irizarry et al. 2009; Fraga et al. 2005a, b; Lister et al. 2009; Doi et al. 2009).

Gene-silencing events were observed to be spanning through a large regions of genome in colorectal cancer. The DNA-methylated and adjoining un-methylated genes were also observed to be coordinately suppressed across the entire chromosome (Frigola et al. 2006). Irizarry et al. (2009) showed that in colon cancer, the epigenetic alterations due to DNA methylation occur in promoters and CpG islands, extending to sequences up to 2Kb distance, termed as 'CpG island shores'. With CpG island hypermethylation, the global hypomethylation across genome has also been a common epigenomic character of cancerous cells (Irizarry et al. 2009). Tumour cells in humans have been attributed with the hallmark of mono-acetylation loss and tri-methylation of H4 histone proteins at global genomic level (Fraga et al. 2005a). A widespread epigenomic difference has been attributed responsible for differential susceptibility to diseases and other anthropomorphic variations observed in the lifetime of monozygotic twins (Fraga et al. 2005b).

While mapping whole genome at single-base resolution, significant differences were observed in pattern and composition of cytosine methylation of embryonic and differentiated cells (Lister et al. 2009). About one-fourth part of total genomic methylations of embryonic stem cells have been identified in a non-CG context reflecting to other methylation mode to regulate gene. Higher degree of non-CG methylations were observed in gene-coding areas, while the same were observed to be depleted in binding sites of proteins and regulatory regions (Lister et al. 2009). There has been suggestion about different mechanisms of epigenetic reprogramming resulting in differential methylation of tissue and disease-specific CpG islands in differentiated cells (fibroblasts), pluripotent stem cells and embryonic stem cells (Doi et al. 2009).

The role of epigenetic mechanism in altered embryonic developments and its manifestations in adulthood diseases, like cardiovascular diseases, is still unknown, but there is a growing volume of evidences supporting the epigenetic regulations (Martinez et al. 2015). Since the early 1980s, several studies have indicated the possible involvement of adverse intrauterine condition to adulthood diseases like cardiovascular diseases (Barker and Osmond 1988), diabetes (type 2), metabolic syndrome, ischemic heart disease, hypertension, etc. (Chen and Zhang 2011). These studies formed the basis of 'thrifty phenotype hypothesis' (Hales and Baker 2001) stating the abnormal intrauterine situations like toxins, hypoxia, undernutrition, chemicals, etc., pushes the developing embryo to undergo irreversible changes to adapt and survive the suboptimal environment, resulting in increased susceptibility of neonate or adult towards developing a disease.

Since the completion of 'Human Genome Project' in the early twenty-first century and next-generation sequencing technology, there has been a flood of genomic and epigenomic data. These developments have resulted in a paradigm shift in our concepts of cell and cellular function. The idea of reprogramming a cell in form of stem cell development has triggered a race for understanding the cellular function beyond genomics, i.e. epigenomics. Several diseases have been and will be associated with epigenetic and epigenomic basis and needs to exploit more to develop therapeutic applications of the same (Kanherkar et al. 2014). After the development of induced pluripotent stem cells (Park et al. 2012), there have been several successful attempts to modulate them epigenetically for generating differential potentials. These have led to many device novel therapies and develop novel disease models for neurodegenerative disorders, cardiovascular diseases, metabolic disorders (PCOS), etc. (Huang and Wu 2013).

5.6 Conclusion

Epigenetic modifications provide a link between nature and nurture. The epigenomics paves the way through a new research that is valuable for predicting and diagnosing various diseases. Epigenomics has the potential to overthrow the disease-causing genes and identify the altered gene expression. Recent research by American Association for Cancer Research (AACR) offers the epigenetic therapy of killing cancer cells by methylating without disrupting its pathway (Jones 2012). The epigenetic tags and their response make it a valuable technology. Various other stocks are in queue for targeting different tissues affected by the disease. The epigenetic data adds a flavour to both the computational- and wet-lab work, and we can say 'epigenomics: an era beckoning'.

5.7 Future Directions

The initial phase of computational epigenetic research got impetus from the ever-progressing experimental ways of data generation. The high-throughput epigenomic data thus generated need proper computational analyses for low-level data processing and quality control. This has led to epigenome predictions as a way out which understand the epigenetic information distributed throughout genome. In conclusion, exciting times are ahead for research in epigenetics with high computational input requirement.

References

Aguilera O et al (2010) Epigenetics and environment: a complex relationship. J Appl Physiol 109 (1):243–251

Altschul SF et al (1990) Basic local alignment search tool. J Mol Biol 215:403–410

Bannister AJ, Kouzarides T (2011) Regulation of chromatin by histone modifications. Cell Res 21 (3):381–395

Barker DJ, Osmond C (1988) Low birth weight and hypertension. BMJ Br Med J 297(6641):134

Bhasin M et al (2005) Prediction of methylated CpG in DNA sequences using a support vector machine. FEBS Lett 579(20):4302–4308

Bock C, Lengauer T (2007) Computational epigenetics. Bioinformatics 24(1):1–10

Brookes E, Shi Y (2014) Diverse epigenetic mechanisms of human disease. Ann Rev Gen 48:237–268

Campagna-Slater V et al (2011) Structural chemistry of the histone methyltransferases cofactor binding site. J Chem Inf Model 51(3):612–623

Chen M, Zhang L (2011) Epigenetic mechanisms in developmental programming of adult disease. Drug Discov Today 16(23):1007–1018

Choi SW, Friso S (2010) Epigenetics: a new bridge between nutrition and health. Am Soc Nut Adv Nutr 1:8–16
Chuang JC, Jones PA (2007) Epigenetics and microRNAs. Pediatr Res 61(5):24R–29R
Collas P (2010) The current state of chromatin immunoprecipitation. Mol Biotechnol 45 (1):87–100
Collins FS et al (2003) A vision for the future of genomics research. Nature 422(6934):835–847
de Pretis S, Pelizzola M (2014) Computational and experimental methods to decipher the epigenetic code. Front Genet 5:335
Dodge JE et al (2002) de novo methylation of MMLV provirus in embryonic stem cells: CpG versus non- CpG methylation. Gene 289(1–2):41–48
Doi A, Park IH, Wen B, Murakami P, Aryee MJ, Irizarry R, Herb B, Ladd-Acosta C, Rho J, Loewer S, Miller J (2009) Differential methylation of tissue-and cancer-specific CpG island shores distinguishes human induced pluripotent stem cells, embryonic stem cells and fibroblasts. Nat Genet 41(12):1350–1353
Eastvaran HP et al (2004) Replication- independent chromatin loading of Dnmt1 during G2 and M phases. EMBO Rep 5(12):118
Espada J, Esteller M (2007) Epigenetic control of nuclear architecture. Cell Mol Life Sci 64 (4):449–457
Fazzari MJ, Greally JM (2004) Epigenomics: beyond CpG islands. Nature 5(6):446–455
Feinberg AP (2010) Epigenomics reveals a functional genome anatomy and a new approach to common disease. Nat Biotechnol 28(10):1049–1052
Fraga MF et al (2005a) Loss of acetylation at Lys16 and trimethylation at Lys20 of histone H4 is a common hallmark of human cancer. Nat Genet 37(4):391–400
Fraga MF et al (2005b) Epigenetic differences arise during the lifetime of monozygotic twins. Proc Natl Acad Sci U S A 102(30):10604–10609
Frigola J et al (2006) Epigenetic remodeling in colorectal cancer results in coordinate gene suppression across an entire chromosome band. Nat Genet 38(5):540–549
Gitan RS et al (2002) Methylation-specific oligonucleotide microarray: a new potential for high-throughput methylation analysis. Genome Res 12(1):158–164
Gonzalgo M, Jones P (1997) Mutagenic and epigenetic effects of DNA methylation. Mutat Res 386(2):107–118
Hales CN, Barker DJ (2001) The thrifty phenotype hypothesis. Br Med Bull 60(1):5–20
Halkidou K et al (2004) Upregulation and nuclear recruitment of HDAC1 in hormone refractory prostate cancer. Prostate 59(2):177–189
Huang C, Wu JC (2013) Epigenetic modulations of induced pluripotent stem cells: novel therapies and disease models. Drug Discov Today Dis Model 9(4):e153–e160
Iizuka M, Smith MM (2003) Functional consequences of histone modifications. Curr Opin Genet Dev 13(2):154–160
Irizarry RA et al (2009) The human colon cancer methylome shows similar hypo-and hypermethylation at conserved tissue-specific CpG island shores. Nat Genet 41(2):178–186
Jones PA (2012) Functions of DNA methylation: islands, start sites, gene bodies and beyond. Nat Rev Gen 13:484–492
Jones PA, Takai D (2001) The role of DNA methylation in mammalian epigenetics. Science 293 (5532):1068–1070
Kaneda M et al (2004) Essential role for de novo methyltransferase Dnmt3a in paternal and maternal imprinting. Nature 429(6994):900–903
Kanherkar RR et al (2014) Cellular reprogramming for understanding and treating human disease. Front Cell Dev Biol 2:67
Kawamura A et al (2010) Development of homogeneous luminescence assays for histone demethylase catalysis and binding. Anal Biochem 404(1):86–93
Kent WJ (2002) BLAT – the BLAST-like alignment tool. Genome Res 12:656–664
Kubota T et al (2012) Epigenetic understanding of gene-environment interactions in psychiatric disorders: a new concept of clinical genetics. Clin Epigenetics 4(1):1

Laird PW (2003) The power and the promise of DNA methylation markers. Nat Rev Cancer 3 (4):253–266
Lister R et al (2009) Human DNA methylomes at base resolution show widespread epigenomic differences. Nature 462(7271):315–322
Lieberman-Aiden E, Van Berkum NL, Williams L, Imakaev M, Ragoczy T, Telling A, Amit I, Lajoie BR, Sabo PJ, Dorschner MO, Sandstrom R (2009) Comprehensive mapping of long-range interactions reveals folding principles of the human genome. Science 326 (5950):289–293
Marchevsky AM et al (2004) Classification of individual lung cancer cell lines based on DNA methylation markers: use of linear discriminant analysis and artificial neural networks. J Mol Des 6(1):28–36
Marmorstein R (2001) Structure and function of histone acetyltransferases. Cell Mol Life Sci 58:693–703
Martinez SR et al (2015) Epigenetic mechanisms in heart development and disease. Drug Discov Today 20(7):799–811
Meissner A et al (2005) Reduced representation bisulfite sequencing for comparative high-resolution DNA methylation analysis. Nucleic Acids Res 33(18):5868–5877
Okano M et al (1998) Cloning and characterization of a family of novel mammalian DNA (cytosine-5) methyltransferases. Nat Genet 19(3):219–220
Okano M et al (1999) DNA methyltransferases Dnmt3a and Dnmt3b are essential for de novo methylation and mammalian development. Cell 99(3):245–257
Park LK, Friso S, Choi SW (2012) Nutritional influences on epigenetics and age-related disease. Proc Nutr Soc 71(1):75–83
Petrossian T, Clarke S (2009) Bioinformatics identification of novel methyltransferases. Epigenomics 1(1):163–175
Pfister S et al (2007) Array-based profiling of reference-independent methylation status (aPRIMES) identifies frequent promoter methylation and consecutive downregulation of ZIC2 in pediatric medulloblastoma. Nucleic Acids Res 35(7):e51
Piplani S et al (2016) Homology modelling and molecular docking studies of human placental cadherin protein for its role in teratogenic effects of anti-epileptic drugs. Comp Biol Chem 60:1–8
Pradhan S, Esteve PO (2003) Mammalian DNA (cytosine-5) methyltransferases and their expression. Clin Immunol 109(1):6–16
Pullirsch D et al (2010) The trithorax group protein Ash 2l and Saf – A are recruited to inactive X-chromosome at the onset of stable X inactivation. Development 137(6):935–943
Putiri EL, Robertson KD (2011) Epigenetic mechanisms and genome stability. Clin Epigenetics 2 (2):299–314
Rivera CM, Ren B (2013) Mapping human epigenomes. Cell 155(1):39–55
Roberts RJ et al (2003) A nomenclature for restriction enzymes, DNA methyltransferases, homing endonucleases and their genes. Nucleic Acids Res 31(7):1805–1812
Robinson MD, Pelizzola M (2015) Computational epigenomics: challenges and opportunities. Front Genet 6(88):1–3
Robinson MD et al. (2014) Statistical methods for detecting differentially methylated loci and regions. Front Genet. 5:324: eCollection-2014
Sato F et al (2011) MicroRNAs and epigenetics. FEBS J 278:1598–1609
Schumacher A et al (2006) Microarray based DNA methylation profiling: technology and application. Nucleic Acids Res 34(2):528–542
Serman A et al (2006) DNA methylation as a regulatory mechanism for gene expression in mammals. Coll Antropol 30(3):665–671
Song J et al (2005) Increased expression of histone deacetylase 2 is found in human gastric cancer. APMIS 113(4):264–268
Tammen SA et al (2013) Epigenetics: the link between nature and nurture. Mol Asp Med 34 (4):753–764

Thompson JD et al (1994) CLUSTAL W: improving the sensitivity of progressive multiple sequence alignment through sequence weighting, position-specific gap penalties and weight matrix choice. Nucleic Acids Res 22(22):4673–4680

Virani S et al (2012) Cancer epigenetics: a brief review. ILAR J 53(3–4):359–369

Wu H et al (2010) Redefining CpG islands using hidden Markov Models. Biostatistics 11 (3):499–514

Yan PS et al (2004) Methylation-specific oligonucleotide microarray. Methods Mol Biol 287:251–260

Part II

Bioinformatics Approaches

Molecular Modeling and Drug Design: A Contemporary Analysis in *Vibrio cholerae*

6

Mobashar Hussain Urf Turabe Fazil, K. Konda Reddy, Haushila Prasad Pandey, and Sunil Kumar

Abstract

Vibrio cholerae causes the diarrheal disease cholera. This microbe inhabits well in human host and aquatic environments. Excessive use of antibiotics has contributed to the emergence of antibiotic resistance in *V. cholerae*. Quorum sensing is one of the lucrative targets presently pursued for drug design in bacteria to encounter virulence. LuxO, a response regulator, is an important part of quorum-sensing machinery in *V. cholerae* contributing in biofilms and virulence machinery. In this chapter, we will concisely discuss the disease, its clinical display, and distinctive methodologies to find drug targets. As a treatise on the method of drug design in *V. cholerae*, we used our in silico model of LuxO in this chapter, to predict probable sites of interference, and then used these sites for in silico drug design.

Keywords

Vibrio cholerae · Drug design · Novel drug targets · Quorum sensing

---

M. H. U. T. Fazil and K. Konda Reddy equally contributed to this chapter.

M. H. U. T. Fazil
Lee Kong Chian School of Medicine, Nanyang Technological University, Singapore, Singapore

K. Konda Reddy
Department of Pharmacy, National University of Singapore, Singapore, Singapore

H. P. Pandey
Department of Biochemistry, Nepalgunj Medical College, Chisapani Campus, Kathmandu University, Nepalgunj, Nepal

S. Kumar (✉)
Bioinformatics Centre, Institute of Life Sciences, Bhubaneswar, Odisha, India

ICAR-NBAIM, Mau, Uttar Pradesh-275103, India

G. Wadhwa et al. (eds.), *Current trends in Bioinformatics: An Insight*,
https://doi.org/10.1007/978-981-10-7483-7_6

6.1 Introduction

The elusive nature of the disease makes the history of cholera fascinating. There are considerable deliberations over the origin of the term cholera and which ancient cultures this disease touched. In India, the Sanskrit word "Visuchika" is believed to denote cholera that has been mentioned in Sushruta Samhita. Cholera has been denoted by different words in almost every country in Europe, every language in India and Arabia, suggesting that cholera may have been known to ancient civilizations in the Mediterranean, Middle East, and Indian subcontinent (Howard Jones 1979). Populations all over the world have sporadically been affected by the devastating outbreaks of cholera. There have been seven pandemics of cholera reported so far. The reasons for increased cholera cases, differentiated spatiotemporally, or decline are not all that well understood.

6.2 The Microbe

Vibrios are a common free-living group of Gram-negative, curved, or straight motile rods that normally inhabit the aquatic environments possessing one or more flagella. In most of its natural environments like estuaries and coastal seawaters, *Vibrios* are located in multispecies biofilm structures (Huq et al. 1984a, b).

Waldor and Mekalanos (1996) reported that the genes encoding for the enterotoxin (CT) responsible for the virulence in *V. cholerae* are harbored and transmitted by a bacteriophage CTXΦ. The CTXΦ element can integrate into *V. cholerae* chromosome and replicate and be transmitted vertically as a plasmid. With the exception of the toxin genes (*ctxAB*), the organization of the core region is similar to other filamentous phage genomes that infect *E. coli* and *V. cholerae*. Several other filamentous phages have also been reported in *V. cholerae* (Albert et al. 1996; Campos et al. 2003; Chang et al. 1998; Ehara et al. 1997; Ikema and Honma 1998; Jouravleva et al. 1998). These phages do not code for any virulence factor.

V. cholerae colonizes in the small intestine. *V. cholerae* can endure, colonize, and produce CT in spite of harsh environmental conditions (Sanchez and Holmgren 2008). CT comprises of one A subunit and five B subunits. The B subunit explicitly binds to GM1 ganglioside receptor of epithelial cells (Van Heyningen et al. 1971; King and Van Heyningen 1973; Pierce 1973; Spangler 1992). Extensive hydrogen-bonding interactions between carbohydrate components of the GM1 and B subunit are critical for interaction. The A subunit is translocated into the host cell cytosol (Gill and Rappaport 1979) and is activated by thiol-dependent reduction oxidoreductases (Moss et al. 1980). ADP-ribosylating activity of the A subunit targets the host cell G protein Gsα. ADP-ribosylated Gsα in turn activates adenylate cyclase, resulting in increased levels of intracellular cAMP (Cassel and Selinger 1977; Kahn and Gilman 1984). Critical organ failure can be caused due to this increase in cAMP leading to passive water loss and eventual death if not treated in time.

Prevention measures for cholera consist of clean drinking water supply and improved sanitation. Treatment with antibiotics, or mass chemoprophylaxis, does not affect the spread of cholera. El Tor-derived protection is more labile against El Tor strains (Levine and Tacket 1994), and infection with the classical biotype protects against different serotypes (Inaba, Ogawa, and Hikojima) of classical strains. Naturally acquired immunity against *Vibrio* was reported to last for 3 years (Levine et al. 1981). Dukoral, the vaccine registered by WHO, consists of killed *V. cholerae* organisms along with the cholera B subunit, and the vaccine therefore stimulates both antibacterial and antitoxic immunity. Two doses are given 1–6 weeks apart (Holmgren et al. 1989). The other oral vaccine Orochol is an avirulent mutant of *V. cholerae*, strain CVD103HgR (Tacket et al. 1999).

Variable antibiotic resistance had been found in *V. cholerae* strains isolated from different cholera epidemics varying in space and time. An increase in multidrug-resistant phenotype of *V. cholerae* strains had been observed by continuous monitoring of epidemic strains (Ramamurthy 2008). The nature of the spread of antibiotic resistance was found to be due to number of agents like transmissible plasmids, bacteriophages, conjugative transposons, transposon-like elements, integrons, integrative and conjugative elements, and presence of mobile DNA or R factor (resistance factor), and the latest addition was a new type of element CONSTIN termed SXT identified for the first time in *V. cholerae* O139. The rapid emergence of multidrug resistance (MDR) in *V. cholerae* was of great concern in public health perspective. Integrons have also been described as vehicles for the acquisition of resistance genes (Stokes and Hall 1989; Waldor et al. 1996).

Considering the case of *V. cholerae* O139, history tells us that there had been severe modifications in the bacteria in resisting various antibiotics over different periods of time. The initial outbreak of cholera due to O139 in 1992 witnessed specific resistance of these bacteria to sulfamethoxazole, trimethoprim, and streptomycin (Waldor et al. 1996). The next round of emergence in 1996 details about susceptibility to the very same antibiotics mentioned above. Genetic analyses had revealed that O139 strains carry a unique integron called the int SXT element that carries the genes for antibiotic resistance for these antibiotics (Waldor et al. 1996). It was assumed that this element was lost owing to genetic recombination events during later outbreaks caused by O139 *Vibrios* (Basu et al. 2000). A series of other events of this kind had been on record for both O1 and O139 *Vibrios* pertaining to the antibiotic resistance phenotype (Faruque et al. 2003). This is a serious issue concerning the failure of provision for a common vaccine against O1 and O139 *V. cholerae* and therapeutics used actually helping in the cause of the disease by generating antibiotic-resistant phenotypes. In this regard novel avenues for drug targeting and vaccine design are the order of the day for the disease cholera apart from many other diseases the tropical countries encounter (Fazil et al. 2012a). Various new methods are thus being considered for design, targeting, or intervention of the disease cholera (Van Dellen and Watnick 2006). These new therapeutics include inhibiting *V. cholerae* quorum sensing, because the most recent of detailed genetic experiments reveal an undeniable role for quorum sensing in *V. cholerae* biofilms and virulence gene regulation. Targeting virulence genes will decrease the dose of toxins and thus the overall severity of infection. Surface attachments prove to be a critical step in establishing infection by the bacteria as nutrients tend to be concentrated at surfaces. Biofilms improves the ability of the bacterium to endure osmosis and chemical stress. The role of surface attachment and biofilm formation is critical in decreasing volume of water required to deliver an infectious dose, thus concentrating the bacteria in addition to forming a physical barrier for proper dispersal and diffusion of antibiotics. Keeping in view the above details, novel approaches to treat *V. cholerae* should include quorum sensing not only for keeping a check on antibiotic resistance but also for the sublime reason that it could lead to answers for a common vaccine or target against diverse toxigenic serogroups of *V. cholerae*. Large amounts of genomic/proteomic data need to be analyzed before target assessment, and this humongous task has been largely relieved due to computational methods adding to the foray of physical determination technology existing in use (Vitkup et al. 2000; Chance et al. 2002).

Approximately 5,553,655 protein sequences in the Swiss-Prot and 18,200,000 protein sequences in TrEMBL protein sequence databases are present. At present approximately more than 1,30,000 protein structures had been deposited in Protein Data Bank. Several structural genomic projects are initiated to solve many proteins particularly from pathogenic organisms. It is expected that around 5000–7000 structures per year will be solved; however only 2–3% of the newly solved structures will have new or novel folds. It has been revealed that proteins adopt a limited number of protein folds approximately 1400–1600. Structure of proteins is more conserved than the protein sequence in the evolution which is well understood. Protein structure prediction tools will play an important role to reduce the gap between protein sequence data and protein structural data, in generating 3D homology models for many protein sequences that help us to predict the protein function from structure. Several molecular modeling tools are developed/available for ab initio structure prediction and secondary structure prediction and fold recognition/threading methods for fold assignment and generating tertiary structures using homology/comparative modeling, energy minimization, and molecular dynamic simulation techniques. The generated 3D models shall be evaluated by various evaluating tools, i.e., Procheck, Whatcheck, Verify3D, and ProSA programs.

The process of designing a new drug and launching it into the market is a very complex task. It takes approximately 6–10 years and 500–800 million dollars for the average new drug to go from the research laboratory to patient use. Three-dimensional structures of drug target proteins are essential to design novel drug compounds using rational structure-based drug design (SBDD) approach. Generally virtual screening is used to find initial lead structures from large chemical/drug databases like PubChem/CMC/NCI/ZINC databases against a given drug target structure. There are three major methodologies applied in virtual screening including label extension, piggy bank extension, and de novo drug discovery (Nwaka and Hudson 2006). If the 3D structures of drug target protein are known, docking algorithms can be applied to virtually search potential hits/lead molecules. The docking of two molecules involves search methods (systematic search methods (incremental construction, conformational search), random search (Monte Carlo genetic algorithm, Tabu Search), simulation methods (molecular simulation, energy minimization for the prediction ligand conformation, and orientation with in the target binding site). Search methods (systematic search methods (incremental construction, conformational search), random search (Monte Carlo genetic algorithm, Tabu Search), simulation methods (molecular simulation) are involved in the docking of two molecules. To predict the binding affinity/biological activity, ranking the ligands, various scoring functions (force field-based scoring functions (G-score, Gold Fitness, DOCK score, etc.), empirical scoring functions (SCORE, X-SCORE, ChemScore), knowledge-based scoring functions (PMF, Drug Score, etc.), and consensus scoring functions (SCORE, X-SCORE, CSCORE, etc.)) are used. The well-known protein-ligand docking tools are DOCK, AUTODOCK, LigandFit, AFFINITY, Gold, Glide, FlexX suite, and ICM DOCK. These docking programs use specific search algorithms, scoring functions, and force fields to predict the best mode binding of two molecules. Most of these docking programs treat target protein structures as rigid and ligands as flexible.

It is sensible to perform in silico analysis to identify the target proteins for experimental follow-up. The parameters include druggability, assayability, specificity/selectivity, and importance of the protein in life cycle stages of the pathogen relevant for human health. Though high-throughput screening techniques and biological assays had certainly provided some of the knowledge on gene sequence and functional gene products to this end, further insights come only through structural information. Approximately 36 protein-coding *V. cholerae* genes were determined to be employed in drug target initiatives (Grover et al. 2010). VC0015, VC2730, and VC2770 locus tags of *V. cholerae* are indicated in the genomic target database for specific drug targets in unique metabolic pathways. Herein, we present experimental results of a small set of experiments done for specific case studies in *Vibrio cholerae*, and methods of drug design are discussed.

6.3 The Target

Environmental clones of *V. cholerae*, possessing characteristics of pathogenic isolates, are known to have been associated with clinical isolates causing cholera outbreaks (Singh et al. 2001). Bacterial survival depends on integration of multicellular responses and acclimatizing to changes that occur in the surrounding environment. This is accomplished through a bacterial cell-cell communication process called quorum sensing. Quorum sensing in *V. cholerae* controls virulence and biofilm formation through synthesis, release, and subsequent detection of signaling molecules called autoinducers (Zhu et al. 2002; Hammer and Bassler 2003). Regulation of virulence gene expression, biofilm formation, and protease production is modulated by three sensory pathways in *V. cholerae* (Miller et al. 2002). Quorum-sensing system 1 is composed of the CAI-1 autoinducer and a two-component sensor kinase Cqs. System 2 is composed of AI-2 (a furanosyl borate diester autoinducer), the periplasmic binding protein LuxP, and two-component sensor kinase LuxQ (Chen et al. 2002). Phosphorelay protein LuxU is central to both of these systems. Autophosphorylation of sensor histidine kinases of the respective quorum-sensing circuitries upon activation leads to transfer of phosphate to phospho-transfer protein

LuxU. LuxU interacts with various proteins and acts both as a phosphodonor and acceptor. LuxU transfers the phosphate to LuxO. Phospho-LuxO triggers the transcription of genes encoding sRNAs that destabilize mRNA encoding transcription factor HapR, the terminal effector (Jobling and Holmes 1997; Lenz et al. 2004). Mutations in conserved aspartate residue of LuxO receiver domain severely inhibit the terminal regulator activities and act as a central switch in coordinate regulation of virulence-related phenotypes such as protease production and biofilm formation (Vance et al. 2003). A *LuxO* mutant is severely defective in colonizing the small intestine and production of cholera toxin (Zhu et al. 2002). The quorum sensing systems of bacteria are lucrative avenues for the development of drugs and vaccines against human pathogens, as bacterial signal transduction has no counterpart in the human host.

As a model for exploration of the methods involved in screening drug-like molecules, we used our in silico-generated LuxO 3D structure (Fig. 6.1) which encompasses three distinct domains (Fazil et al. 2012b). We used two sites for drug binding studies: site 1 is the receiver domain and site 2 the ATPase domain; the methods are described in detail in the next section.

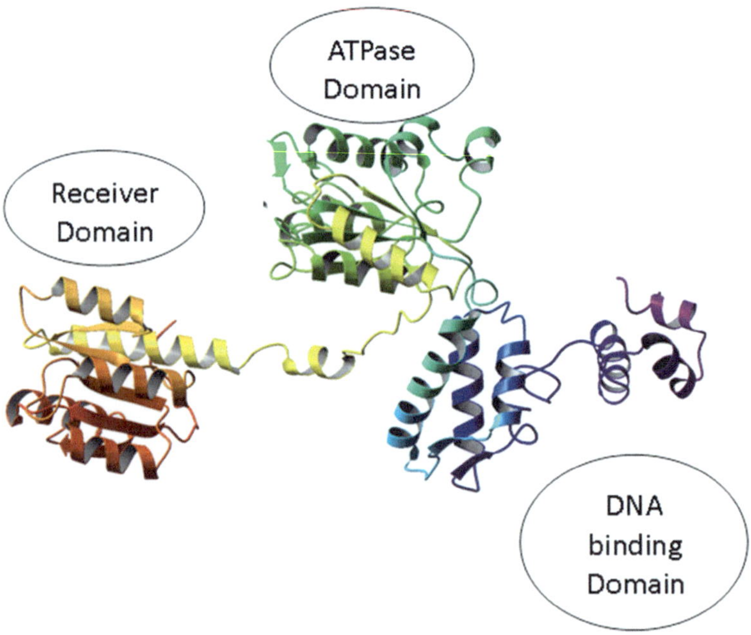

Fig. 6.1 Homology model of LuxO from *Vibrio cholerae*

6.4 The Database

A total of 56,000 molecules from Maybridge database were selected for structure-based virtual screening against the LuxO protein. These ligands were prepared using the LigPrep 2.4 module of the Schrödinger suite of tools. Initially all hydrogen atoms were added to ligand molecules as they had implicit hydrogen atoms. The bond orders of these ligands were fixed. The ionization states of the ligands were generated in the pH range of 5.0–9.0. Tautomers and stereoisomers were generated to study the activity of individual stereotypes of each ligand. Later on compounds were minimized with OPLS-2005 force field.

6.5 Protein Preparation

In in silico docking simulation, the role of protein preparation is critical to get correct interactions with ligands (Lionta et al. 2014). Protein Preparation Wizard of the Schrödinger suite was used in protein structure preparation of LuxO. Protein was subjected to energy minimization using Schrödinger implementation of OPLS-2005 force field with implicit solvation after bond orders in the protein and hydrogen atoms were assigned. Root mean square deviation (RMSD) of the heavy atoms was crucial in determination of energy minimization and was terminated if the RMSD in the minimized structure relative to the homology model structure exceeded 0.3 Å.

6.6 Virtual Screening

Structure-based virtual in silico screening method was used to identify potential ligand molecules that interact with LuxO. In these docking simulations, we used semiflexible docking protocols (Spyrakis and Cavasotto 2015). An arbitrary number of torsional degrees of freedom were explored keeping the ligand docking flexible. The ligand poses created were subjected to a series of hierarchical filters that calculated the ligands interactions with the LuxO protein. The three phases of virtual screening involved three different docking protocols of varying precisions and computational intensities. Virtual screening against LuxO protein was performed each time discarding the molecules based on the Glide HTVS, SP, and XP docking scores. Out of the 56,000 ligands from Maybridge database, a total of 20 potential hits were identified and studied in detail. Primarily, all virtual hits were docked using high-throughput virtual screening (HTVS) protocol. About 10% ligands were selected for standard precision docking based on docking score followed by extra precision docking. Finally, we identified a total of four compounds targeted against binding site 1 and six compounds targeted against binding site 2 using extra precision molecular docking and visual inspection (Tables 6.1a and 6.1b). Analysis of the binding mode of identified compounds revealed interactions with conserved

Table 6.1a The docking results and QikProp properties of screened compounds against binding site 1

S. No	Ligand ID[a]	Interacting residues	Glide score	QPlog P^b	QP log S^c	QP $P_{Caco}{}^d$
1	ligand_57047	Arg63, Ser93, and Glu99	−5.131	1.021	2.514	0.915
2	ligand_45243	Asp18 and Arg63	−4.246	2.285	−3.319	1.006
3	ligand_97998	Leu62 and Leu64	−4.085	1.174	−4.152	87.086
4	ligand_16384	Leu64, Pro65, and Glu99	−3.902	−0.206	−3.459	13.697

[a]Ligand ID
[b]Predicted octanol/water log P from QikProp
[c]Predicted aqueous solubility S in mol/L from QikProp
[d]Predicted Caco-2 cell permeability in nm/s from QikProp

Table 6.1b The docking results and QikProp properties of screened compounds against binding site 2

S. No	Ligand ID[a]	Interacting residues	Glide score	QPlog P^b	QP log S^c	QP $P_{Caco}{}^d$
1	ligand_45214	Gly179, Lys180, Glu181, Gln367	−11.822	1.356	−4.163	2.203
2	ligand_45221	Gly179, Lys180, Glu181, and Asn202	−11.403	1.472	−4.483	0.331
3	ligand_64313	Gly177, Gly179, Lys180, Asp245, and Glu246	−10.988	−2.157	−0.75	70.667
4	ligand_45201	Gly177, Gly179, Lys180, Glu181, Glu185, Asn202	−10.798	1.572	−4.138	0.513
5	ligand_25753	Ile147, Lys180, and Gln367	−10.772	6.064	−7.954	905.946
6	ligand_13590	Gly177, Gly179, Lys180, Glu181, Gln367	−10.712	4.088	−7.298	12.725

[a]Ligand ID
[b]Predicted octanol/water log P from QikProp
[c]Predicted aqueous solubility S in mol/L from QikProp
[d]Predicted Caco-2 cell permeability in nm/s from QikProp

active site residues of LuxO (Figs. 6.2a and 6.2b). The top hit compounds were further subjected to Lipinski's rules and ADME properties. The ADME properties of the final compounds were in acceptable range.

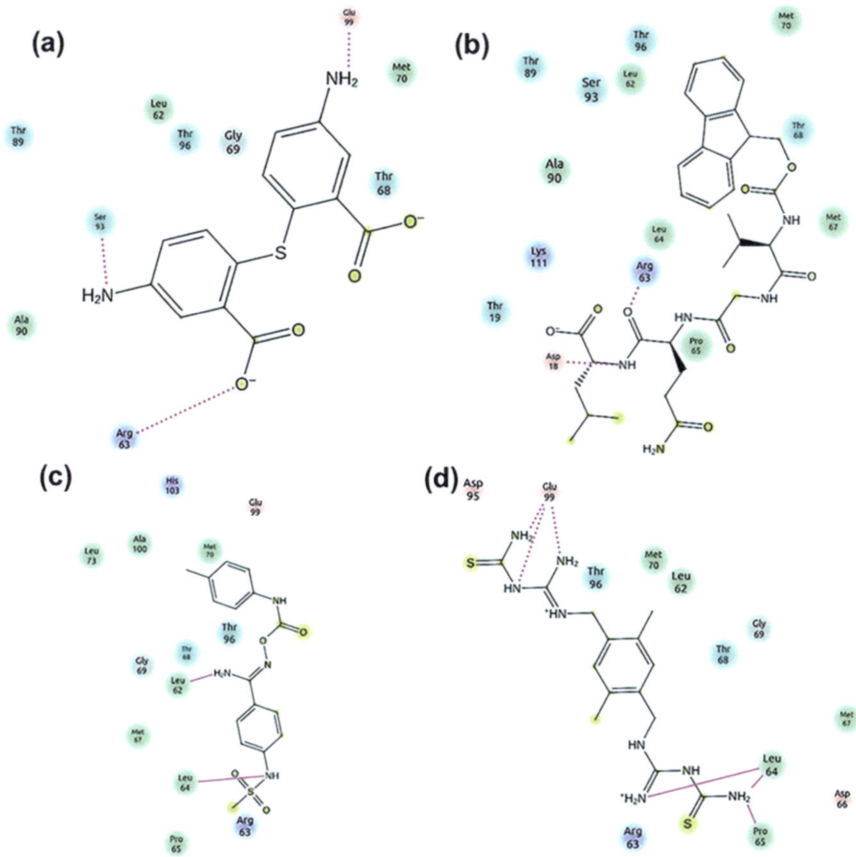

Fig. 6.2a Ligand interaction diagram of top 4 compounds in the binding site 1

6.7 Conclusion

With increasing reports of MDR phenotype in pathogenic bacteria, the antimicrobial armamentarium might soon be obsolete. The knowledge regarding infectious disease processes and the efficiency of novel drugs and available drug targets in *V. cholerae* puts us in a difficult spot in the event of rapid spread of an epidemic. Elucidation of potential sites of ligand binding in quorum-sensing proteins could give a deeper insight into methodologies to subdue this pathogen. Given the modulations in climatic conditions and the bacterium's ability to survive and spread in harsh environments, a rapid breakthrough in drug discovery research against cholera must include computational approaches to target identification and structure-based drug design. Targeted disruption of positions involved in signal

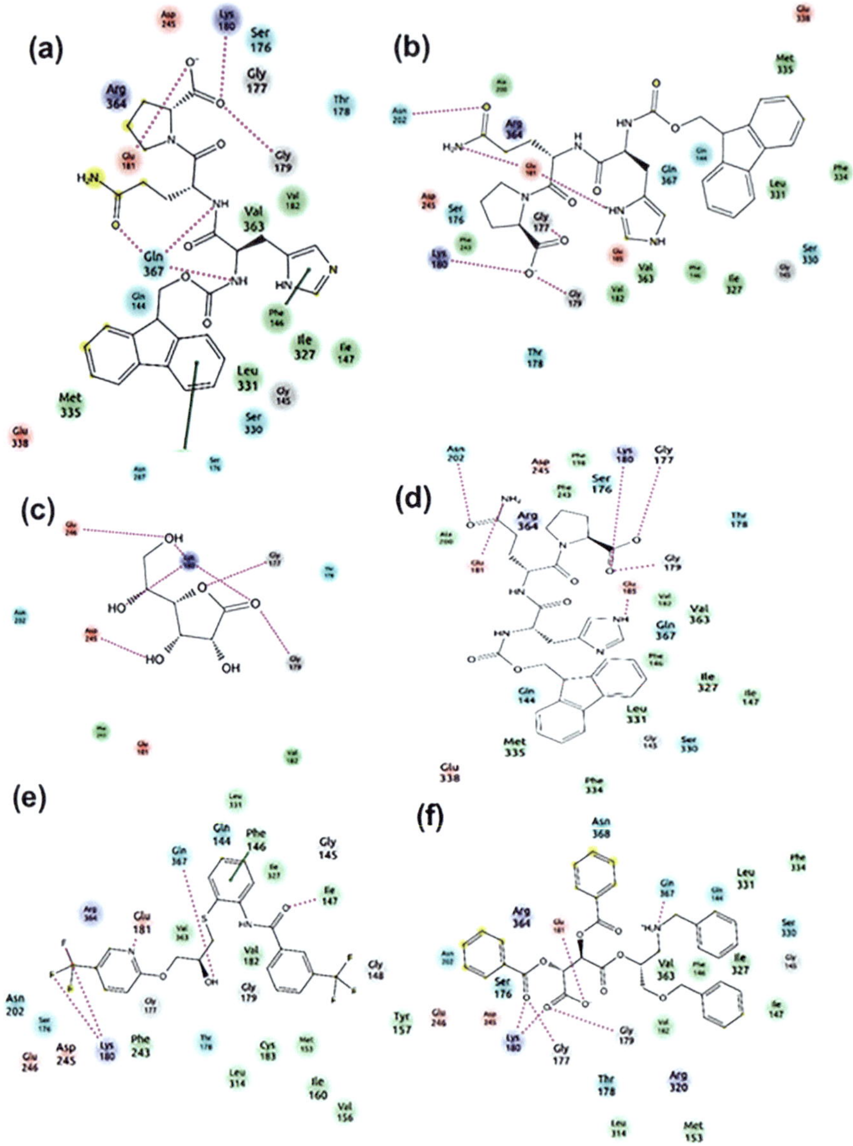

Fig. 6.2b Ligand interaction diagram of top 6 compounds in the binding site 2

propagation and activation of LuxO could result in probable loss of function in these quorum sensors. Elaboration of binding sites for ATP and phosphorylation site in LuxO provided us with an opportunity to utilize binding pockets in designing targeted inhibitors, and thus potential molecules for repression of virulence. Further evaluation of these compounds would be necessary to understand precise mode of action.

References

Albert MJ et al (1996) Phage specific for *Vibrio cholerae* O139 Bengal. J Clin Microbiol 34:1843–1845

Basu A et al (2000) Vibrio cholerae O139 in Calcutta, 1992–1998: incidence, antibiograms, and genotypes. Emerg Infect Dis 6:139–147

Campos J et al (2003) VGJΦ, a novel filamentous phage of *Vibrio cholerae*, integrates into the same chromosomal site as CTXΦ. J Bacteriol 185:5685–5696

Cassel D, Selinger Z (1977) Mechanism of adenylate cyclase activation by cholera toxin: inhibition of GTP hydrolysis at the regulatory site. Proc Natl Acad Sci U S A 74:3307–3311

Chance MR et al (2002) Structural genomics: a pipeline for providing structures for the biologist. Protein Sci 11:723–738

Chang B et al (1998) Filamentous bacteriophages of *Vibrio parahaemolyticus* as a possible clue to genetic transmission. J Bacteriol 180:5094–5101

Chen X et al (2002) Structural identification of a bacterial quorum-sensing signal containing boron. Nature 415:545–549

Ehara M et al (1997) Characterization of filamentous phages of *Vibrio cholerae* O139 and O1. FEMS Microbiol Lett 154:293–301

Faruque SM et al (2003) CTXphi-independent production of the RS1 satellite phage by *Vibrio cholerae*. Proc Natl Acad Sci U S A 100:1280–1285

Fazil MHUT, Kumar S, Farmer R, Pandey HP, Singh DV (2012a) Binding efficiencies of carbohydrate ligands with different genotypes of cholera toxin B: molecular modeling, dynamics and docking simulation studies. J Mol Model 18(1):1–10

Fazil MH et al (2012b) Comparative structural analysis of two proteins belonging to quorum sensing system in *Vibrio cholerae*. J Biomol Struct Dyn 30(5):574–584

Gill DM, Rappaport RS (1979) Origin of the enzymatically active A1 fragment of cholera toxin. J Infect Dis 139:674–680

Grover PA, Kuntal H, Sharma V (2010) In silico prediction of drug targets in Vibrio cholerae. Protoplasma 248(4):799–804

Hammer BK, Bassler BL (2003) Quorum sensing controls biofilm formation in *Vibrio cholerae*. Mol Microbiol 50:101–104

Hammer BK, Bassler BL (2009) Distinct sensory pathways in Vibrio cholerae El Tor and classical biotypes modulate cyclic dimeric GMP levels to control biofilm formation. J Bacteriol 191:169–177

Holmgren J et al (1989) New cholera vaccines. Vaccine 7:94–96

Howard Jones N (1979) Cholera nomenclature and nosology; a historical note. Bull WHO 51:317–324

Huq A et al (1984a) Influence of water temperature, salinity, and pH on survival and growth of toxigenic Vibrio cholerae serovar 01 associated with live copepods in laboratory microcosms. Appl Environ Microbiol 48:420–424

Huq A et al (1984b) The role of planktonic copepods in the survival and multiplication of *Vibrio cholerae* in the aquatic environment. In: Colwell RR (ed) Vibrios in the environment. Wiley, New York, pp 521–534

Ikema M, Honma Y (1998) A novel filamentous phage, fs-2, of *Vibrio cholerae* O139. Microbiology 144:1901–1906
Jobling MG, Holmes RK (1997) Characterization of hapR a positive regulator of the *Vibrio cholerae* HA/protease gene hap and its identification as a functional homologue of the *Vibrio harveyi* luxR gene. Mol Microbiol 26:1023–1034
Jouravleva EA et al (1998) Characterization and possible functions of a new filamentous bacteriophage from *Vibrio cholerae* O139. Microbiology 144:315–324
Kahn RA, Gilman AG (1984) ADP-ribosylation of Gs promotes the dissociation of its alpha and beta subunits. J Biol Chem 259:6235–6240
King CA, Van Heyningen WE (1973) Deactivation of cholera toxin by a sialidase-resistant monosialosylganglioside. J Infect Dis 127:639–647
Lenz DH et al (2004) The small RNA chaperone Hfq and multiple small RNAs control quorum sensing in *Vibrio harveyi* and *Vibrio cholerae*. Cell 118:69–82
Levine MM, Tacket CO (1994) Recombinant live cholera vaccines. In: Wachsmuth IK, Blake PA, Olsvik O (eds) *Vibrio cholerae* and cholera: molecular to global perspectives. American Society for Microbiology, Washington, DC, pp 39–415
Levine MM et al (1981) Duration of infection-derived immunity to cholera. J Infect Dis 143:818–820
Lionta E et al (2014) Structure-based virtual screening for drug discovery: principles, applications and recent advances. Curr Top Med Chem 14(16):1923–1938
Miller MB et al (2002) Parallel quorum sensing systems converge to regulate virulence in *Vibrio cholerae*. Cell 110:303–314
Moss J et al (1980) Activation of choleragen by thiol: protein disulfide oxidoreductase. J Biol Chem 255:11085–11087
Nwaka S, Hudson A (2006) Innovative lead discovery strategies for tropical diseases. Nat Rev Drug Discov 5:941–955
Pierce NF (1973) Differential inhibitory effects of cholera toxoids and ganglioside on the enterotoxins of *Vibrio cholerae* and *Escherichia coli*. J Exp Med 137:1009–1023
Ramamurthy T (2008) Antibiotic resistance in *Vibrio cholera*. In: Faruque SM, Nair GB (eds) *Vibrio cholerae*: genomics and molecular biology. Caister Academic Press, London, pp 191–207
Sanchez, Holmgren J (2008) Cholera toxin structure, gene regulation and pathophysiological and immunological aspects. Cell Mol Life Sci 65:1347–1136
Singh DV et al (2001) Molecular analysis of *Vibrio cholerae* O1, O139, non-O1 and non-O139 strains: clonal relationships between clinical and environmental isolates. Appl Environ Microbiol 67:910–921
Spangler BD (1992) Structure and function of cholera toxin and the related Escherichia coli heat-labile enterotoxin. Microbiol Rev 56:622–647
Spyrakis F, Cavasotto CN (2015) Open challenges in structure-based virtual screening: receptor modeling, target flexibility consideration and active site water molecules description. Arch Biochem Biophys 583:105–119
Stokes HW, Hall RM (1989) A novel family of potentially mobile DNA elements encoding site-specific gene-integration functions: integrons. Mol Microbiol 3:1669–1683
Tacket CO et al (1999) Randomized, double-blind, placebo-controlled, multicentered trial of the efficacy of a single dose of live oral cholera vaccine CVD 103-HgR in preventing cholera following challenge with *Vibrio cholerae* O1 El tor Inaba three months after vaccination. Infect Immun 67:6341–6345
Van Dellen KL, Watnick PI (2006) The Vibrio cholerae biofilm: a target for novel therapies to prevent and treat cholera. Drug Discov Today Dis Mech 3:261–266
Van Heyningen WE et al (1971) Deactivation of cholera toxin by ganglioside. J Infect Dis 124:415–418

Vance RE, Zhu J, Mekalanos JJ (2003) A constitutively active variant of the Quorum-Sensing regulator LuxO affects protease production and biofilm formation in *Vibrio cholerae*. Infect Immun 71(5):2571–2576

Vitkup D et al (2000) Solvent mobility and the protein 'Glass' transition. Nat Struct Biol 7:34–38

Waldor MK, Mekalanos JJ (1996) Lysogenic conversion by a filamentous phage encoding cholera toxin. Science 272:1910–1914

Waldor MK, Tschape H, Mekalanos JJ (1996) A new type of conjugative transposon encodes resistance to sulfamethoxazole, trimethoprim, and streptomycin in Vibrio cholerae O139. J Bacteriol 178:4157–4165

Zhu J et al (2002) Quorum-sensing regulators control virulence gene expression in *Vibrio cholerae*. Proc Natl Acad Sci U S A 99:3129–3134

Modelling Polyketide Synthases and Similar Macromolecular Complexes

7

Rohit Farmer, Christopher M. Thomas, and Peter J. Winn

Abstract

Science is slowly unlocking the secrets of the exquisite chemical synthesis capabilities of polyketide synthases (PKSs), as well as other secondary metabolites' biosynthesis pathways, and learning to re-engineer such pathways to synthesize novel chemical compounds. Research over the last 30 years has involved innovative experiments and bioinformatics focused on a wide range of medicinal compounds ranging from antibiotics to anticholesterol agents. Furthermore, it has been possible to manipulate PKSs to produce novel compounds for pharmaceutical use. However, despite great progress, our knowledge is still sketchy, and experiments continue to be time-consuming and difficult. PKSs, and secondary metabolite biosynthetic pathways in general, provide model systems for developing and testing experimental and bioinformatic tools for synthetic biology application. Bioinformatic and molecular modelling are important for making sense of existing and future experimental data. Bioinformatic and structural modelling can help in several ways: by predicting how manipulations of protein domains might yield viable novel biosynthetic pathways to generate variants of existing chemicals/pharmaceuticals of high value or to allow the synthesis of totally novel compounds, by assisting the discovery of novel gene clusters in genomic and metagenomic data, by

R. Farmer (✉)
School of Biosciences, University of Birmingham, Edgbaston, Birmingham B15 2TT, UK

Department of Computational Biology and Bioinformatics, Jacob Institute of Biotechnology and Bioengineering, Sam Higginbottom University of Agriculture, Technology and Sciences, Allahabad 211007, Uttar Pradesh, India
e-mail: rohit.farmer@shiats.edu.in

C. M. Thomas · P. J. Winn (✉)
School of Biosciences, University of Birmingham, Edgbaston, Birmingham B15 2TT, UK
e-mail: p.j.winn@bham.ac.uk

G. Wadhwa et al. (eds.), *Current trends in Bioinformatics: An Insight*,
https://doi.org/10.1007/978-981-10-7483-7_7

predicting the metabolites synthesized by novel gene clusters and by interpreting experimental data to elucidate the rules governing polyketide synthase function, which feeds back into the others on this list.

Keywords
Polyketide synthases · Pharmaceuticals · Structural modelling

7.1 Introduction

Science is slowly unlocking the secrets of the exquisite chemical synthesis capabilities of polyketide synthases (PKSs), as well as other secondary metabolites' biosynthesis pathways, and learning to re-engineer such pathways to synthesize novel chemical compounds (Weissman and Leadlay 2005). Research over the last 30 years has involved innovative experiments and bioinformatics focused on a wide range of medicinal compounds ranging from antibiotics to anticholesterol agents. Furthermore, it has been possible to manipulate PKSs to produce novel compounds for pharmaceutical use (Weissman and Leadlay 2005; Challis 2008; Park et al. 2010). However, despite great progress, our knowledge is still sketchy, and experiments continue to be time-consuming and difficult. In particular, decrypting how the different components of these mega-dalton complexes interact and how their spatial organization affects their function is important for developing a PKS tool box for the production of novel therapeutic and other compounds of high commercial value. Bioinformatic and molecular modelling are important for making sense of existing and future experimental data, for the rational design of secondary metabolite biosynthesis de novo and for the efficient discovery of existing but unknown secondary metabolites of potential therapeutic or commercial value.

Of particular interest is adapting PKS proteinaceous assembly lines to re-engineer pharmaceuticals derived from existing natural products to improve their drug-like properties. For example, the antibiotic mupirocin is one of the few compounds currently effective against methicillin-resistant *Staphylococcus aureus*, but it can only be used topically since it contains an ester linkage that rapidly degrades in the body (Thomas et al. 2010). Remodelling the mupirocin biosynthetic pathway to replace the ester linkage with something more stable is currently underway and might produce a mupirocin derivative with greatly enhanced pharmacological properties. Developing our ability to routinely manipulate the mupirocin or other similar PKS pathways would thus be of clear benefit. Indeed, with many of our current antibiotics being derived from PKS pathways and related systems, re-engineering PKS provides one possibility for fighting back against increasing bacterial resistance to current treatments.

Bioinformatic and structural modelling can help in several ways: by predicting how domain manipulations might yield viable novel synthetic pathways to generate variants of existing chemicals/pharmaceuticals of high value, or to allow the synthesis of totally novel compounds; by assisting the discovery of novel gene clusters in genomic and metagenomic data; by predicting the metabolites synthesized by novel gene clusters; and by interpreting experimental data to elucidate the rules governing polyketide synthase function, which feeds back into the others on this list. In addition to predicting what manipulations should be made, there is much ongoing research into the genetic tools needed to manipulate these pathways (Weissman and Leadlay 2005; Challis 2008; Park et al. 2010; Kwon et al. 2012) something that we do not discuss further here.

7.1.1 Polyketides

Polyketides are a large set of secondary metabolites derived from natural sources such as bacteria, fungi and plants (Fig. 7.1). They display a myriad of clinical applications such as antibiotic (erythromycin, tetracycline and mupirocin), antifungal (amphotericin), immunosuppressant (FK506 and rapamycin), antitumor (doxorubicin, geldanamycin and epothilone B), antituberculosis (rifamycin B) and anticholesterol (lovastatin). Such structurally diverse and therapeutically useful compounds are synthesized by the addition of simple keto units catalyzed by polyketide synthases (PKSs), while additional tailoring enzymes add diversity during and post-polyketide synthesis.

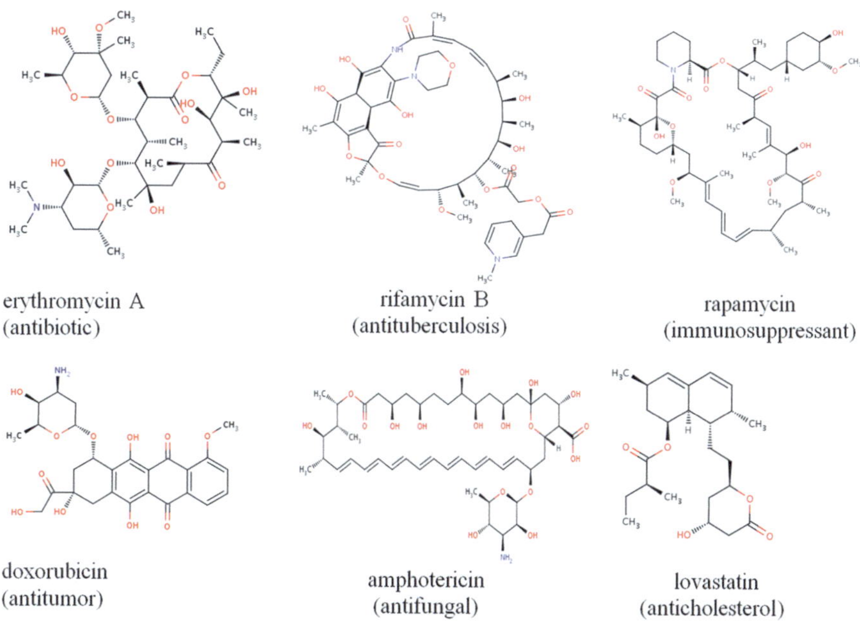

Fig. 7.1 Some examples of polyketides (Knox et al. 2011)

7.1.2 Polyketide Synthases

Polyketide synthases are related through their evolution and biochemical function to the very well-understood fatty acid synthases (FASs), which catalyze a two-carbon backbone extension by the addition of a keto moiety followed by biochemical reduction; when iterated this process produces a saturated fatty acid. PKSs differ from FASs in that they may only partially reduce the keto group or may not reduce it at all. Like FASs, PKSs are classified as type I or type II but also have a type III. Type III PKSs are much simpler than types I and II, consisting of homodimeric ketosynthase (KS) domains. They are responsible for chalcones, stilbenes and similar compounds and are often referred to as chalcone synthases (CHSs) or stilbene synthases. They have primarily been observed in plants but in recent years also in some bacteria (Austin and Noel 2003) and fungi (Hertweck 2009). The full intricacies of PKSs are discussed in detail elsewhere (Austin and Noel 2003; Hertweck 2009), with a brief introduction to type I and type II given below such that the reader can understand the bioinformatic and molecular modelling problems that they pose and the possibilities for commercial and therapeutic benefit that they proffer.

Type I PKSs (and FASs) are composed of diverse catalytic domains covalently fused to form a large multifunctional complex (Hertweck 2009). A minimal type I PKS must have at least three domains, a KS, an acyltransferase (AT) and an acyl carrier protein (ACP), which together form a module capable of extending the backbone of a polyketide synthase by two carbons. The AT domain loads an extender unit, i.e. the basic biochemical building block of a polyketide, onto the ACP. The KS domain is loaded with a substrate that is either a starter unit, if it is the first elongation step of the pathway, or an intermediate from the previous step of the biosynthetic pathway. The KS domain catalyzes a Claisen condensation creating a new carbon-carbon bond between the extender unit on the ACP and the substrate attached to the KS. The ACP holds the extended product, which is subsequently passed to the next module, as shown in the example in Fig. 7.2, or passed back to the KS of the same module for further extension, as shown in Fig. 7.3. A PKS can accept a diverse range of starter and extender units comprising two, three or four carbon building blocks, such as acetyl-CoA, propionyl-CoA and butyryl-CoA malonyl-CoA, methylmalonyl-CoA, and ethylmalonyl-CoA (Yinyan Tang et al. 2006; Khosla et al. 1999; Staunton and Weissman 2001).

Apart from the minimal essential domains, a variety of other domains add functionality to the growing polyketide chain; some of them are shown in Fig. 7.2. Ketoreductase (KR) domains catalyze beta-keto reduction. Dehydratase (DH) domains catalyze the dehydration at the beta hydroxyl. Enoylreductase (ER) domains further reduce a double bond to a fully saturated carbon-carbon bond, such in fatty acids. Methyltransferase (MT) domains add a methyl group at the alpha carbon (Fig. 7.3) using S-adenosyl methionine as the methyl donor. Terminal thioesterase (TE) domains catalyze the final hydrolysis of the polyketide chain to release the end product from the ACP.

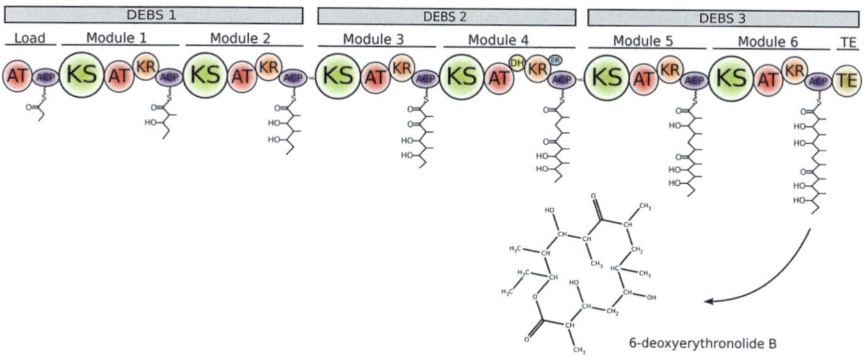

Fig. 7.2 An example of a type I modular PKS with an AT *in cis*: DEBS biosynthesis. The 6-deoxyerythronolide B synthase (DEBS) system comprises three polypeptides that have between them a loading didomain, six extension modules and a terminal thioesterase (*TE*) domain. The order of processing is determined by docking domains at the polypeptide termini. These ensure the specific interaction of the C-terminus of the polypeptide DEBS1 with the N-terminus of DEBS2 and the C-terminus of DEBS2 with the N-terminus of DEBS3. The loading domain loads the KS of module 1 with a propionyl starter unit, which is then elongated by a Claisen condensation with a methyl-malonyl extender group held on the ACP. One molecule of CO_2 is released during this process. The KR of module 1 then reduces the beta-keto group to OH, to give the product shown attached to the ACP of module 1. The ACP then passes this substrate to the KS of the next module, where it is condensed with another methyl malonyl, with the same pattern repeated along the biosynthetic pathway until the product is released by the thioesterase (*TE*). *Abbreviations*: *KS* ketosynthase, *AT* acyltransferase, *ACP* acyl carrier protein, *KR* ketoreductase, KR^0 inactive ketoreductase, *DH* dehydratase, *ER* enoylreductase (Khosla et al. 2007)

Tailoring enzymes may add functionality to the end product of a PKS extending their structural diversity (Bender et al. 1999), for example, cyclization in the DEBS system (6-deoxyerythronolide B synthase), formation of a pyran ring and beta methylation in mupirocin biosynthesis (Chan et al. 2009) and addition of sugar domains by glucosyl transfer domains in macrolides (Lairson et al. 2008).

There a number of variations around the type I PKS described above. Fungal PKSs of type I tend to be iterative (Fig. 7.3), i.e. the same multienzyme module performs a substrate extension reaction multiple times, whereas bacterial PKSs of type I tend to undertake one synthetic cycle per module (Fig. 7.2), with multiple modules involved in polyketide chain extension and tailoring. Type I PKSs can have AT domains that exist as separate polypeptide chains that act *in trans*, e.g. in mupirocin biosynthesis (Gurney and Thomas 2011; Piel 2010); these are referred to as AT-less or *trans-AT* type I PKS and are interesting for their ability to incorporate additional modifications *in trans*, such as the HCS cassette mentioned later. Type I PKS where the AT is part of the PKS polypeptide chain, e.g. as shown in Fig. 7.2, are referred to as *cis*-AT PKSs.

Type II PKSs are complexes of individual enzymes performing the same functions as in type I PKSs, but they might not be encoded in a linear fashion on the genome and may dissociate in vivo and typically lack an acyl transferase (AT) domain. Decarboxylase (KS^Q or KS_β), KS and ACP domains are the minimal

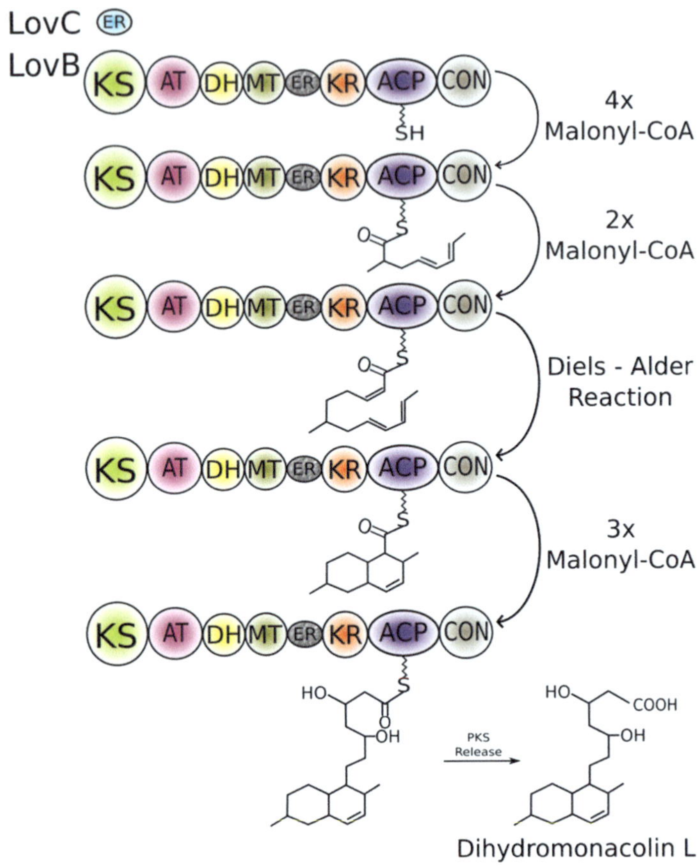

Fig. 7.3 An example of a type I iterative PKS: the action of LovB in lovastatin biosynthesis. Lovastatin is synthesized by two megasynthases, the lovastatin nonaketide synthase (*LovB*) and lovastatin diketide synthase (LovF) along with LovC (a dissociated *ER*) and numerous accessory enzymes. This figure shows LovB which is a single module used *iteratively* to synthesize the nonaketide decaline core from nine malonyl-CoA units. The ER domain in LovB is found to be non-functional, this function being provided by LovC. The malonyl-CoA:ACP acyltransferase (*AT*) selects and transfers the extender units in the form of malonic esters, and the methyl transferase (*MT*) incorporates the methyl group from S-adenosyl methionine (SAM). The LovB also contains a C-terminal domain similar to a condensation domain from a nonribosomal peptide synthase (labelled CON in the figure) with unknown function. Four iterations of LovB-catalyzed elongation are followed by the action of the MT domain. Two further iterations of LovB-catalyzed elongation are followed by a Diels-Alder reaction and then a further three elongations by LovB (Staunton and Weissman 2001; Ma and Tang 2007)

unit for a type II PKS (Austin and Noel 2003), and these are often accompanied by KR, cyclase and aromatase domains (Austin and Noel 2003; Hertweck 2009). Type II PKSs are typically involved with the production of aromatic compounds (Hutchinson et al. 2000; Hertweck 2009).

Tailoring enzymes may also act *in trans* during polyketide elongation reactions, notably *in trans*-AT type I PKSs, e.g. the HCS (hydroxyl methylglutaryl coenzyme A synthase) cassette. HCS cassettes can introduce β-methyl, cyclopropane and vinyl chloride moieties, depending on the exact domains present in a system. The HCS cassette has at least four enzymatic domains: an ACP, a hydroxymethyl glutaryl-CoA synthase, one or more dehydratases belonging to the crotonase superfamily and a decarboxylase, in some cases supplemented with other functions, e.g. a halogenase domain (Busche et al. 2012). HCS cassettes act on the type I modules which lack KR, DH or ER domains but have tandemly repeated ACPs (Wu et al. 2007). For example, methylation at the beta carbon position of the C14 monic acid moiety in mupirocin biosynthesis is hypothesized to be carried out by an HCS cassette. Similarly an HCS cassette together with a halogenase in the curacin A biosynthetic pathway is responsible for catalyzing cryptic chlorination, subsequently leading to cyclopropane ring formation. A mutagenesis study characterized the molecular interactions between the ACP and halogenase domains in the curacin A PKS (Khare et al. 2010; Busche et al. 2012). Such studies combined with computational modelling can help decipher the mechanisms of molecular interaction essential for modification of polyketides by tailoring enzymes, as we discuss later. This in turn would benefit our ability to both predict the metabolites produced by gene clusters and re-engineer PKS/NRPS clusters to our own advantage.

A number of groups have made contributions to our current understanding of PKS pathways and how to re-engineer them, reviewed in detail elsewhere (Weissman and Leadlay 2005; Challis 2008; Kwon et al. 2012; Piel 2010) with the macrolide systems being particularly popular systems for study (Park et al. 2010). Much work has been focused on the type I PKS. For example, Khosla's group has extensively manipulated and modelled the DEBS system (Khosla et al. 2007) with recent work using a number of chimaeric constructs of ACPs, crystal structure determination and computational protein-protein docking to reveal how an ACP has specificity for elongation of the DEBS chain in its own module in the DEBS pathway (Kapur et al. 2010), how an ACP specifically passes the processed substrate onto the next module (Kapur et al. 2012, 2010) and why a PKS dimer might be required for function.

This insight into the DEBS system, which is shown in Fig. 7.2, was achieved by considering a number of chimaeras using the sequences of ACP3 and ACP6 from deoxyerythronolide B synthase, the ACPs from modules 3 and 6 of the synthetic pathway. The chimaeras indicated that the loop between helix I and helix II was critical during the elongation process of synthesis for the interaction of the chimaeric or WT ACPs with KS5-AT5, the ketosynthase-acetyl transferase didomain from module 5. Computer docking of the ACP onto the crystal structure of the KS5-AT5 homodimer indicated two residues in the loop that appeared critical for electrostatic complementarity (D44 and R45). Further modelling showed electrostatic complementarity between the KS-AT didomain in each module and the equivalent residues in their cognate ACP, ACP3 (R44, R45), ACP4 (44R, 45 K), ACP5 (44D, 45R) and ACP6 (44D, 45Q), while R44A/R45A mutations in ACP3 confirmed the importance of the residues at these positions. They also found residues at the N-terminus of helix II to be important. The authors noted that these key residues of the ACP interact with the linker region connecting the KS and AT modules, as well as the AT module, and thus this mechanism cannot be the same for PKSs that have the AT acting *in trans*. In contrast to the elongation mechanism, for the transfer of the substrate from an ACP to the KS on the next module, they found a different mechanism, one that relied on the residues in helix I of the ACP. While this work emphasizes the benefit of computational modelling working together with well-designed experiments to elucidate the engineering principles underlying the PKS, a recent structure of the whole PikAIII module from the pikromycin biosynthetic pathway obtained by electron microscopy (Dutta et al. 2014) suggests that the KS-AT structure used for the docking may not be in a physiologically relevant conformation, as discussed later.

7.1.3 NRPS

NRPSs are also highly prevalent in secondary metabolite biosynthesis and often synthesize compounds in conjunction with PKSs. The NRPS and PKS systems have similar modular structures and often combine in the same biosynthetic pathway; thus NRPSs warrant some discussion here. Nonribosomal peptides are the product of the sequential addition of amino acid monomers catalyzed by NRPSs, involving domains similar in function to that of PKS systems. Amino acid monomers are selected and activated by an adenylation (A) domain; peptide bond formation is catalyzed by a condensation (C) domain and a thiolation domain, also known as a peptide carrier protein, (T or PCP) with a phosphopantetheine prosthetic group that facilitates the transfer of growing peptide chain/monomers to the various active sites. The C, A, T domains form the minimal set of domains required to carry out the NRP biosynthesis. In both the modular PKS and NRPS systems, the individual domains are linked together by a polypeptide linker region which has also been found to be responsible for functional communication within the domains (Gokhale and Khosla 2000).

7.2 Knowledge-Based Modelling of Polyketide Synthases

It is clear from the previous sections that PKSs are complex machines that produce equally complex products of great therapeutic and commercial value. The above description only scratches the surface of what we know of their complexity, and there is much that is still to be understood, particularly if we are to be able to successfully modify existing pathways or use them to make de novo compounds. Beyond their own intrinsic complexities, PKSs can also fuse with other biosynthetic machinery, such as the NRPSs, adding further complexity to the natural products produced. Computational tools are needed to assist in interpreting experimental data, in the discovery of novel compounds and, in the longer term, to dassist in the design of de novo biosynthesis pathways.

The term knowledge-based modelling here refers to the methods of protein structure/complex and secondary metabolite prediction based on the prior knowledge of similar systems, knowledge derived via various data mined from the literature or via novel experiments, e.g. mutagenesis experiments, spectrometry, crystallography and genomics. With such data various computational algorithms can be trained to identify prediction rules and signatures to classify or identify similar features in the novel pathways or genomes.

Below we review some of the resources available for identifying novel clusters, domain organization in the cluster, predicting the products produced by the clusters and predicting the 3D structure of the PKS complexes, with an emphasis on the use of structural modelling.

7.2.1 Bioinformatic Approaches in PKS Research

Research into PKS pathways has been carried out in two major areas: firstly, in identification and experimental characterization of new polyketide natural products and, secondly, in developing a synthetic biology tool box for the design and synthesis of novel "natural" products by the re-engineering of naturally occurring polyketide biosynthetic machinery, both of which might provide the basis for novel drugs. Bioinformatic analysis of PKS clusters has played a major role in guiding experiments via a variety of tools; examples are PKSDB (Yadav et al. 2003) a database of PKS domains developed over a decade ago, which was one of the first tools available, and SBSPKS (Anand et al. 2010) and antiSMASH (Medema et al. 2011), which are recent advanced sequence analysis tools guided by protein structure analysis. Computational analysis of experimentally characterized PKS clusters can enhance our understanding of polyketide biosynthesis and also help formulate rules for classification of PKSs into modular and iterative types, the order of substrate channelling for modular PKSs and the number of chain extension reactions catalyzed by iterative PKSs (Yadav et al. 2009).

Most work in the field has focused on the core PKS functions, particularly type I PKS; thus there is still little known about the mechanisms of the tailoring enzymes acting *in trans*. Such tailoring enzymes are an essential part of any PKS system since they provide additional chemical diversity to the polyketide chain. An ability to predict the function of tailoring enzymes and their compatibility with other PKS modules would provide much additional functionality to the synthetic biology tool box. Similarly *trans*-AT systems are less well studied, and since software tools are primarily targeted at *cis*-AT systems, they tend to completely fail or have limited capabilities in predicting *trans*-AT systems.

7.2.2 Computational Tools Available for the PKS and NRPS Researcher

PKSDB/SEARCHPKS and NRPS-PKS were the first available web-based tools for identifying PKS/NRPS domains in an unknown sequence as well as relating PKS/NRPS sequences to their corresponding secondary metabolites, and these are reviewed in detail elsewhere (Jenke-Kodama and Dittmann 2009; Bachmann and Ravel 2009). Following similar lines, resources like ASMPKS, ClustScan, CLUSEAN, NP.searcher, NRPSpredictor, NRPSsp and antiSMASH have been developed for the discovery of secondary metabolites through genome analysis. All these servers primarily utilize sequence information either for domain identification or to correlate the various PKS domains to their corresponding metabolic products. However, many of them also utilize structural information for predicting the most likely starter and extender units picked by the AT domains and SBSPKS models the 3D structure of PKS modules. Table 7.1 gives a summary of the resources available for the PKS/NRPS pathway analysis.

Comparing all the above-mentioned resources, the various tasks that can be performed can be divided into four major categories: (1) obtaining a well-curated PKS/NRPS cluster database for further analysis, (2) protein domain detection for various modular PKSs and NRPs, (3) predicting substrate specificity for various starter and extender units and (4) correlating identified clusters to their corresponding metabolites.

The NRPS-PKS database serves as an amalgamated source for PKS clusters (type I and type III) and NRPS clusters. Data in this database is derived from the PKSDB (19 modular PKS cluster), NRPSDB (17 NRPS clusters and 5 hybrid PKS +NRPS clusters), ITERDB (21 iterative PKSs) and CHSDB (11 plant chalcone PKSs and 3 bacterial chalcone PKSs) databases. Another database, ASMPKS, presently contains 41 characterized PKS pathways including everything in PKSDB, with more entries being added. A user may also add or delete their own entries. NORINE is a database for nonribosomal peptides (Caboche et al. 2008). At the time of writing, July 2016, it contained 1184 peptide products and over 500 monomers; however it does not provide information on biosynthesis.

Why is there a need for specialized databases for PKS/NRPS research when databases like CDD (Marchler-Bauer et al. 2011) and InterPro (Hunter et al. 2012) or domain finding software like SMART (Letunic et al. 2012) exist? It was observed by Yadav et al. (2003) during the construction of PKSDB/SEARCHPKS that in spite of CDD and InterPro being a vast source of protein domains, they suffer from being general and not tailored for a specific purpose. Comparing the CDD results from their domain identification program, they found at that time that CDD identified no DH domains in the modular PKS clusters and could not distinguish between KR and ER domains. However, over time CDD has improved and in our analyses has been generally able to detect PKS domains, e.g. with MmpD from the mupirocin pathway, it is able to predict all the domains except the catalytically inactive DH domain in module 1. Nonetheless, PKS-specific tools continue to be

Table 7.1 Resources available for secondary metabolite prediction

Resources	Clusters	Prediction tools	Domain specifities analysed	Back-end/training data source	Hyperlink
SEARCHPKS/ PKSDB (database)	PKS	BLAST	AT	PKSDB	http://www.nii.res.in/searchpks.html, http://www.nii.res.in/pksdb.html
NRPS-PKS (database)	NRPS, PKS	BLAST	AT, A	PKSDB, NRPSDB, ITERDB, CHSDB	http://www.nii.res.in/nrps-pks.html
ASMPKS	PKS	GLIMMER, BLAST	AT	PKSDB, more	http://gate.smallsoft.co.kr:8008/~hstae/asmpks/genome.pl
NRPS predictor/ NRPSpredictor2	NRPS	SVM/TSVM	A	Training data amalgamated from various sources	http://nrps.informatik.uni-tuebingen.de/Controller?cmd=SubmitJob
CLUSTSCAN/ CompGen (homologous recombination module)	PKS/NRPS/PKS-NRPS hybrid	HMM	KR, AT	Pfam, specialized	http://bioserv.pbf.hr/cms/ http://bioserv.pbf.hr/cms/index.php?page=compgen
SBSPKS	NRPS, PKS	BLAST, 3D structure modelling	At, A	PKSDB, NRPSDB, ITERDB, CHSDB	http://www.nii.ac.in/~pksdb/sbspks/master.html
NORINE (database)	NRPS products, monomers				http://bioinfo.lifl.fr/norine/
CLUSEAN (Perl module framework)	NRPS, PKS	BLAST, HMM	A	NCBI NR, Pfam, specialized	http://redmine.secondarymetabolites.org/projects/clusean

(continued)

Table 7.1 (continued)

Resources	Clusters	Prediction tools	Domain specifities analysed	Back-end/training data source	Hyperlink
antiSMASH (metaserver and standalone)	NRPS, PKS, terpenes, aminoglycosides, aminocoumarins, indolocarbazoles, lantibiotics, bacteriocins, nucleosides, beta-lactams, butyrolactones, siderophores, melanins and others	NCBI BLAST+, HMMER 3, Muscle 3, Glimmer 3, FastTree, TreeGraph 2	AT, A, KR	Amalgamated from various previously published works	http://antismash.secondarymetabolites.org/
NRPSsp	NRPS	HMM	A	Specialized	http://www.nrpssp.com
SMURF (secondary metabolite unique regions finder)	PKSs, NRPSs, hybrid PKS-NRPSs and prenyltransferases	HMM		Pfam, TIGRFAM	www.jcvi.org/smurf/

developed and are likely to outperform generic domain finding algorithms by, e.g. having more complete training sets and taking into account the context of co-occurrence of genes when undertaking gene cluster annotation.

Tailored NRPS/PKS domain prediction has been achieved by either a BLAST (Altschul et al. 1990) search of the query sequence against a back-end database or through querying a database of profile hidden Markov models (HMMs) trained on domains from PKS/NRPS clusters. The domain detection algorithms in NRPS-PKS, ASMPKS and SBSPKS use BLAST, whereas ClustScan, NRPSsp, CLUSEAN and antiSMASH use profile HMMs. HMMs have also been used by Ansari and coworkers (2008) to identify MT domains present in type I PKS and NRPS megasynthases and to sub-group them into N-MT, C-MT and O-MT groups based on the prediction of their site of methylation. In another study, Foerstner et al. (2008) used HMMs to screen eight metagenomics shotgun data sets in order to estimate the frequency of type I PKSs using HMMs for eight domains. They also incorporated analysis of maximum likelihood phylogenetic trees to increase the reliability and resolution of the data set and to discriminate true PKS I domains from evolutionarily related but functionally different ones.

ClustScan (Starcevic et al. 2008) also utilizes profile HMMs, both extracted from Pfam and specifically constructed profiles. ClustScan works as a client server application with the main program running on a Linux server and a Java client running on the user's computer, making it compatible with all operating systems. It uses Glimmer or GeneMark for gene prediction followed by HMMs for the domain identification in PKS/NRPS/PKS-NRPS hybrid clusters. At time of writing, the ClustScan database contains data for 57 PKS clusters, 51 NRPS clusters and 62 PKS-NRPS hybrid clusters. Apart from domain identification and substrate specificity detection, ClustScan can also export the chemical structures of predicted products in a SMILES/SMARTS format for further analysis by standard chemistry programs. ClustScan also acts as a graphical interface to CompGen, a tool that undertakes in silico homologous recombination of PKS gene clusters, predicting whether a particular recombination is likely to be functional and the polyketide product to be expected (Starcevic et al. 2012). ClustScan can be accessed using a 30-day evaluation license, and the database behind ClustScan, ClustScanDB, is freely available via a web interface (http://csdb.bioserv.pbf.hr/csdb/ClustScanWeb.html).

To facilitate the systematic mapping of secondary metabolites in fungal genomes, Khaldi et al. (2010) developed the SMURF (Secondary Metabolite Unknown Region Finder). SMURF is a web-based tool which relies on hidden Markov model searches against Pfam and TIGRFAM (Haft et al. 2001) domains to detect genes that are central to the production of secondary metabolites, which they refer to as backbone genes. The analysis of fungal genomes is unusual, with most tools having focused on prokaryotic genomes. The backbone genes include PKSs, NRPSs, hybrid polypeptides including both NRPSs and PKSs, terpene cyclases and prenyltransferases. The clustering and specific domain content of these backbone genes are used to identify putative secondary metabolite biosynthetic gene clusters, e.g. an NRPS cluster requires at least one polypeptide with C, A and T domains. The location of genes for tailoring enzymes is also determined using HMMs, and their proximity to the backbone genes is used as part of the cluster identification process, and 27 secondary metabolite defining domains were identified by analysis of existing genome annotations. Khaldi et al. (2010) used SMURF to catalogue putative clusters in 27 publically available fungal genomes. They also compared the predicted results with genetically characterized clusters from six fungal species and demonstrated that SMURF was capable of predicting accurately all of these clusters and identify novel uncharacterized clusters.

Many research groups have used structural data to determine conserved residues lining the active site, which they then use for the prediction of substrate specificity. In the case of PKSs, this is typically specificity for extender or starter units and in the case of NRPs, specificity for amino acids. Stachelhaus and coworkers (1999) found ten key active site-lining residues from the crystal structure of the A-domain of the NRPS gramicidin S synthetase 1 (GrsA, PDB ID 1AMU). By looking at the variation of residue types at these 10 positions in different A-domains, they could predict substrate specificity for 20 substrate types. Shortly after that work, Challis et al. (2000), adopting a similar strategy, extended the number of predictable NRPS substrates to 33. Challis et al. used eight amino acids within the binding pocket combined with phylogenetic clustering.

The idea that a few residues can define the substrate specificity of an A-domain has since been developed by several groups, using a variety of machine learning techniques (e.g. HMMs, support vector machines) for correlating active site residues or whole sequence with substrate specificity, reviewed in Weber and Kim (2016). For example, NRPSpredictor2 (Röttig et al. 2011) used an innovative method employing a machine learning algorithm called a transductive support vector machine (SVM). The method is based on previous work (Rausch et al. 2005) from the same group although the new version outperforms the previous version by predicting the specificity of adenylation domains at four hierarchical levels, ranging from gross physicochemical properties of an A-domain's substrate to single amino acid substrates as well as predicting A-domain specificity in fungal systems, which was not achieved in the previous version. The NRPSpredictor2 utilizes the active site-lining residues within 8 Å of the bound phenylalanine ligand in the crystal structure (PDB ID 1AMU) of the peptide synthetase gramicidin S synthetase 1 (GrsA). These 34 extracted positions were then located in the A-domain sequences of the training data set, and this data was input into the SVM to train predictors of substrate specificity. The NRPSpredictor2 also has a larger database of training data including the sequences from fungal counterparts, as compared to its previous version, thus enabling a wider and more accurate prediction rate.

However, the above machine learning methods are limited to categorizing data within the already experimentally characterized parameter space. Recently, Lee et al. (2015) used in silico molecular docking techniques to try and predict A-domain specificity based on ligand and protein physical properties, without including prior knowledge of binding specificity. The cognate ligands were among the top-ranking docked ligands and top-ranking hits tended to have similar properties to the cognate ligand, thus showing some promise; however it was not possible to definitively specify the cognate ligand from the docking results. The limits of accuracy of protein structure modelling and the inherent difficulty of modelling ligand binding are likely to restrict the application of this type of method for the foreseeable future, e.g. it is difficult to correctly model the entropic contribution to ligand binding, since this depends on changes in the dynamics of the system and is thus a non-local property; whether water molecules are favourably bound in an active site and assist ligand binding is also difficult to predict.

Substrate specificity-determining residues in AT domains were found by analysis of the structure of the AT from *E* coli FAS (PDB ID 1MLA), combined with experimental data from the literature and sequence conservation (Yadav et al. 2003). Of 13 active site residues, they found that 9 were conserved across AT sequences, suggesting that the 4 variable residues might define specificity. An additional 11 non-active site residues were highlighted based on experimental work in the literature. Modelling malonate and methylmalonate extender units into the active site of 1MLA identified position 200 as being a key determinant of substrate specificity. Combining sequence conservation data with known AT specificity data led to the conclusion that F/P at position 200 would indicate malonate specificity, and a serine at this position would indicate methylmalonate specificity, while any other residue would be indicative of some unusual extender unit. This analysis was developed for the first polyketide synthase database (PKSDB) and associated domain prediction program SEARCHPKS (Yadav et al. 2003). SEARCHPKS includes the ability to search for AT domains and define their substrate specificity by looking in PKSDB for AT examples with identical active site residues and known substrate or, in the absence of an identical active site composition, by looking at the residue at position 200, as described above. Curiously, although providing a basis for classifying AT domains, it is still the case that it is difficult to use site-directed mutagenesis to alter the substrate specificity of AT domains (Dunn et al. 2013 and references therein).

More recent phylogenetic analysis of AT domains reconfirms their separation into distinct clades with Musiol and Weber (2012) finding that *trans*-acting ATs primarily separate into two evolutionarily distinct clades, "clade A" and "clade B". They found separate clades for *cis*-acting ATs, which split into a clade for ATs binding a malonyl extender unit and a clade for ATs binding a methyl-malonyl extender unit. Further distinct clades were found for FAS-derived ATs and for lipopeptide biosynthetic ATs. Curiously the essential AT for a *trans*-AT biosynthetic pathway seems to be found in clade B, excepting kirromycin synthesis, where KirII is found in the *cis*-AT methylmalonyl-CoA specific clade. Clade A seems to contain a second, non-essential AT that is present in many *trans*-AT pathways, which potentially has some function that is yet to be fully understood.

Yadav et al. (2009) again combined sequence and structural analysis, this time applying it to the KS domains from iterative PKSs. They found that KS domains that carry out similar numbers of elongation iterations and reductions cluster together in a phylogenetic tree. Homology models of different KS domains showed that the cavity size of the active site would change as the side chains lining the cavity changed, without a need for backbone conformational changes, and the number of iterations performed by the KS correlated with the size of the active site cavity. Thus the phylogenetic relationship might in part be explained by the adaptations required in the active site, and Yadav et al. noted that a relationship between KS and chain length had been seen previously in experimental work on iterative PKSs (Yadav et al. 2009 and references therein).

In the same work, Yadav et al. (2009) also utilized profile HMMs to distinguish between KS domains from modular PKSs and iterative PKSs. They also observed that HMMs are not only capable of broadly classifying KSs as modular or iterative but also of distinguishing *trans*-AT from *cis*-AT modular KSs and enediyne from non-enediyne iterative PKSs as well as identifying KSs from PKS-NRPS hybrid systems. It should be added that, in the preceding year, Nguyen et al. (2008) reported that the KS of *trans*-AT systems formed a phylogenetic tree such that KSs formed clades that grouped KSs according to the biosynthetic transformations in the module immediately upstream, whereas the KSs of *cis*-AT systems did not. This was a particularly important breakthrough since *trans*-AT PKSs often have tailoring enzymes that act *in trans*, which are hard to predict, and thus the metabolite synthesized by the cluster is hard to predict. In contrast the products of *cis*-AT systems can be predicted directly from the domains present in each module, excepting ambiguities in AT substrate specificity and misidentifying domains. Both Yadav et al. (2009) and Nguyen et al. (2008) proposed that such methods can help sequencing projects because just by analysing the KS domains of a novel PKS cluster, one can identify its type and subtype and decide whether sequencing the entire cluster would be of interest.

Coordinates of crystal structures of various type I PKS catalytic domains (Khosla 2009; Khosla et al. 2007; Tang et al. 2007; Keatinge-Clay and Stroud 2006; Keatinge-Clay 2008; Tsai and Ames 2009) and an almost complete module of the homodimeric mammalian FAS protein (Maier et al. 2008) allow the modelling of PKS domains in a homodimeric modular context, assuming the PKSs have a similar structure to the mammalian FAS (Tsai and Ames 2009; Gokhale et al. 2007). Based on such available structural data, Mohanty's group developed SBSPKS (Anand et al. 2010). SBSPKS is a web-based tool and probably the first which can model the 3D structures of a PKS module in a biologically active dimeric conformation. SBSPKS consists of three main components MODEL_3D_PKS, DOCK_DOM_ANAL and an updated version of NRPS-PKS. There are however now questions over the validity of using the mammalian FAS structure as a template following the publication of a structure of the whole PikAIII module from the pikromycin biosynthetic pathway (Dutta et al. 2014). This differs substantially from the FAS structure and from the X-ray-determined structure of the KS-AT dimer from DEBS module 3 (Tang et al. 2007), the latter having highly similar structure to the mammalian FAS. These disagreements need resolving if we are to understand the structures of these systems and be able to use structural data to model by homology novel, structurally uncharacterized systems.

Currently, MODEL_3D_PKS models the dimeric 3D structure of individual modules from type I modular PKSs, using the mammalian FAS structure as a template. Query sequence and template sequence for homology modelling are aligned by BLAST and side chains remodelled using SCRWL (Canutescu et al. 2003). The templates reported for the various domains of PKSs were those with PDB IDs 2HG4, 3LE6, 1IZ0 and 2FR0 along with nine threading models based on 2FR0 for modelling structural sub-domains of KR in the cases where DH-KR and DH-ER linkers lacked homology to 2FR0. The modelled domains are then superimposed onto the corresponding mammalian FAS module to provide the relative orientation of the PKS domains in dimeric state. MODEL_3D_PKS can model 3D structures for any of the four typical combinations of modular PKS, i.e. KS-AT-ACP, KS-AT-KR-ACP, KS-AT-DH-KR-ACP and KS-AT-DH-ER-KR-ACP, excluding the ACP domain as there is no experimental information available to provide the relative orientation of ACP domains to the rest of the domains in the module. Although MODEL_3D_PKS can model the typical combinations of domains found in modular PKSs, it is not designed to model modules from *trans*-AT systems, although it may be possible to "trick" the system into modelling this. The MODEL_3D_PKS web interface also has an embedded JMOL applet for quick visualization of the modelled structure.

The DOCK_DOM_ANAL module of the SBSPKs analyses the docking domains between the related subunits in a modular PKS. The docking domains are the inter-subunit linker region characterized by a four helical bundle; one helix is contributed by the C-terminal linker of the preceding ORF and three helical stretches from the N-terminus linker of the succeeding ORF. In previous studies (Broadhurst et al. 2003; Weissman and Müller 2008; Weissman 2006), a "docking code" was proposed, in which two crucial electrostatic residue pairs in a docking domain structure form inter-subunit contacts during substrate channelling between two ORFs from a modular PKS cluster. DOCK_DOM_ANAL has implemented the protocol developed by Yadav et al. (2009) to estimate the crucial inter-subunit contacts and predict the preferred order of substrate channelling, thus improving prediction of the likely product. As a point of terminology, these inter-subunit docking domains should not be confused with intra-module segments seen *in trans*-AT PKS I systems, which show sequence similarity to *cis*-AT domains and may thus be remnants of such domains, although their role is still poorly understood (Gurney and Thomas 2011; Lohman et al. 2015).

The NRPS-PKS component of SBSPKS is essentially an updated version of NRPS-PKS database developed by the same group. The most important updated feature seems to be a wider range of substrate specificity detection for the AT domain. The initial version of the AT domain specificity protocol was only able to discriminate between the malonate and methylmalonate selectivity, while the new version can now detect specificity for a total of 13 substrates. Another enhanced feature is its integration into the SBSPKS interface, thus providing links for automated input of its results to various other programs in the site.

Apart from the easy-to-use web servers and client server-based applications like ClustScan, initiatives by Weber and coworkers resulted in CLUSEAN which is a computer-based framework for the automated analysis of bacterial secondary metabolite biosynthetic gene clusters (Weber et al. 2009). CLUSEAN (CLUster SEquence ANalyzer) is an open source resource for semiautomatic analysis of secondary metabolite gene clusters. It is a modular framework of Bioperl (Stajich et al. 2002) programs that are compatible with Linux, Unix or MS Windows systems. It currently includes BLAST and HMMER as the tools for sequence annotation and domain identification, respectively. CLUSEAN scripts search the NCBI nonredundant protein database with BLASTp and use HMMER to search the following hidden Markov models: Pfam domains, PKS/NRPS domains and motifs and C-domain types and NRPS adenylation domain models developed by Rausch and coworker (2007, 2005). CLUSEAN currently offers one of the most comprehensive analyses by including a full genome annotation, but it seems to be difficult to operate by researchers who are not comfortable with scripts, and requires manual analysis of the output.

Another software pipeline called antiSMASH (antibiotics and Secondary Metabolite Analysis Shell) (Medema et al. 2011) has been developed for rapid identification, annotation and analysis of secondary metabolite biosynthesis gene clusters in bacterial and fungal genome sequences. It serves as a metaserver which amalgamates the data and methods available from various sources (de Jong et al. 2010; Weber et al. 2009; Finn et al. 2010; Letunic et al. 2009; Yadav et al. 2009; Ansari et al. 2008; Rausch et al. 2007). It is capable of analysing not only PKS/NRPS clusters as most of the software mentioned above do but also for the identification of the biosynthetic loci for terpenes, aminoglycosides, aminocoumarins, indolocarbazoles, lantibiotics, bacteriocins, nucleosides, beta-lactams, butyrolactones, siderophores, melanins and others.

AntiSMASH can be accessed via a web server, or it can be run as a stand-alone Java graphical user interface. It utilizes Glimmer3/GlimmerHMM for the gene prediction in the input sequence data and HMMER3 for the prediction of biosynthetic gene clusters using both existing profile HMMs as well as new profile HMMs from seed alignments. The substrate specificity of acyltransferase (AT) and adenylation (A) domains was performed using the methods proposed by Yadav et al. (2003) and Röttig et al. (2011), respectively, along with the method proposed by Minowa et al. (2007) for both. Finally the predictions from all the methods are integrated into a consensus by a majority vote. The stereochemistry predictions for PKSs based on the ketoreductase (KR) domain were carried out using the method used in the program ClustScan (Starcevic et al. 2008). To predict the biosynthetic order of PKS/NRPS modules, antiSMASH uses the same method as SBSPKS to match the docking domain residues in the ORFs of type I modular PKSs and otherwise assumes colinearity in the biosynthetic gene cluster. It also generates the SMILES string for the final predicted core chemical structure along with its picture. antiSMASH can also annotate the accessory genes surrounding the detected core signature genes in the various types of secondary metabolite biosynthesis gene clusters by utilizing the HMMs constructed on smCOG (secondary metabolite clusters of orthologous groups). It also provides features like ClusterBlast which can be used for comparative gene cluster analysis between the queried cluster and the clusters in the database and genome-wide BLAST and Pfam analysis using the modules from the CLUSEAN framework. The latest release, at the time of writing, is antiSMASH 3.0 (Weber et al. 2015), which has improved algorithms for assigning clusters and for predicting chemical structures produced by clusters. Currently, antiSMASH seems to be the most commonly used software for secondary metabolite gene cluster discovery and annotation, presumably because it is so comprehensive.

7.2.3 Application of Bespoke Structural Modelling to Elucidate Details of Molecular Specificity in PKS and NRPS Clusters

Computational modelling of the protein structures of PKSs and NRPSs has the potential to give insights into the specificity of protein-protein interactions, e.g. the previously mentioned study of Kapur et al. (2012) into the specificity of ACP elongation and downstream transfer interactions in the DEBS system and the previously mentioned work on determining the specificity of different adenylation domains in NRPSs.

Our interest in PKSs has been focused on modelling tailoring enzymes, which have to date been largely ignored by the bioinformatic and molecular modelling communities. We have investigated the mechanism of recognition in HCS function using docking techniques and HMMs in conjunction with biochemical and genetic experiments (Haines et al. 2013). Taking a similar approach to Anand and Mohanty (2012), we used HADDOCK (de Vries et al. 2010) to investigate protein-protein as well as protein-ligand docking with enforced distance restraints between the interacting pairs. However, in contrast to Anand and Mohanty's method, instead of docking the proteins first and then docking the ligands as several separate parts, our first step was to dock the whole ligand in the active site. We used structural information available from X-ray and NMR experiments and applied our knowledge of the enzymatic reaction by putting distance restraints between the catalytic active site residues and appropriate sites on the ligand. In the second step, the tail end of the ligand protruding from the active site (e.g. phosphate in case of phosphopantetheine) was used to define a distance restraint between the two proteins to be docked, e.g. between the phosphate of phosphopantetheine protruding from the enzyme and the catalytic serine of the ACP, which is covalently bound to the phosphopantetheine. Thus, information about the interaction between protein and substrate can be used to guide the protein-protein docking process, the former being typically experimentally better characterized than the latter. We further verified the predicted interaction using PIER (Kufareva et al. 2007), a software that predicts likely protein interaction sites from the physicochemical properties of the surface, and evolutionary trace analysis (Wilkins et al. 2012), which analyses sequence conservation patterns with the context of the predicted phylogenetic tree, which when mapped to the protein surfaces highlight important functional sites, e.g. protein-protein interaction sites. Further confirmation was given by mutagenesis experiments (Haines et al. 2013).

Computational docking is a promising method for investigating the interactions between the modules of polyketide synthases, but the basic methods still suffer from limitations in their conformational search algorithms, scoring functions and force fields. Therefore, molecular modelling should be augmented with data from sources such as literature searching, mutagenesis experiments, phylogenetics, evolutionary trace and amino acid covariance analysis. Based on such techniques, we focused experiments onto the features by which the HCS cassette recognizes where and when to modify the growing polyketide chain (Haines et al. 2013). The interaction of PKS domains with enzymes such as the HCS cassette in mupirocin and halogenase in curacin that function *in trans* are still complicated to model; however, the incentives involved in developing novel therapeutics through synthetic biology and its impact on healthcare industry remain a prime motivator for PKS researchers. Further insight might be obtained by other structural analysis techniques such as molecular dynamic techniques, and we are also following this direction.

7.3 Conclusions

Polyketide synthases and nonribosomal peptide synthases are large mega-dalton multidomain enzyme complexes that synthesize a wide range of natural products of medicinal interest. The structure, dynamics and organization that governs these multidomain proteins during secondary metabolite biosynthesis are still poorly understood. Routinely re-engineering secondary metabolite systems for the biosynthesis of novel compounds is a long-standing goal of much of the secondary metabolite research community, as is developing efficient methods for discovering and characterizing gene clusters producing natural products. Progress on both these goals will aid in the development of novel therapies such as anticancer drugs and antibiotics and of other compounds of high commercial value.

Re-engineering biosynthetic gene clusters requires a better understanding of how the different proteins in a biosynthetic pathway interact to efficiently produce the correct product, with current re-engineering work being very "hit and miss" and, on the occasions when successfully re-engineered, pathways tending to be inefficient. While much experimental work needs to be done to fill in the gaps in our knowledge, bioinformatic and molecular modelling also have had and will continue to have an important role in making sense of the experimental data and accelerating the rate of progress of the field. Various specialized databases, web servers and computational algorithms have been designed and are being used with increasing success to decipher the domain organization and substrate specificity of the PKS/NRPS systems from genome sequences, and these continue to be developed. The secondary metabolite portal (SMBP; http://www.secondarymetabolites.org) aims to summarize and link all secondary metabolite bioinformatic resources, past, present and future (Weber and Kim 2016). Recent developments have also shown progress towards computational prediction and understanding of 3D structure and dynamics of the PKS complexes.

Acknowledgements RF thanks the Darwin Trust of Edinburgh for financial support and Sam Higginbottom University of Agriculture, Technology and Sciences, India, for study leave and financial support. PJW and CMT thank the BBSRC/EPSRC for support via grant BB/F014570/1.

References

Altschul SF et al (1990) Basic local alignment search tool. J Mol Biol 215(3):403–410

Anand S, Mohanty D (2012) Modeling holo-ACP:DH and holo-ACP:KR complexes of modular polyketide synthases: a docking and molecular dynamics study. BMC Struct Biol 12(1):10

Anand S et al (2010) SBSPKS: structure based sequence analysis of polyketide synthases. Nucleic Acids Res 38:W487–W496

Ansari MZ et al (2008) In silico analysis of methyltransferase domains involved in biosynthesis of secondary metabolites. BMC Bioinforma 9:454

Austin MB, Noel JP (2003) The chalcone synthase superfamily of type III polyketide synthases. Nat Prod Rep 20(1):79–110

Bachmann BO, Ravel J (2009) Methods for in silico prediction of microbial polyketide and nonribosomal peptide biosynthetic pathways from DNA sequence data. Methods Enzymol 458:181–217

Bender C, Rangaswamy V, Loper J (1999) Polyketide production by plant-associated pseudomonads. Annu Rev Phytopathol 37:175–196

Broadhurst RW et al (2003) The structure of docking domains in modular polyketide synthases. Chem Biol 10(8):723–731

Busche A et al (2012) Characterization of molecular interactions between ACP and halogenase domains in the Curacin A polyketide synthase. ACS Chem Biol 7(2):378–386

Caboche S et al (2008) NORINE: a database of nonribosomal peptides. Nucleic Acids Res 36: D326–D331

Canutescu AA, Shelenkov AA, Dunbrack RL (2003) A graph-theory algorithm for rapid protein side-chain prediction. Protein Sci Publ Protein Soc 12(9):2001–2014

Challis GL (2008) Mining microbial genomes for new natural products and biosynthetic pathways. Microbiology (Reading, England) 154(6):1555–1569

Challis GL, Ravel J, Townsend C (2000) Predictive, structure-based model of amino acid recognition by nonribosomal peptide synthetase adenylation domains. Chem Biol 7 (3):211–224

Chan YA et al (2009) Biosynthesis of polyketide synthase extender units. Nat Prod Rep 26 (1):90–114

de Jong A et al (2010) BAGEL2: mining for bacteriocins in genomic data. Nucleic Acids Res 38: W647–W651

de Vries SJ, Van Dijk M, Bonvin AMJJ (2010) The HADDOCK web server for data-driven biomolecular docking. Nat Protoc 5(5):883–897

Dunn BJ, Cane DE, Khosla C (2013) Mechanism and specificity of an acyltransferase domain from a modular polyketide synthase. Biochemistry 52(11):1839

Dutta S et al (2014) Structure of a modular polyketide synthase. Nature 510:512–517

Finn RD et al (2010) The Pfam protein families database. Nucleic Acids Res 38:D211–D222

Foerstner KU et al (2008) A computational screen for type I polyketide synthases in metagenomics shotgun data. PLoS One 3(10):3515

Gokhale RS, Khosla C (2000) Role of linkers in communication between protein modules. Curr Opin Chem Biol 4(1):22–27

Gokhale RS, Sankaranarayanan R, Mohanty D (2007) Versatility of polyketide synthases in generating metabolic diversity. Curr Opin Struct Biol 17(6):736–743
Gurney R, Thomas CM (2011) Mupirocin: biosynthesis, special features and applications of an antibiotic from a gram-negative bacterium. Appl Microbiol Biotechnol 90(1):11–21
Haft DH et al (2001) TIGRFAMs: a protein family resource for the functional identification of proteins. Nucleic Acids Res 29(1):41–43
Haines AS et al (2013) A conserved motif flags acyl carrier proteins for β-branching in polyketide synthesis. Nat Chem Biol 9(11):685–692
Hertweck C (2009) The biosynthetic logic of polyketide diversity. Angew Chem Int Ed Eng 48 (26):4688–4716
Hunter S et al (2012) InterPro in 2011: new developments in the family and domain prediction database. Nucleic Acids Res 40:D306–D312
Hutchinson CR et al (2000) Aspects of the biosynthesis of non-aromatic fungal polyketides by iterative polyketide synthases. Antonie Van Leeuwenhoek 78(3–4):287–295
Jenke-Kodama H, Dittmann E (2009) Bioinformatic perspectives on NRPS/PKS megasynthases: advances and challenges. Nat Prod Rep 26(7):874–883
Kapur S et al (2010) Molecular recognition between ketosynthase and acyl carrier protein domains of the 6-deoxyerythronolide B synthase. Proc Natl Acad Sci U S A 107(51):22066–22071
Kapur S et al (2012) Reprogramming a module of the 6-deoxyerythronolide B synthase for iterative chain elongation. Proc Natl Acad Sci U S A 109(11):4110–4115
Keatinge-Clay A (2008) Crystal structure of the erythromycin polyketide synthase dehydratase. J Mol Biol 384(4):941–953
Keatinge-Clay AT, Stroud RM (2006) The structure of a ketoreductase determines the organization of the beta-carbon processing enzymes of modular polyketide synthases. Structure 14 (4):737–748
Khaldi N et al (2010) SMURF: genomic mapping of fungal secondary metabolite clusters. Fungal Genet Biol 47(9):736–741
Khare D et al (2010) Conformational switch triggered by alpha-ketoglutarate in a halogenase of curacin A biosynthesis. Proc Natl Acad Sci U S A 107(32):14099–14104
Khosla C (2009) Structures and mechanisms of polyketide synthases. J Org Chem 74 (17):6416–6420
Khosla C et al (1999) Tolerance and specificity of polyketide synthases. Annu Rev Biochem 68:219–253
Khosla C et al (2007) Structure and mechanism of the 6-deoxyerythronolide B synthase. Annu Rev Biochem 76:195–221
Knox C et al (2011) DrugBank 3.0: a comprehensive resource for "omics" research on drugs. Nucleic Acids Res 39:D1035–D1041
Kufareva I et al (2007) PIER: protein interface recognition for structural proteomics. Proteins 67 (2):400–417
Kwon SJ et al (2012) Expanding nature's small molecule diversity via in vitro biosynthetic pathway engineering. Curr Opin Chem Biol 16(1–2):186–195
Lairson LL et al (2008) Glycosyltransferases: structures, functions, and mechanisms. Annu Rev Biochem 77:521–555
Lee TV, Johnson RD, Arcus VL, Lott JS (2015) Prediction of the substrate for nonribosomal peptide synthetase (NRPS) adenylation domains by virtual screening. Proteins 83:2052–2066
Letunic I, Doerks T, Bork P (2009) SMART 6: recent updates and new developments. Nucleic Acids Res 37:D229–D232
Letunic I, Doerks T, Bork P (2012) SMART 7: recent updates to the protein domain annotation resource. Nucleic Acids Res 40:D302–D305
Lohman JR et al (2015) Structural and evolutionary relationships of "AT-less" type I polyketide synthase ketosynthases. PNAS 112(41):12693–12698
Ma SM, Tang Y (2007) Biochemical characterization of the minimal polyketide synthase domains in the lovastatin nonaketide synthase LovB. FEBS J 274(11):2854–2864

Maier T, Leibundgut M, Ban N (2008) The crystal structure of a mammalian fatty acid synthase. Science 321(5894):1315–1322

Marchler-Bauer A et al (2011) CDD: a conserved domain database for the functional annotation of proteins. Nucleic Acids Res 39:D225–D229

Medema MH et al (2011) antiSMASH: rapid identification, annotation and analysis of secondary metabolite biosynthesis gene clusters in bacterial and fungal genome sequences. Nucleic Acids Res 39:W339–W346

Minowa Y, Araki M, Kanehisa M (2007) Comprehensive analysis of distinctive polyketide and nonribosomal peptide structural motifs encoded in microbial genomes. J Mol Biol 368 (5):1500–1517

Musiol EW, Weber T (2012) Discrete acyltransferases involved in polyketide biosynthesis. Med Chem Commun 3:871–886

Nguyen T et al (2008) Exploiting the mosaic structure of trans-acyltransferase polyketide synthases for natural product discovery and pathway dissection. Nat Biotechnol 26(2):225–233

Park SR et al (2010) Genetic engineering of macrolide biosynthesis: past advances, current state, and future prospects. Appl Microbiol Biotechnol 85(5):1227–1239

Piel J (2010) Biosynthesis of polyketides by *trans*-AT polyketide synthases. Nat Prod Rep 27 (7):996–1047

Rausch C et al (2005) Specificity prediction of adenylation domains in nonribosomal peptide synthetases (NRPS) using transductive support vector machines (TSVMs). Nucleic Acids Res 33(18):5799–5808

Rausch C et al (2007) Phylogenetic analysis of condensation domains in NRPS sheds light on their functional evolution. BMC Evol Biol 7:78

Röttig M et al (2011) NRPS predictor2 – a web server for predicting NRPS adenylation domain specificity. Nucleic Acids Res 39:W362–W367

Stachelhaus T, Mootz HD, Marahiel MA (1999) The specificity-conferring code of adenylation domains in nonribosomal peptide synthetases. Chem Biol 6(8):493–505

Stajich JE et al (2002) The Bioperl toolkit: Perl modules for the life sciences. Genome Res 12 (10):1611–1618

Starcevic A et al (2008) ClustScan: an integrated program package for the semi-automatic annotation of modular biosynthetic gene clusters and in silico prediction of novel chemical structures. Nucleic Acids Res 36(21):6882–6892

Starcevic A et al (2012) Recombinatorial biosynthesis of polyketides. J Ind Microbiol Biotechnol 39(3):503–511

Staunton J, Weissman KJ (2001) Polyketide biosynthesis: a millennium review. Nat Prod Rep 18 (4):380–416

Tang Y et al (2006) The 2.7-Angstrom crystal structure of a 194-kDa homodimeric fragment of the 6-deoxyerythronolide B synthase. Proc Natl Acad Sci U S A 103(30):11124–11129

Tang Y et al (2007) Structural and mechanistic analysis of protein interactions in module 3 of the 6-deoxyerythronolide B synthase. Chem Biol 14(8):931–943

Thomas CM et al (2010) Resistance to and synthesis of the antibiotic mupirocin. Nat Rev Microbiol 8(4):281–289

Tsai SCS, Ames BD (2009) Structural enzymology of polyketide synthases. Methods Enzymol 459(09):17–47

Weber T, Kim HU (2016) The secondary metabolite bioinformatics portal: computation tools to facilitate synthetic biology of secondary metabolite production. Synth Syst Biotechnol 1:69–79

Weber T et al (2009) CLUSEAN: a computer-based framework for the automated analysis of bacterial secondary metabolite biosynthetic gene clusters. J Biotechnol 140(1–2):13–17

Weber T et al (2015) antiSMASH 3.0—a comprehensive resource for the genome mining of biosynthetic gene clusters. Nucleic Acids Res 43:W237–W243

Weissman KJ (2006) The structural basis for docking in modular polyketide biosynthesis. Chembiochem Eur J Chem Biol 7(3):485–494

Weissman KJ, Leadlay PF (2005) Combinatorial biosynthesis of reduced polyketides. Nat Rev Microbiol 3(12):925–936

Weissman KJ, Müller R (2008) Protein-protein interactions in multienzyme megasynthetases. Chembiochem Eur J Chem Biol 9(6):826–848

Wilkins A et al (2012) Evolutionary trace for prediction and redesign of protein functional sites. Methods Mol Biol 819:29–42

Wu J et al (2007) Mupirocin H, a novel metabolite resulting from mutation of the HMG-CoA synthase analogue, mupH in Pseudomonas fluorescens. Chem Commun 8(20):2040–2042

Yadav G, Gokhale RS, Mohanty D (2003) Computational approach for prediction of domain organization and substrate specificity of modular polyketide synthases. J Mol Biol 328 (2):335–363

Yadav G, Gokhale RS, Mohanty D (2009) Towards prediction of metabolic products of polyketide synthases: an in silico analysis. PLoS Comput Biol 5(4):1000351

In Silico Studies on Colon Cancer

8

Sharad Singh Lodhi, Manish Sinha, Yogesh K. Jaiswal, and Gulshan Wadhwa

Abstract

The colon cancer (CC) is one of the most common malignancies, and more than one million people become the prey of colon cancer every year worldwide. The CC initiates, develops, and progresses in various ways. But all these ways cannot be understood fully to till date. So, any effort taken in the CC research is a step toward prevention and cure of this devastating disease. In the present scenario, there are various in silico approaches as well as computational strategies available for gene expression analysis for CC like SAGEmap, X Profiler, digital gene expression displayer (DGED), digital differential display (DDD), and Digital Extractor. The drug discovery and drug development are very cumbersome, intense, and interdisciplinary processes. There are various methods available for the in silico drug designing and development such as homology modeling, molecular docking, virtual high-throughput screening, quantitative structure-activity relationship, hologram quantitative structure-activity relationship (HQSAR), comparative molecular field analysis (CoMFA), comparative molecular similarity indices analysis (CoMSIA), pharmacophore mapping, microarray analysis, conformational analysis, and Monte Carlo simulation. With the help of in silico approaches, many novel drug targets for CC like cytochrome P450 2A7, Rab3A, SFRP1, TLR4, MLH1, MSH6, survivin, FGFR-4, and ras oncogene products (H-ras, K-ras, and N-ras) have been identified.

S. S. Lodhi · Y. K. Jaiswal
School of Studies in Biochemistry, Jiwaji University, Gwalior, India

M. Sinha
Laureate Institute of Pharmacy, Kangra, Himachal Pradesh, India

G. Wadhwa (✉)
Department of Biotechnology, Apex Bioinformatics Centre, Ministry of Science & Technology, New Delhi, India
e-mail: gulshan@dbt.nic.in

G. Wadhwa et al. (eds.), *Current trends in Bioinformatics: An Insight*,
https://doi.org/10.1007/978-981-10-7483-7_8

Although these tools can act at good starting points in disease gene discovery, there is a need for experimental validation of in silico-derived differential expression and drug design results.

Keywords
Colon cancer · Tumor suppressor gene · In silico studies · Gene expression

8.1 Introduction

The colon cancer (CC) is one of the most common malignancies and is also the leading cause of mortality in cancer-related cases according to WHO report. More than one million people become the prey of colon cancer every year worldwide. The CC initiates, develops, and progresses in various ways (The Cancer Genome Research Atlas Network 2012). But all these ways cannot be understood fully to till date. So, any effort taken in the CC research is a step toward prevention and cure of this devastating disease.

It is categorized under carcinoma and more specifically adenocarcinoma. The mechanism of colon carcinogenesis is complex and influenced by various genetic and environmental factors. The CC progression occurs through many clinical and histopathological stages varying from small benign tumors to carcinomas. The genetic instability is the major cause for the progression of a normal cell to a colon cancer neoplasm (Fearon and Vogelsteins 1990). The cause of CC is the accumulating significant mutations in K-ras, adenomatous polyposis coli (APC), tumor protein P53 (TP53), and deleted in colorectal carcinoma (DCC) genes. However, these genes are not totally accountable for all colon cancers.

In silico approaches are not only helpful in disclosing the complex nature of changes in genetic activity during colon carcinogenesis but also aid in the information database which may be useful for identifying novel therapeutic targets.

8.2 Colon Cancer: An Overview

The colon cancer can be differentiated as two major types depending upon treatment and prognosis: the proximal colon cancer and distal colon cancer. The histology and physiology of these parts are very different, and this leads to differentiation at molecular level. There are three ways of transmission of colon carcinoma:

1. Sporadic: This is the major way of transmission of colon cancer usually used to differentiate cancers occurring in individuals who do not carry a mutation conferring tumor susceptibility from cancers occurring in individuals who carry a known mutation associated with this disease.

2. Familial: Familial transmission occurs in 10–30% cases where first degree relatives have sporadic cancer in the colon. Till now no gene associated to familial cancers has been identified.
3. Hereditary: This form and hamartomatous syndromes are least common types. The hereditary actors express only under the influence of environmental factors. Familial adenomatous polyposis (FAP) and hereditary nonpolyposis colorectal cancer (HNPCC) are the commonest forms of hereditary CC.

The pathogenic pathways leading to CC are heterogeneous in nature. There are two major tumorigenic pathways known for colon carcinogenesis. The first pathway involves chromosomal instability (CIN) which includes both oncogenes and tumor suppressor genes residing on chromosomes 5q, 17p, and 18q (Delattre et al. 1989; Gervaz et al. 2001). Chromosome 5q resides the APC gene, 17p resides the TP53 gene, and 18q resides the DCC or SMAD4 genes. K-ras is the most common oncogene responsible for CC. All these genes play important roles in the transformation of adenoma to carcinoma. CIN pathway increases the rate of loss of heterozygosity (LOH), which causes the inactivation of tumor suppressor genes. In an experiment, around 100 genes are made mutated in yeast cells which cause CIN. Many of these genes have several homologues in humans (Jallepalli and Lengauer 2001) and are involved in chromosome metabolism, spindle assembly and dynamics, cell cycle regulation, and mitotic checkpoint control mechanism (Cahill et al. 1998). The first pathway is prevalent in the distal colon.

The second pathway involves microsatellite instability (MSI) which includes the alterations in mismatch repair (MMR) genes (Thibodeau et al. 1993; Miyakura et al. 2001). It has been evident that MMR enzymes like hMSH2, HMLH1, hPMS1, hPMS2, and hMSH6 are responsible for MSI. The second pathway is predominant in proximal colon (Thibodeau et al. 1993; Miyakura et al. 2001). Epigenetics has also been reported to play an important role in sporadic CCs with MSI which is regulated by methylation (Anacleto et al. 2005).

Recently it is shown that miRNAs play a crucial role in cancer initiation and progression by negative regulation of their target genes. miR-15a and miR-16 target multiple oncogenes including BCL2, MCL1, CCND1, and WNT3A (Aqeilan et al. 2010). Both miR-135a and miR-135b are oncomiRs, and they directly target 3′ untranslated region of the APC gene suppressing its expression and inducing downstream activity in Wnt signaling pathway (Nagel et al. 2008).

8.2.1 Environmental and Genetic Factors Causing Colon Cancer

There are various environmental and genetic factors which are supposed to cause CC. Aging is an important risk factor for CC, and around 50% of western industrialized population developed adenomas in the gastrointestinal tract (Kinzler and Vogelstein 1996). But the presence of one or more adenomas does not surely cause CC (Jemal et al. 2002). Epidemiological studies show that populations migrating from low- to high-risk geographical areas have similar incidence of CC in comparison to already existing population in that area (Tomatis and Bartsch 1990). Diet is the major environmental factor, and studies have shown that fiber- and nutrient-rich diet is protective against CC (WCRF and AICR 1997). It is shown that CC risk can be reduced by 40% if the dietary fibers are doubled in diet of populations having less average fiber consumption (Bingham et al. 2003). Also nondigestible fructo-oligosaccharides and high starch content diets contribute to reduce CC risk (WCRF and AICR 1997). The other environmental factors are high intake of alcohol, animal fat, red and processed meat, smoking, and low physical activity (WCRF and AICR 1997) (Giovannucci and Martinez 1996). Some of the inflammatory diseases of gastrointestinal tract like inflammatory bowel disease and Crohn's disease are also the risk factors for CC (Labianca et al. 2000; Chapelle 2004). The mechanism of protective or predisposing effects modulated by diet, lifestyle, and other nongenetic factors is still not known.

On the contrary, the mechanism of genetic factors is better elucidated in the last decade. The patients having familial colorectal cancer risk are 20–25% of all CC patients (Chapelle 2004). Majority of CC syndromes are risk factors of CC which include familial adenomatous polyposis (FAP) caused due to APC mutations, MYH-associated polyposis due to MYH gene, hereditary nonpolyposis colorectal cancer (HNPCC) caused by the mismatch repair genes, Peutz Jegher syndrome (PJS) due to STK11 gene, and juvenile polyposis syndrome (JPS) caused due to mutations in SMAD4, BMPR1A, and ENG gene. The contribution of these CC syndromes to the total CC burden is around ~5% of the cases. The remaining genetic factors predisposing to CC need to be revealed.

8.3 Colon Cancer at Molecular Level

The conversion of normal epithelial cells into benign adenomas and then to malignancy is a multi-step process of histological stages accompanying concurrent genetic and epigenetic changes (Fearon and Vogelstein 1990). Due to which the neoplastic cells undergo clonal expansion and cell contact inhibition is lost, parenchyma cells evade apoptotic and cell cycle arrest mechanisms and attain stem cell-like features. With the advancing histological stages of CC progression, the genetic alterations become more intense in comparison to early adenoma stages.

Mutations cause the activation of oncogenes with inactivation of tumor suppressor genes in the tumorigenic pathway which occur in a definite sequence specific for CC stages. The small lesions of irregular glandular architecture known as aberrant crypt focus (ACF) initiate the adenoma-carcinoma sequence. ACF is characterized by hyperproliferation, increased size, expanded pericryptal zones, and elongated or slit-like crypt lumina (Rosenberg and Liu 1995; Papanikolaou et al. 1998, 2000; Bird 1987). The exact potential of this lesion is not clear. There are two types of ACFs (Nucci et al. 1997): the heteroplastic ACF which is associated with nonmalignant intestinal mucosa and the dysplastic or unicryptal ACF type which has more likelihood for cancer progression. The heteroplastic ACFs may be caused by mutational activation of the oncogene (KRAS2), whereas latter ones arise due to biallelic inactivation of the tumor suppressor gene (adenomatous polyposis coli (APC)) (Lamlum et al. 1999). The loss of APC gene function deregulates the WNT/β-catenin signaling pathway which initiates CC (Fodde et al. 2001).

The majority of sporadic CCs are triggered by the constitutive activation of WNT pathway. The activation of WNT pathway is caused by mutation either in APC or in CTNNB1 gene, due to which the intestinal epithelial cell gets enhanced cell renewal and decreased differentiation potential leading to formation and growth of dysplastic ACFs to give rise to adenomas (Reya and Clevers 2005). On the contrary, colon tumors which carry intact APC have got mutations in β-catenin that affect relevant phosphorylation sites, and so colon tumors become resistant to proteolytic degradation (Sparks et al. 1998; Ilyas et al. 1997). Some other genes of WNT pathway like AXIN2 are also found to be implicated in CC initiation (Lammi et al. 2004).

Ras-Raf MAPK pathway is a GDP/GTP controlled signaling cascade which modulates cell growth and survival. KRAS2 oncogene and BRAF are important components of this pathway. In 50% of CC cases, the mutational activation of the KRAS2 has been reported (Forrester et al. 1987; Andreyev et al. 2001), whereas in 20% CC cases, the mutational activation of BRAF is seen in the absence of KRAS2 (Rajagopalan et al. 2002; Davies et al. 2002). Mutational activation of both genes is associated with growth and progression of adenoma (Rajagopalan et al. 2002), and it may occur in earlier stages and hyperplastic ACFs (Takayama et al. 2001).

The alterations in transcription growth factor-β (TGFβ) signal transduction pathway cause progression from adenoma to carcinoma. The mutational inactivation of its components like TGFβ receptor 2 (TGFβR2) (Markowitz et al. 1995), SMAD2, and SMAD4 genes (Eppert et al. 1996) causes altered angiogenesis, cell proliferation, and differentiation (Blobe et al. 2000). The majority of CC exhibit mutational inactivation in at least one of its components.

The TP53 protein is a transcription factor which maintains the genomic integrity of intestinal cells by inhibiting the cell growth and stimulating cell death induced by cellular stress. In 45% of CC cases, the transformation from adenoma to carcinoma occurs due to biallelic inactivation of the TP53 tumor suppressor gene (Delattre et al. 1989; Baker et al. 1989).

The studies have also reported that the mutational alterations in some additional pathways including PI3K and receptor tyrosine kinases and phosphatases are also implicated in causing CC (Samuels et al. 2004; Parsons et al. 2005).

8.4 Current Therapies for Colon Cancer

There are four major approaches to cure CC which include surgery, chemotherapy, radiotherapy, and targeted therapy. Out of which, surgery is the commonest approach, but the decision of implementation of an approach or their combination depends on the stage of CC. In first stage of CC, only surgical excision is enough because of less recurrence rate of CC at that stage (Kobayashi et al. 2011). In stage III, adjuvant therapy is recommended to patients for better survival. But in stage II of CC, the benefit of adjuvant therapy is skeptical. Only the patients who have high risk for stage II are offered that type of therapy. However, chemotherapy or targeted therapies in appropriate combination with surgery are offered to patients with stage IV. Although surgery, chemotherapy, and radiation therapy are key approaches for the cure of CC, the targeted therapies are also emerging as a potential cure of CC.

8.4.1 Targeted Therapies in the Treatment of CC

With the advances in the understanding of progression of CC at molecular level, various targeted agents are developed to cure CC. These targeted agents are administered along with chemotherapy. Common targeted agents available for CC are the monoclonal antibodies, which include bevacizumab (Avastin™, Genentech/Roche) targeting vascular endothelial growth factor (VEGF) (Hurwitz et al. 2004) and anti-EGFR monoclonal antibodies cetuximab (Erbitux™, Imclone Systems) (Van Cutsem et al. 2009) and panitumumab (Vectibix™, Amgen, CA, USA) (Douillard et al. 2010). There is an urgent need to develop other effective targeted agents to cure CC.

8.5 In Silico Approaches of Gene Expression Analysis for Colon Cancer

The gene expression profiles are helpful in delineating the gene functions and identifying diagnostic markers as well as novel drug targets. In the present scenario, there are various in silico approaches and computational strategies available for gene expression analysis for CC. Most of the strategies rely on the utilization of expressed sequence tags (ESTs) collections. The ESTs are the best source for profiling of gene expression as they express at the same level as that of normal gene. The EST libraries of various organ- and disease-derived cDNAs are available with various databases. Such as an EST database (dbEST) has more than 28 million public entries. The various methods for in silico gene expression profiling of CC are as follow:

8.5.1 SAGEmap

The serial analysis of gene expression (SAGE) method is used to quantify and compare the groups of transcripts [Velculescu et al.]. There are various repositories for SAGE data. To interpret SAGE data, one of the algorithm-based online tools is SAGEmap. This tool facilitates the submission of SAGE data from any source and thus enables to study the SAGE data from various sources. It consists of the collection of SAGE data from bulk tissues, cell lines from various species, different tissue types, depositors, and tissues which have undergone any treatment. But this tool excludes the ESTs with low counts. The SAGEmap is available at the URL http://www.ncbi.nlm.nih.gov/projects/SAGE/.

8.5.2 X Profiler

X Profiler is a tool which is used to compare the expression between two libraries or a group of two or more libraries or within a library. For example, a comparison of colon cancer tissue and a healthy colon tissue or two colon cancer tissues can be performed by the user with the help of X Profiler. X Profiler has the list of genes of each pool, and it also categorizes these genes as unique or nonunique and known or unknown. X Profiler is available at the URL http://cgap.nci.nih.gov/Tissues/xProfiler.

8.5.3 Digital Gene Expression Displayer (DGED)

DGED compares the gene expression between the two pools of the libraries. Either cDNA libraries or SAGE tag libraries are used for analysis by DGED. It counts the gene in the library pool as degree and compares the amount of a gene in different pools. In this tool, the parameters may vary statistically and results can be linked to microarray data. Moreover, it has an ability to select origin or type of tissue, and the genes with low abundance may also be considered. But the comparison of genes is based on odds ratio. It is available at the URL http://cgap.nci.nih.gov/SAGE/SDGED.

8.5.4 Digital Differential Display (DDD)

The DDD identifies the differential gene expression. DDD is a powerful web-based bioinformatic tool which uses the EST profiles of normal and disease cDNA libraries available in NCBI UniGene database. The comparison of the number of assignments of ESTs in various libraries or their pools to a specific UniGene cluster is carried out. In DDD, conservative test like Fisher's exact test is employed to determine significance. Both absolute and relative counts are given in DDD, but the libraries with low EST count get excluded by analysis, and limited number of normal tissue libraries is available. DDD is available at URL http://www.ncbi.nlm.nih.gov/UniGene/ddd.cgi.

8.5.5 Digital Extractor

The data obtained from DDD is further processed through Digital Extractor which performs automated annotation of the output clusters. Digital Extractor does the high-throughput processing of DDD output. It compiles the profiles of known differentially expressed genes as well as annotates the clusters containing cDNAs with no homology to known genes. It is available at the URL http://cgap.nci.nih.gov/SAGE/SDGED.

8.6 In Silico Drug Designing Approaches for Colon Cancer

The drug discovery and drug development are very cumbersome, intense, and interdisciplinary processes. There are various approaches which can be applied to get the most suitable and effective drugs against CC. Among various approaches, in silico drug designing approaches can also be used for designing the suitable and effective drug for CC treatment. In the recent time, a trend has been started toward the use of in silico chemistry. There are various methods available for the in silico drug development which can be explained as:

8.6.1 Homology Modeling

Homology modeling technique is also called as protein comparative modeling. In homology modeling method, an unknown atomic resolution model of the target protein gets generated from its amino acid sequence, and thus an experimental three-dimensional structure of a related homologous protein is obtained. Homology modeling resembles the structure of the query sequence with one or more already identified protein structures by making aligning the maps residues in the query sequence with the residues in the template sequence.

8.6.2 Molecular Docking (Interaction Networks)

Molecular docking is a technique, in which the most favored orientation of one molecule with the second one is predicted, while these molecules bind with each other and form a stable complex. Molecular docking denotes binding of ligand to its target or receptor which is a protein. Molecular docking recognizes and optimizes the potential drug molecules by analyzing and modeling molecular interactions between ligand and target molecules. By this method, the multiple ligand conformations and orientations get generated, and then the most appropriate candidates are selected.

8.6.3 Virtual High-Throughput Screening

The drug discovery process and research include virtual screening. Virtual screening is one of the computational techniques which evaluate the huge libraries of compounds for their potential to bind specific sites on target molecules. Those compounds which match with the target as per requirement get selected. With the help of virtual screening, the rapid exploration of large libraries of chemical structures can be achieved, and the structures which are most appropriate to bind to a drug target are identified.

8.6.4 Quantitative Structure-Activity Relationship

With the help of quantitative structure-activity relationship (QSAR) methods, the relationship of structure or property descriptors of compounds with their biological activities is described. The steric, topologic, electronic, and hydrophobic properties of large number of molecules are described by these descriptors, which are earlier determined through empirical methods. But in the recent time, these are determined by computational methods like QSAR.

8.6.5 Hologram Quantitative Structure-Activity Relationship (HQSAR)

In hologram quantitative structure-activity relationship method, the molecule gets broken into a molecular fingerprints, by which the frequency of occurrence of various kinds of molecular fragments is encoded. The minimum and maximum length of the obtained fragment is based on the size of the fragment which is included in the hologram fingerprint.

8.6.6 Comparative Molecular Field Analysis (CoMFA)

By comparative molecular field analysis (CoMFA), the values of ClogP can be calculated. This method determines the steric and electrostatic values of the ligands.

8.6.7 Comparative Molecular Similarity Indices Analysis (CoMSIA)

Comparative molecular similarity indices analysis (CoMSIA) determines the steric and electrostatic characteristics of the ligand. The hydrogen bond acceptors, hydrogen bond donor, and hydrophobic fields are also revealed with the help of this method.

8.6.8 Pharmacophore Mapping

The functional groups are arranged in a 3-D manner within a molecular framework. Those functional groups are known as pharmacophore which are important for attaching to an active site of an enzyme or binding to a macromolecule. Recognizing a pharmacophore is considered as the first step for understanding the ligand interaction with a receptor. As soon as the recognition of pharmacophore takes place, the 3-D database search tools can be utilized for retrieving the novel compounds which are appropriate for pharmacophore mapping. The pharmacophore mapping is a very crucial, dynamic, and simple technique, which recognizes the lead compounds together with a preferred target.

8.6.9 Microarray Analysis

Microarray technology is an advanced technique, which involves chips of properly arranged sets of DNA molecules known sequence. These chips are mostly rectangular in shape and can consist of hundreds or thousands sets, on which each single feature of the array is found at the accurate position on the substrate. So that the identity of DNA molecule associated to each feature cannot be changed. Microarray technology has the ability to perform the gene expression profiling of whole genome in one go.

8.6.10 Conformational Analysis

Conformation analysis involves various calculation methods to evaluate deformable molecules and their minimum energy configurations. By conformational analysis, molecular receptor sites of molecules are compared, and most satisfactory 3-D conformation on the basis of minimum energy is calculated.

8.6.11 Monte Carlo Simulation

With the help of Monte Carlo simulation, different conformations of a system are produced by computer simulation. Thermodynamic, structural, and numerical properties are calculated for these conformations with the help of statistical mechanics principles.

8.7 Recent In Silico Studies on Colon Cancer

Sharad et al. compared the gene expression profiles of colorectal cancer cells from normal colonic cells using publicly available microarray data of human genes. The J5 test was used for significance analysis and Naive Bayes Classifier Algorithm for testing defined classification of samples. Correlation distance was analyzed by Pearson's correlation distance method. Some of the upregulated genes including vasopressin-neurophysin 2-copeptin preproprotein, cytochrome P450 2A7 isoform 1, major centromere autoantigen B, myelin-associated glycoprotein, and bone morphogenetic protein 1 isoform 3 precursor have not been reported for their overexpression in colon cancer cells before; however, their overexpression was reported in other cancers such as lung cancer, breast cancer, etc. (Lodhi et al. 2012). The precise function and structure of cytochrome P450 2A7 and Rab3A proteins are not known. So, these genes were further studied in silico for their structure and function by in silico tools. The binding sites for the inhibitors were also identified which help in the targeted inactivation of these proteins for would further research (Lodhi et al. 2014, 2015).

In a study, SFRP1 has been reported to have a differentially methylated pattern at each methylation site, and so SFRP1 may be a potential biomarker for colon cancer survival (Jongbum and Sangsoo 2014). A Toll-like receptor (TLR4) is overexpressed early during the colon cancer initiation and can be used as a drug target to inhibit colon cancer neoplasia (Li et al. 2014). Some non-synonymous single nucleotide polymorphisms (SNPs) are reported in various inherited diseases. In a study, 16 non-synonymous SNPs have been identified in MLH1 gene which is known to cause colon cancer (Joy et al. 2014). It has been recently found that the variants of MSH6 are present in colon cancer cells, and these variants are formed due to mutations in MSH6 gene (Berends et al. 2002). Also, the somatic mutations in TGFBR2 and SMAD genes cause colon cancer. By preventing selected mutations, a reduction in SMAD3 transcriptional activity and SMAD2-SMAD4 complex formation is seen (Fleming et al. 2013).

Survivin is identified as one of the potential target for the treatment of colon cancer and by using virtual screening approach and docking; inhibitors have been identified (Groner and Weiss 2014). The mutational activation of the ras oncogene products (H-ras, K-ras, and N-ras) is frequently observed in colon cancer, and these genes are considered as the promising anticancer drug targets. No effective strategy has been available for the development of Ras inhibitors, partly owing to the absence of well-defined surface pockets suitable for drug binding. Recently, small molecule of Ras inhibitors has been discovered (Shima et al. 2013). Fibroblast growth factor receptor-4 (FGFR4) is a tyrosine kinase, and its frequent mutations are reported in various types of cancers. The FGFR4 inhibitors have been developed as anticancer agent (Ho et al. 2013).

8.8 Conclusion

The in silico data mining strategies provide a platform for the initial identification of genes which are differentially expressed in CC tissue. The data generated through these approaches helps to set a starting point to delineate the molecular basis of colon cancer. The different types of in silico gene expression methods such as EST gene profiling strategy, SAGEmap, X Profiler, digital gene expression displayer, digital differential display, and the Digital Extractor are the latest tools of research helping the researchers to get an insight of molecular level of colon cancer pathology. New approaches of in silico drug targeting are providing easy, fast, and low-cost methods to target desired genes. Molecular docking, virtual high-throughput screening, and computer-aided drug design when accompanied with the in silico gene expression methods and in silico drug targeting methods are giving wings to the new drug discovery and research. Although these tools can act at good starting points in disease gene discovery, there is a need for experimental validation of in silico-derived differential expression and drug design results.

References

Anacleto C et al (2005) Colorectal cancer "methylator phenotype": fact or artifact? Neoplasia 7 (4):331–335

Andreyev et al (2001) Kirsten ras mutations in patients with colorectal cancer: the 'RASCAL II' study. Br J Cancer 85(5):692–696

Aqeilan RI et al (2010) miR-15a and miR-16-1 in cancer: discovery, function and future perspectives. Cell Death Differ 17(2):215–220

Baker SJ et al (1989) Chromosome 17 deletions and p53 gene mutations in colorectal carcinomas. Science 244:217–221

Berends MJW et al (2002) Molecular and clinical characteristics of *MSH6* variants: an analysis of 25 index carriers of a germline variant. Am J Hum Genet 70(1):26–37

Bingham SA et al (2003) Dietary fibre in food and protection against colorectal cancer in the European prospective investigation into cancer and nutrition (EPIC): an observational study. Lancet 361:1496–1501

Bird RP (1987) Observation and quantification of aberrant crypts in the murine colon treated with a colon carcinogen: preliminary findings. Cancer Lett 37:147–151

Blobe GC et al (2000) Role of transforming growth factor beta in human disease. N Engl J Med 342:1350–135810

Cahill DP et al (1998) Mutations of mitotic checkpoint genes in human cancers. Nature 392:300–303

Cancer Genome Atlas Network (2012) Comprehensive molecular characterization of human colon and rectal cancer. Nature 487:330–337

Chapelle A (2004) Genetic predisposition to colorectal cancer. Nat Rev Cancer 4:769–780

Davies et al (2002) Mutations of the BRAF gene in human cancer. Nature 417(6892):949–954

Delattre O et al (1989) Multiple genetic alterations in distal and proximal colorectal cancer. Lancet 334(8659):353–356

Douillard JY et al (2010) Randomized, phase III trial of panitumumab with infusional fluorouracil, leucovorin, and oxaliplatin (FOLFOX4) versus FOLFOX4 alone as first-line treatment in patients with previously untreated metastatic colorectal cancer: the PRIME study. J Clin Oncol 28:4697–4705

Eppert K et al (1996) MADR2 maps to 18q21 and encodes a TGF beta-regulated MAD-related protein that is functionally mutated in colorectal carcinoma. Cell 86:543–552

Fearon ER, Vogelstein B (1990) A genetic model for colorectal tumorigenesis. Cell 61(5):759–767

Fleming NI et al (2013) SMAD2, SMAD3 and SMAD4 mutations in colorectal cancer. Cancer Res 73(2):725–735

Fodde R et al (2001) APC, signal transduction and genetic instability in colorectal cancer. Nat Rev Cancer 1(1):55–67

Forrester et al (1987) Detection of high incidence of K-ras oncogenes during human colon tumorigenesis. Nature 327:298–303

Gervaz P et al (2001) Dukes B colorectal cancer: distinct genetic categories and clinical outcome based on proximal or distal tumor location. Dis Colon Rectum 44(3):364–372

Giovannucci E, Martinez ME (1996) Tobacco, colorectal cancer, and adenomas: a review of the evidence. J Natl Cancer Inst 88:1717–1730

Groner B, Weiss A (2014) Targeting Survivin in cancer: novel drug development approaches. BioDrugs 28(1):27–39

Ho HK et al (2013) Developing FGFR4 inhibitors as potential anti-cancer agents via in silico design, supported by in vitro and cell-based testing. Curr Med Chem 20(10):1203–1217

Hurwitz H et al (2004) Bevacizumab plus irinotecan, fluorouracil, and leucovorin for metastatic colorectal cancer. N Engl J Med 350:2335–2342

Ilyas M et al (1997) β-catenin mutations in cell lines established from human colorectal cancers. Proc Natl Acad Sci U S A 94:10330–10334

Jallepalli PV, Lengauer C (2001) Chromosome segregation and cancer: cutting through the mystery. Nat Rev Cancer 1:109–117

Jemal A et al (2002) Cancer statistics. CA Cancer J Clin 52:23–47

Jongbum K, Sangsoo K (2014) In silico identification of SFRP1 as a Hypermethylated gene in colorectal cancers. Genomics Inform 12(4):171–180

Joy A et al (2014) MLH1 gene: an in silico analysis. J Comput Biol Bioinforma Res 5(1):1–5

Kinzler KW, Vogelstein B (1996) Lessons from hereditary colorectal cancer. Cell 87:159–170

Kobayashi H et al (2011) Characteristics of recurrence after curative resection for T1 colorectal cancer: Japanese multicenter study. J Gastroenterol 46(2):203–211

Labianca R et al (2000) Colon cancer. Crit Rev Oncol Hematol 51:145–170

Lamlum H et al (1999) The type of somatic mutation at APC in familial adenomatous polyposis is determined by the site of the germline mutation: a new facet to Knudson's 'two-hit' hypothesis. Nat Med 5:1071–1075

Lammi L et al (2004) Mutations in AXIN2 cause familial tooth agenesis and predispose to colorectal cancer. Am J Hum Genet 74:1043–1050

Li TT et al (2014) Toll-like receptor signaling in colorectal cancer: carcinogenesis to cancer therapy. World J Gastroenterol 20(47):17699–17708

Lodhi SS et al (2012) Statistical analysis of differential gene expression profile for colon cancer. Indian J Biotechnol 11:396–403

Lodhi SS et al (2014) 3D structure generation, virtual screening and docking of human Ras-associated binding (Rab3A) protein involved in tumourigenesis. Mol Biol Rep 41 (6):3951–3959

Lodhi SS et al (2015) In silico structural, virtual screening and docking studies of human cytochrome P450 2A7 protein. Interdiscip Sci 7(2):129–135

Markowitz S et al (1995) Inactivation of the type II TGF-ß receptor in colon cancer cells with microsatellite instability. Science 268:1336–1338

Miyakura Y et al (2001) Extensive methylation of hMLH1 promoter region predominates in proximal colon cancer with microsatellite instability. Gastroenterology 121(6):1300–1309

Nagel R et al (2008) Regulation of the adeno- matous polyposis coli gene by the miR-135 family in colorectal cancer. Cancer Res 68:5795–5802

Nucci MR et al (1997) Phenotypic and genotypic characteristics of aberrant crypt foci in human colorectal mucosa. Hum Pathol 28:1396–1407

Papanikolaou A et al (1998) Azoxymethane-inducedcolon tumors and aberrant crypt foci in mice of different genetic susceptibility. Cancer Lett 14:29–34

Papanikolaou A et al (2000) Sequential and morphological analyses of aberrant crypt foci formation in mice of differingsusceptibility to azoxymethane-induced colon carcinogenesis. Carcinogenesis 21:1567–1572

Parsons DW et al (2005) Colorectal cancer: mutations in a signalling pathway. Nature 436:792

Rajagopalan H et al (2002) Tumorigenesis: RAF/RAS oncogenes and mismatch-repair status. Nature 418:934

Reya T, Clevers H (2005) Wnt signalling in stem cells and cancer. Nature 434:843–850

Rosenberg DW, Liu Y (1995) Induction of aberrant crypts in murine colon with varyingsensitivity to colon carcinogenesis. Cancer Lett 92:209–214

Samuels Y et al (2004) High frequency of mutations of the PIK3CA gene in human cancers. Science 304:554

Shima F et al (2013) In silico discovery of small-molecule Ras inhibitors that display antitumor activity by blocking the Ras-effector interaction. Proc Natl Acad Sci U S A 110(20):8182–8187

Sparks AB et al (1998) Mutational analysis of the APC/β-catenin/Tcf pathway in colorectal cancer. Cancer Res 58:1130–1134

Takayama T et al (2001) Analysis of K-ras, APC and β-catenin in aberrant crypt foci in patients with adenoma and cancer, and familial adenomatous polyposis. Gastroenterology 121:599–611

Thibodeau SN et al (1993) Microsatellite instability in cancer of the proximal colon. Science 260 (5109):816–819

Tomatis L, Bartsch H (1990) The contribution of experimental studies to risk assessment of carcinogenic agents in humans. Exp Pathol 40:251–266

Van Cutsem E et al (2009) Cetuximab and chemotherapy as initial treatment for metastatic colorectal cancer. N Engl J Med 360:1408–1417

WCRF and AICR Food (1997) In: AIoC Research (ed) Nutrition and prevention of cancer: a global perspective. WCRF and AICR. World Cancer Research Fund and American Institute for Cancer Research, Washington

Tools, Databases, and Applications of Immunoinformatics

9

Namrata Tomar and Rajat K. De

Abstract

A large volume of data relevant to immunology research has accumulated due to sequencing of the human and other model organism genomes. At the same time, huge amounts of clinical and epidemiologic data are being deposited in various scientific literature and clinical records. This accumulation of the information is like a gold mine for researchers looking for mechanisms of immune function and disease pathogenesis. Thus the need to handle this rapidly growing immunological resource has given rise to the field known as immunoinformatics. Immunoinformatics, otherwise known as computational immunology, is the interface between computer science and experimental immunology. It represents the use of computational methods and resources for the understanding of immunological information. It not only helps in dealing with huge amount of data but also plays a great role in defining new hypotheses related to immune responses. This article reviews classical immunology, different databases, and prediction tools. Further, it describes applications of immunoinformatics in designing in silico vaccination and immune system modeling, cancer diagnosis, and therapy. It also explores the idea of integrating immunoinformatics with Systems Biology for the development of personalized medicine.

N. Tomar (✉)
Department of BioMedical Engineering, Medical College of Wisconsin, Milwaukee, WI 53226, USA

Machine Intelligence Unit, Indian Statistical Institute, Kolkata, India
e-mail: ntomar@mcw.edu

R. K. De
Machine Intelligence Unit, Indian Statistical Institute, Kolkata, India
e-mail: rajat@isical.ac.in

G. Wadhwa et al. (eds.), *Current trends in Bioinformatics: An Insight*,
https://doi.org/10.1007/978-981-10-7483-7_9

Keywords
Systems biology · Immunomics · In silico models · T cells · B cells · Allergy · Reverse vaccinology · Personalized medicine

9.1 Introduction

The term "immunity" comes into consideration through individuals who had recovered from certain infectious disease and got protected from the same disease when it would be encountered in future. Thus there exists an immune system and associated biological processes within these individuals, which are responsible for developing "immunity." The role of an immune system is to protect against diseases by identifying and killing pathogens. An immune system includes innate and adaptive components. According to traditional dogma of immunology, vertebrates have both innate and adaptive immune systems whereas invertebrates possess only innate immune system (Kimbrell and Beutler 2001; Tomar and De 2010).

An immune system may be considered as a network of thousands of molecules, which leads to many intertwined responses. It is found to be structurally and functionally diverse. This diversity is both temporal and varies over the individuals. Thus huge amount of data related to immune systems is being generated. Immunologists have been using high-throughput experimental techniques for quite a long time, which have generated a vast amount of functional, clinical, and epidemiological data. So the development of new computational approaches to store and analyze these data is needed. Recently immunology focused resources and softwares are coming up, which help in understanding the properties of whole immune system (Gardy et al. 2009). This gives rise to a new field, called immunoinformatics. Immunogenomics, immunoproteomics, epitope prediction, and in silico vaccination are different areas of computational immunological research. Recently, Systems Biology approaches are being applied to investigate the properties of dynamic behavior of an immune system network (De and Tomar 2014; Tomar and De 2010, 2014).

The information provides immunoinformatics domain, an immunologist can explore the potential binding sites, which, in turn, leads to development of new vaccines. This methodology is termed as "reverse vaccinology" which analyzes the pathogen genome in order to identify potential antigenic proteins (Davies and Flower 2007). These tools are also helpful to identify virulence genes and surface-associated proteins. Immunomics itself is a new discipline, with high-throughput techniques to get the immune system mechanism (De Groot 2006; Grainger 2004) as mentioned in Fig. 9.1.

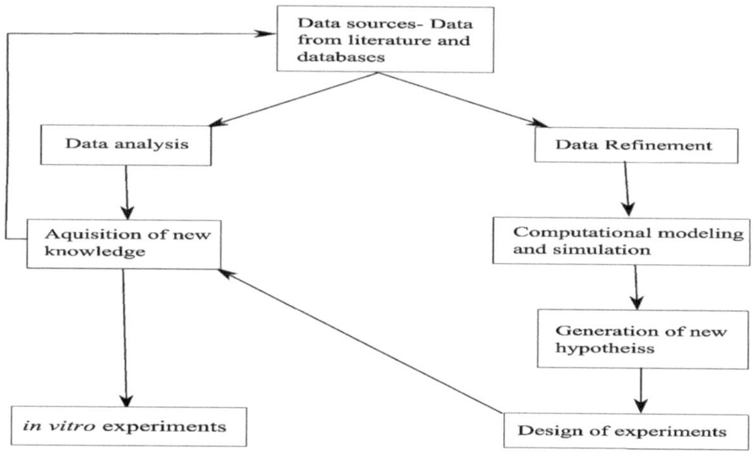

Fig. 9.1 Shows the workflow in immunomics

9.2 Data Sources

This section provides information on immune system-related datatypes and databases. The detailed version of information on data sources is available from our previously published articles and book on immunoinformatics (De and Tomar 2014; Tomar and De 2010, 2014).

9.2.1 Data from Lab Experiments

Immunology has a vast amount of experimental data due to the high-throughput molecular biology techniques. These techniques help in finding the structure and function of immune genes and their products (Yates et al. 2001). Experiments involve many immunological techniques to understand the underlying mechanism of an immune system and its responses to various infections, diseases, and drugs, viz., affinity chromatography (Kaplan et al. 1974), flow cytometry (Davey 2003), radioimmunoassay (Mari et al. 2006) (Durkin et al. 1997), enzyme-linked immunosorbent assay (ELISA) (Durkin et al. 1997; Ma et al. 2006), competitive inhibition assay (Levine et al. 1980), and Coombs test (Nishimaki et al. 1987).

Purification techniques like affinity chromatography are used to purify MHC peptide from membrane MHC molecules, which can be analyzed by capillary high-pressure liquid chromatography electrospray ionization-tandem mass spectrometry (Admon et al. 2003).

9.2.2 Immunomic Microarray Technology

The similar technology, like microarray one, is used in functional immunomics and is referred to as "immunomic microarray" that includes dissociable antibody microarray (Wang 2004), serum microarray (Magdalena et al. 2005), and serological analysis of cDNA expression library (SEREX) (Sahin et al. 1997). Antibody microarray is used to measure concentration of antigen for a specific antibody probes and thereby consists of antibody probes and antigen targets. Peptide microarray uses antigen peptides as fixed probes and serum antibodies as targets. Peptide-MHC microarray or artificial antigen-presenting chip technique has recombinant peptide-MHC complexes and co-stimulatory molecules, which are immobilized on a surface. The T cell spots act as artificial antigen-presenting cells (Oelke et al. 2003) containing a defined MHC-restricted peptides. One can measure two or more signals simultaneously determined by a single feature, i.e., epitope in immunomic microarray (Braga-Neto and Marques 2006; Nahtman et al. 2007).

9.2.3 Immunomic Databases

Epitope information-related databases, bioinformatics tools, and prediction algorithms are very crucial for basic immunological studies, diagnosis, and vaccine research (Peters et al. 2005). InnateDB (Lynn et al. 2008) (www.innatedb.ca) has been created to understand complete network of pathways and interactions of innate immune system responses. The newer version is Cerebral (Barsky et al. 2007) and has a Java plug-in for the Cytoscape biomolecular interaction viewer version 2.8.2 (Shannon et al. 2003) for automatically generating layouts of biological pathways. Table 9.1 has some of the databases for B cell epitopes, T cell epitopes, allergy prediction, and evolution of immune system genes and proteins (Tomar and De 2010).

9.2.4 B Cell Epitope Databases

Epitome (Schlessinger et al. 2006) (http://www.rostlab.org/services/epitome/) contains all known antigen-antibody complex structures. More details are available in our previously published article (Tomar and De 2010, 2014).

9.2.5 T Cell Epitope Databases

Some recent investigations include finding and mapping of potential epitopes. Epitope mapping leads to design effective vaccines. SYFPEITHI database (Rammensee et al. 1999) (www.syfpeithi.de) has information on MHC class I and II anchor motifs and their bindings. IEDB (Sathiamurthy et al. 2005) (http://www.

Table 9.1 Databases on B cell epitopes, T cell epitopes, allergen, and molecular evolution of immune system components (Tomar and De 2010, 2014)

Databases	Names	URLs
B cell epitopes	CED	http://immunet.cn/ced/log.html
	Bcipep	http://www.imtech.res.in/raghava/bcipep
	Epitome	http://www.rostlab.org/services/epitome/
	IEDB	http://www.immuneepitope.org/
	IMGT®	http://imgt.org
T cell epitopes	JenPep	Ask d.r.flower@aston.ac.uk
	SYFPEITHI	http://www.syfpeithi.de
	IEDB	http://www.immuneepitope.org/
	FRED	http://abi.inf.uni-tuebingen.de/Software/FRED
	IMGT®	http://imgt.org
Allergen	Database of IUIS	http://www.allergen.org
	SDAP	http://fermi.utmb.edu/SDAP/
Information related to molecular evolution of immune system components	ImmTree	http://bioinf.uta.fi/ImmTree
	Immunome database	http://bioinf.uta.fi/Immunome/
	ImmunomeBase	http://bioinf.uta.fi/ImmunomeBase
	Immunome Knowledge Base	http://bioinf.uta.fi/IKB/

immuneepitope.org/) and ontology-related information (http://ontology.iedb.org/) are specifically designed to get intrinsic, chemical, and biochemical information on immune epitopes and their interactions with host molecules.

FRED (Feldhahn et al. 2009) (http://abi.inf.uni-tuebingen.de/Software/FRED) deals with the methods for data processing and to compare the performance of the prediction methods considering experimental values. IMGT® (Lefranc et al. 2009) (the international ImMunoGeneTics information system®) (http://imgt.org) has 5 databases and 15 interactive online tools for sequence, genome, and 3D structure analysis. The IMGT/HLA Database (Robinson et al. 2011) (http://www.ebi.ac.uk/imgt/hla/) provides a specialist database as a part of the international ImMunoGeneTics project (IMGT). This information is also available in our previously published article (Tomar and De 2010).

9.3 Immunomic Tools and Algorithms

The main objective of B cell epitope prediction is to design a molecule that can replace an antigen in the process of either antibody production or antibody detection. Such a molecule can be synthesized, or, in case of a protein, its gene can be cloned into an expression vector. Designed molecules are preferable to use because they are inexpensive and noninfectious in contrast to viruses or bacteria, which may be harmful to a researcher or experimental animal. Epitopes are important for disease understanding, host-pathogen interaction analyses, antimicrobial target discovery, and vaccine design. The experimental techniques are found to be difficult and time consuming. Due to this reason, several in silico methodologies are being developed and used to identify epitopes. Table 9.2 lists some of the tools that deal with B and T cell epitope prediction, allergy prediction, and in silico vaccination. Here, we describe different methodologies for epitope and allergy prediction and the process of in silico vaccination briefly.

9.3.1 B Cell Epitope Prediction

B cell epitopes are classified as continuous/linear and discontinuous/conformational. A synthetic peptide may correspond to a short continuous stretch from a protein sequence and bind an antibody raised against a protein; such a peptide is called a continuous epitope of the protein. Both sequence- and structure-based prediction tools are available; however, prediction tools are less available for discontinuous B cell epitopes (Tong and Ren 2009; Saha et al. 2005).

9.3.2 Prediction of Continuous B Cell Epitopes

9.3.2.1 Sequence Based Methods

The majority of the sequence-based methods assume that epitopes must be accessible for antibody binding, and, thus, these methods used epitope properties related to surface exposure. These methods are limited to the prediction of continuous epitopes. Sequence-based methods have been tested on prediction of two protective epitopes known in influenza A virus hemagglutinin HA1 (Bui et al. 2007). The first continuous epitope is the 91–108 epitope (SKAFSNCYPYDVPDYASL), which is a protective epitope in a rabbit, able to elicit antibodies neutralizing infectivity of influenza viruses (Muller et al. 1982). The second continuous epitope is the 127–133 epitope (WTGVTQN) protective against the influenza strain A/Achi/2/68 (H3N2) in mouse (Naruse et al. 1994).

Table 9.2 Web servers and tools for prediction of B cell epitopes, T cell epitopes, allergy, and for in silico vaccination (De and Tomar 2014; Tomar and De 2010, 2014)

Web servers and tools	Names	URLs
B cell epitope prediction	ABCpred	http://www.imtech.res.in/raghava/abcpred
	BEPITOPE	jlpellequer@cea.fr
	COBEpro	http://scartch.proteomics.uci.edu
	BepiPred	http://www.cbs.dtu.dk/services/BepiPred
	IMGT®	http://imgt.org
	Bcepred	http://www.imtech.res.in/raghava/bcepred/
	Discotope	http://www.cbs.dtu.dk/services/DiscoTope/
	CEP	http://115.111.37.205/cgi-bin/cep.pl
	AgAbDb	http://115.111.37.206:8080/agabdb2/home.jsp
	MIMOP	Request from franck.molina@cpbs.univ-montp1.fr
	MIMOX	http://immunet.cn/mimox/
	Pepitope	http://pepitope.tau.ac.il/
	3DEX	http://www.schreiber-abc.com/3dex/
	IEDB	http://www.immuneepitope.org
T cell epitope prediction	MMBPred	http://www.imtech.res.in/raghava/mmbpred/
	NetCTL	http://www.cbs.dtu.dk/services/NetCTL/
	NetMHC 3.0	http://www.cbs.dtu.dk/services/NetMHC/
	TAPPred	http://www.imtech.res.in/raghava/tappred/
	Pcleavage	http://www.imtech.res.in/raghava/pcleavage/
	ElliPro	http://tools.immuneepitope.org/tools/ElliPro
	MHCPred	http://www.ddg-pharmfac.net/mhcpred/MHCPred/
	ProPred	http://www.imtech.res.in/raghava/propred1/
	EpiToolKit	http://www.epitoolkit.org
	SYFPEITHI	http://www.syfpeithi.de
	IMGT®	http://imgt.org
	IEDB	http://www.immuneepitope.org/
	EpiJen v 1.0	http://www.ddg-harmfac.net/epijen/EpiJen/EpiJen.htm
Allergy prediction	AlgPred	http://www.imtech.res.in/raghava/algpred
	Allermatch	http://www.allermatch.org
	APPEL	http://jing.cz3.nus.edu.sg/cgi-bin/APPEL
	EVALLER	http://www.slv.se/en-gb/Group1/Food-Safety/e-Testing-of-protein-allergenicity/
In silico vaccination	VaxiJen	http://www.ddg-pharmfac.net/vaxijen/
	DyNAVacS	http://miracle.igib.res.in/dynavac/
	NERVE	http://www.bio.unipd.it/molbinfo
	VIOLIN	http://www.violinet.org
	Vaxign	http://www.violinet.org/vaxign/

9.3.2.2 Prediction Using Amino Acid Propensity Scale

Amino acid scale-based methods apply amino acid scales to compute the scores of a residue *i* in a given protein sequence. The *i-(n-1)/2* neighboring residues on each side of residue *i* are used to compute the score for residue *i* in a window of size *n*. The final score for residue *i* is the average of the scale values for *n* amino acids in the window. Amino acid propensity scales such as hydrophilicity and characteristic flexibility can be used to identify epitopes.

9.3.2.3 Prediction Using Machine Learning Methodologies

Several researchers used machine learning algorithms and tools to retrieve characteristics of an epitope through learning a dataset. For example, Saha and Raghava used ANN in ABCpred (Saha and Raghava 2006) (www.imtech.res.in/raghava/abcpred). Sweredoski and Baldi (2009) presented COBEpro using SVM. Saha et al. (2005) used feed forward and recurrent neural networks to predict continuous B cell epitopes. COBEpro (Sweredoski and Baldi 2009) that is a two-step system for prediction of continuous B cell epitopes. For BepiPred (Larsen et al. 2006), (http://www.cbs.dtu.dk/services/BepiPred), three datasets of linear B cell epitopes were constructed.

9.3.2.4 Mimotope-Based Methodology

Phage display library has a large number (more than 109) of random peptides (Mayrose et al. 2007a, b). These pools of peptides are called as mimotopes (Moreau et al. 2006). MIMOP tool (Moreau et al. 2006) has been developed. MIMOP predicts linear and conformational epitopes based on two algorithms, viz., MimAlign uses degenerated alignment analyses, and MimCons is based on consensus identification. MIMOX (Huang and Honda 2006) (http://web.kuicr.kyoto-u.ac.jp/~hjian/mimox) comes in the same category, which maps a single mimotope or a consensus sequence of a set of mimotopes, on to the corresponding antigen structure.

9.3.2.5 T Cell Epitope Prediction

There exist several methodologies for prediction of MHC binding peptides, which are based on the idea of quantitative matrices, hidden Markov model (HMM), artificial neural networks (ANN), support vector machine (SVM), and structure of the (De and Tomar 2014; Tomar and De 2010, 2014).

9.3.2.6 Prediction Through Matrix-Driven Methods

Huang and Dai (2006) first investigated a new encoding scheme of peptides. This scheme has BLOSUM matrix with the amino acid indicator vectors for direct prediction of T cell epitopes. It replaced each nonzero entry in the amino acid indicator vector by the corresponding value appeared in the diagonal entries in BLOSUM matrix. MMBPred (Bhasin and Raghava 2003) (www.imtech.res.in/raghava/mmbpred/) server is one such example, which predicts the mutated promiscuous and high affinity MHC binding peptide.

9.3.2.7 Prediction Through Hidden Markov Model

Zhang et al. developed PRED^{TAP} (Zhang et al. 2006) for the prediction of peptide binding to hTAP, also mentioned in Tomar and De (2010).

9.3.2.8 Prediction Through Artificial Neural Networks

MHC class I molecule motifs are well defined; however, MHC class II binding peptide prediction is found to be difficult. The reasons are variable length of reported binding peptides, undetermined core region for each peptide, and number of amino acids as primary anchor. Brusic et al. developed PERUN (Brusic et al. 1998), a hybrid method for the prediction of MHC class II binding peptide. The use of PlaNet package version 5.6 (Miyata 1991) to design and train a three-layered fully connected feed forward artificial neural network has provided the needful impact. The whole process of MHC class I ligands' degradation and presentation has been modeled in EpiJen (Doytchinova et al. 2006) (http://www.ddg-pharmfac.net/epijen/EpiJen/EpiJen.htm), which uses a multistep algorithm based on quantitative matrices.

9.4 Applications of Immunoinformatics

This section focuses on applications of immunoinformatics that includes cancer diagnosis and therapy, along with the idea of integrating Systems Biology with immunoinformatics.

9.4.1 Immunoinformatics for Cancer Diagnosis and Therapy

Antigen presentation plays a central role in the immune response and as a result also in immunotherapeutic methods like antitumor vaccination. There is a need to rapidly screen the antigens and to design specific types of expression constructs for immunotherapy of cancer. Competent immune responses to cancer are likely to be restricted to the immunome of a specific cancer, including the set of antigens that drive successful immune responses. However, it is still difficult to find the set of antigens that varies between different tumors. Antitumor vaccination takes advantage of in vivo processes, and it harnesses the full power of the immune system, unlike, the more artificial ex vivo expansion of T cells.

Changes in the cancer diagnosis and prevention are being supported by informatics (Hu et al. 2004). For example, the Cancer Biomedical Informatics Grid (caBIG) connects a network of 500 individuals and 50 institutions who share data and analyze tools to speed up the development of innovative approaches for the prevention and treatment of cancer (Sanchez et al. 2004). The 2005 database issue of Nucleic Acids Research lists 14 cancer-related molecular databases, which mainly focus on cancer-related genes and gene expression (Galperin 2005). Listings of tumor antigens are also available (Novellino et al. 2005). This list includes antigens that have defined T cell epitopes. Tumor-associated antigens (TAA) have played a vital role in both diagnosis and treatment of human carcinomas, such as *prostate-specific antigen* (PSA) in the diagnosis of prostate cancer. Despite of this, the process of TAA identification has often been hampered by the complicated lab procedures. To fasten the process of tumor antigen discovery, and improve diagnosis and treatment of human carcinoma, a publicly available database Human Potential Tumor Associated Antigen database (HPtaa) (http://www.hptaa.org) has been established (Wang et al. 2006). Systems Biology approaches target identification of a small number of antigens expressed by cancer cells that are suitable targets of immune responses against cancer. A proteomic mapping of in vivo targets for antibodies in the lungs, and solid tumors in experimental animals, defines aminopeptidase-P and annexin A1 as targets of anticancer immune responses (Oh et al. 2004). Informatic methods have also been used for classification of tumors into subtypes, which supports decisionmaking for the selection of therapeutic approaches; however, such applications in cancer immunology are yet to come (Camp et al. 2004).

9.4.2 Vaccine Against Tumors

Reliable predictions of immunogenic T cell epitope peptides are crucial for rational vaccine design and represent a key problem in immunoinformatics. Computational approaches have been developed to facilitate the process of epitope detection and show potential applications to the immunotherapeutic treatment of cancer. Epitope-driven vaccine design employs these bioinformatics algorithms to identify potential targets of vaccines against cancer (Rosa et al. 2010). The development of epitope-based DNA vaccines and their antitumor effects in preclinical research against B-cell lymphoma has been described (Iurescia et al. 2012).

Most immunotherapeutic approaches work on the induction of antitumor $CD8^+$ T cells, which exhibit cytolytic activity toward tumor cells expressing tumor-specific or tumor-associated Ags. But the immunization strategies that focus solely on $CD8^+$ T cell immunity might prove to be insufficient because they will be unable to provide long-term protective immunity (Khanolkar et al. 2007). It has been shown that the peptides predicted to bind MHC can elicit a tumor-killing cytotoxic T lymphocytes (CTL) response (Lu and Celis 2000). Although CTLs have been found to be the key player in the generation of antitumor therapeutic effects, sometimes it also remains as suboptimal. $CD4^+$ T cells are critical for the generation and maintenance of CTLs response through providing cytokines or by major pathway, i.e., dendritic cell licensing (Smith et al. 2004; Wan and Flavell 2009). Class II MHC-bound epitopes activate $CD4^+$ T cells and maintain effective CTL response that plays an important role in the antitumor response (Hung et al. 1998; Kalams and Walker 1998).

$CD4^+$ T cells determine the functional status of both innate and adaptive immune responses; thus, the inclusion of appropriate $CD4^+$ T cell epitopes may be essential for vaccine efficacy. Idiotypic immunoglobulin M (IgM) expressed by B-cell lymphoma is a clonal marker and a tumor-specific antigen. Thus, it can be used as an immune target. Specific immunogenic epitopes identified from these tumor antigens can be used as vaccines to activate an immune response against tumors cells (Houot and Levy 2009). Concerning to lymphoproliferative malignancies, TTFrC (tetanus toxin fragment C) fusion vaccine design was able to activate anti-Id antibody responses and to suppress tumor growth in murine models (King et al. 1998; Thirdborough et al. 2002) as well as was effective in inducing $CD8^+$ CTL in several tumor model et al. 2001).

9.4.3 Immunoinformatics and Systems Biology for Personalized Medicine

The idea to integrate immunoinformatics with Systems Biology approaches is for the better understanding of immune-related diseases at various systems levels. This integration can open the path of several translational studies for better clinical practices. The association between a disease and genetic variations is one of the most important aspects in pharmacogenomics and the development of personalized medicine. Figure 9.2 shows the integration that leads to the development of personalized medicine. The information about allele frequencies of immune molecules in a human population is especially important as different patient subgroups can be identified with different vaccine or drug responses (Yan 2010). For example, a SNP (S427T) in the innate immune gene interferon regulatory factor 3 (IRF3) has been associated with increased risk of human papillomavirus (HPV) persistence and cervical cancer (Wang et al. 2009). Genomic variation databases such as HapMap (http://snp.cshl.org/) and dbSNP (http://www.ncbi.nlm.nih.gov/SNP/) provide information on individual genotype data. The Allele Frequencies Database can be used to search for polymorphic regions of various populations on histocompatibility and immunogenetics (http://www.allelefrequencies.net/). This includes polymorphism information on HLA, cytokines, and killer-cell immunoglobulin-like receptors (KIR). Thus, there is a scope of the development of optimized vaccines and drugs tailored to personalized prevention and treatment through the integration of Systems Biology and immunoinformatics.

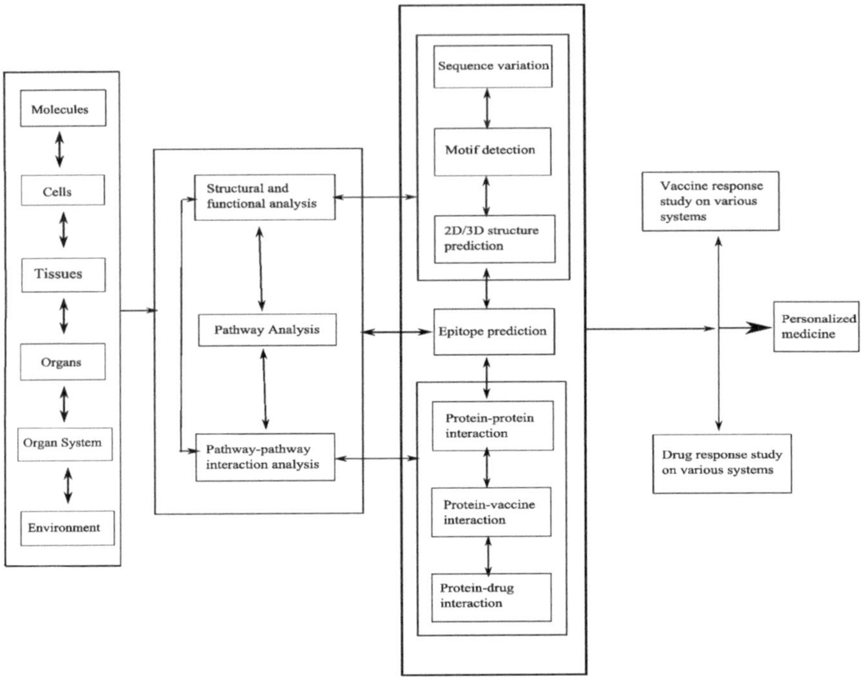

Fig. 9.2 Shows an integration of immunoinformatics and Systems Biology and how it leads to the development of a personalized medicine (Idea inspired from Yan 2010)

9.5 Conclusions

High-throughput experimental techniques are combined with immunoinformatics which result in explosive growth of immunology. This is as similar as the event that has transformed genetics into genomics as domain immunoinformatics can help in reducing time and cost for traditional immunology lab practices. This review article contains online immunological databases, tools and web servers, and the application of immunoinformatics.

Immunoinformatics models simulate the real behavior of immune system reactions and kinetics; therefore, these are engineered in a way that it can be interpreted and/or modified. These mathematical models take over the systems' uncertainty as compared to lab experiments. However, these cannot be directly compared to real biological data and it is a limitation. Immune system modeling capabilities take us toward designing a drug with no side effects. Therefore, integrating Systems Biology with immunoinformatics can lead to better clinical trials.

References

Admon A, Barnea E, Ziv T (2003) Tumor antigens and proteomics from the point of view of the major histocompatibility complex peptides. Mol Cell Proteomics 2(6):388–398

Barsky A, Gardy JL, Hancock RE, Munzner T (2007) Cerebral: a Cytoscape plugin for layout of and interaction with biological networks using subcellular localization annotation. Bioinformatics 23(8):1040–1042

Bhasin M, Raghava G (2003) Prediction of promiscuous and high-affinity mutated MHC binders. Hybrid Hybridomics 22(4):229–234

Braga-Neto UM, Marques ET Jr (2006) From functional genomics to functional immunomics: new challenges, old problems, big rewards. PLoS Comput Biol 2(7):e81

Brusic V, Rudy G, Honeyman M, Hammer J, Harrison L (1998) Prediction of MHC class II-binding peptides using an evolutionary and artificial neural network. Bioinformatics 14:121–130

Bui HH, Peters B, Assarsson E, Mbawuike I, Sette A (2007) Ab and T cell epitopes of influenza A virus, knowledge and opportunities. Proc Natl Acad Sci USA 104(1):246–251

Camp RL, Dolled-Filhart M, Rimm DL (2004) X-tile: a new bio-informatics tool for biomarker assessment and outcome-based cut-point optimization. Clin Cancer Res 10(21):7252–7259

Davey HM (2003) Flow cytometric techniques for the detection of microorganisms. Advanced flow cytometry: applications in biological research. Springer, pp 91–97

Davies MN, Flower DR (2007) Harnessing bioinformatics to discover new vaccines. Drug Discov Today 12(9):389–395

De Groot AS (2006) Immunomics: discovering new targets for vaccines and therapeutics. Drug Discov Today 11(5):203–209

De RK, Tomar N (2014) Immunoinformatics. Humana Press, New York

Durkin MM, Connolly PA, Wheat LJ (1997) Comparison of radioimmunoassay and enzyme-linked immunoassay methods for detection of Histoplasma capsulatum var. capsulatum antigen. J Clin Microbiol 35(9):2252–2255

Doytchinova IA, Guan P, Flower DR (2006) EpiJen: a server for multistep T cell epitope prediction. BMC Bioinform 7:131

Feldhahn M, Dönnes P, Thiel P, Kohlbacher O (2009) FRED—a framework for T-cell epitope detection. Bioinformatics 25(20):2758–2759

Galperin MY (2005) The molecular biology database collection: 2005 update. Nucleic Acids Res 33(Database issue):D5–24

Gardy JL, Lynn DJ, Brinkman FS, Hancock RE (2009) Enabling a systems biology approach to immunology: focus on innate immunity. Trends Immunol 30(6):249–262

Grainger DJ (2004) Immunomics: principles and practice. IRTL 2:1–6

Houot R, Levy R (2009) Vaccines for lymphomas: idiotype vaccines and beyond. Blood Rev 23 (3):137–142

Hu H, Brzeski H, Hutchins J, Ramaraj M, Qu L, Xiong R, Kalathil S, Kato R, Tenkillaya S, Carney J, Redd R, Arkalgudvenkata S, Shahzad K, Scott R, Cheng H, Meadow S, McMichael J, Sheu SL, Rosendale D, Kvecher L, Ahern S, Yang S, Zhang Y, Jordan R, Somiari SB, Hooke J, Shriver CD, Somiari RI, Liebman MN (2004) Biomedical informatics: development of a comprehensive data warehouse for clinical and genomic breast cancer research. Pharmacogenomics 5(7):933–941

Huang L, Dai Y (2006) Direct prediction of T-cell epitopes using support vector machines with novel sequence encoding schemes. J Bioinforma Comput Biol 4(01):93–107

Huang J, Honda W (2006) CED: a conformational epitope database. BMC Immunol 7(1):1

Hung K, Hayashi R, Lafond-Walker A, Lowenstein C, Pardoll D, Levitsky H (1998) The central role of CD4(+) T cells in the antitumor immune response. J Exp Med 188(12):2357–2368

Iurescia S, Fioretti D, Fazio VM, Rinaldi M (2012) Epitope-driven DNA vaccine design employing immunoinformatics against B-cell lymphoma: a biotech's challenge. Biotechnol Adv 30(1):372–383

Kalams SA, Walker BD (1998) The critical need for CD4 help in maintaining effective cytotoxic T lymphocyte responses. J Exp Med 188(12):2199–2204
Kaplan NO, Everse J, Dixon JE, Stolzenbach FE, Lee C-Y, Lee C-LT, Taylor SS, Mosbach K (1974) Purification and separation of pyridine nucleotide-linked dehydrogenases by affinity chromatography techniques. Proc Natl Acad Sci 71(9):3450–3454
Khanolkar A, Badovinac VP, Harty JT (2007) CD8 T cell memory development: CD4 T cell help is appreciated. Immunol Res 39:94–104
Kimbrell DA, Beutler B (2001) The evolution and genetics of innate immunity. Nat Rev Genet 2 (4):256–267
King CA, Spellerberg MB, Zhu D, Rice J, Sahota SS, Thompsett AR, Hamblin TJ, Radl J, Stevenson FK (1998) DNA vaccines with single-chain Fv fused to fragment C of tetanus toxin induce protective immunity against lymphoma and myeloma. Nat Med 4(11):1281–1286
Larsen JEP, Lund O, Nielsen M (2006) Improved method for predicting linear B-cell epitopes. Immunome Res 2(1):1
Lefranc M-P, Giudicelli V, Ginestoux C, Jabado-Michaloud J, Folch G, Bellahcene F, Wu Y, Gemrot E, Brochet X, Lane JM (2009) IMGT®, the international ImMunoGeneTics information system®. Nucleic Acids Res 37(suppl 1):D1006–D1012
Levine AM, Thornton P, Forman SJ, Van Hale P, Holdorf D, Rouault CL, Powars D, Feinstein D, Lukes R (1980) Positive coombs test in Hodgkin's disease: significance and implications. Blood 55(4):607–611
Lu J, Celis E (2000) Use of two predictive algorithms of the world wide web for the identification of tumor-reactive T-cell epitopes. Cancer Res 60(18):5223–5227
Lynn DJ, Winsor GL, Chan C, Richard N, Laird MR, Barsky A, Gardy JL, Roche FM, Chan TH, Shah N (2008) InnateDB: facilitating systems-level analyses of the mammalian innate immune response. Mol Syst Biol 4(1):218
Ma H, Hongbao K, Lee SL (2006) Study of ELISA technique. Nat Sci 4(2):36
Magdalena J, Odling J, Qiang PH, Martenn S, Joakin L, Uhlen M, Hammarstrom L, Nilsson P (2005) Serum microarrays for large scale screening of protein levels. Mol Cell Proteomics 4:1942–1947
Mari A, Scala E, Palazzo P, Ridolfi S, Zennaro D, Carabella G (2006) Bioinformatics applied to allergy: allergen databases, from collecting sequence information to data integration. The Allergome platform as a model. Cell Immunol 244(2):97–100
Mayrose I, Penn O, Erez E et al (2007a) Pepitope: epitope mapping from affinity-selected peptides. Bioinformatics 23:3244–3246
Mayrose I, Shlomi T, Rubinstein ND, Gershoni JM, Ruppin E, Sharan R, Pupko T (2007b) Epitope mapping using combinatorial phage-display libraries: a graph-based algorithm. Nucleic Acids Res 35:69–78
Miyata J (1991) A user's guide to PlaNet Version 5.6
Moreau V, Granier C, Villard S, Laune D, Molina F (2006) Discontinuous epitope prediction based on mimotope analysis. Bioinformatics 22(9):1088–1095
Muller GM, Shapira M, Arnon R (1982) Anti-influenza response achieved by immunization with a synthetic conjugate. Proc Natl Acad Sci USA 79(2):569–573
Nahtman T, Jernberg A, Mahdavifar S, Zerweck J, Schutkowski M, Maeurer M, Reilly M (2007) Validation of peptide epitope microarray experiments and extraction of quality data. J Immunol Methods 328(1):1–13
Naruse H, Ogasawara K, Kaneda R, Hatakeyama S, Itoh T, Kida H, Miyazaki T, Good RA, Onoe K (1994) A potential peptide vaccine against two different strains of influenza virus isolated at intervals of about 10 years. Proc Natl Acad Sci U S A 91(20):9588–9592
Nishimaki T, Sagawa K, Motogi S, Saito K, Morito T, Yoshida H, Kasukawa R (1987) A competitive inhibition test of enzyme immunoassay for the anti-nRNP antibody. J Immunol Methods 100(1):157–160

Novellino L, Castelli C, Parmiani G (2005) A listing of human tumor antigens recognized by T cells: March 2004 update. Cancer Immunol Immunother 54(3):187–207
Oelke M, Maus MV, Didiano D, June CH, Mackensen A, Schneck JP (2003) Ex vivo induction and expansion of antigen-specific cytotoxic T cells by HLA-Ig–coated artificial antigen-presenting cells. Nat Med 9(5):619–625
Oh P, Li Y, Yu J, Durr E, Krasinska KM, Carver LA, Testa JE, Schnitzer JE (2004) Subtractive proteomic mapping of the endothelial surface in lung and solid tumours for tissue-specific therapy. Nature 429(6992):629–635
Rammensee H-G, Bachmann J, Emmerich NPN, Bachor OA, Stevanović S (1999) SYFPEITHI: database for MHC ligands and peptide motifs. Immunogenetics 50(3–4):213–219
Robinson J, Mistry K, McWilliam H, Lopez R, Parham P, Marsh SG (2011) The IMGT/HLA database. Nucleic Acids Res 39(Database issue):D1171–D1176
Rosa DS, Ribeiro SP, Cunha-Neto E (2010) CD4+ T cell epitope discovery and rational vaccine design. Arch Immunol Ther Exp 58(2):121–130
Saha S, Raghava GP (2006) Prediction of continuous B-cell epitopes in an antigen using recurrent neural network. Proteins 65(1):40–48
Saha S, Bhasin M, Raghava GPS (2005) Bcipep: a database of B-cell epitopes. BMC Genomics 6:79
Sahin U, Türeci Ö, Pfreundschuh M (1997) Serological identification of human tumor antigens. Curr Opin Immunol 9(5):709–716
Sanchez W, Gilman B, Kher M, Lagou S, Covitz P (2004) caGRID white paper (cancer biomedical informatics grid prototype project)
Sathiamurthy M, Peters B, Bui H-H, Sidney J, Mokili J, Wilson SS, Fleri W, McGuinness DL, Bourne PE, Sette A (2005) An ontology for immune epitopes: application to the design of a broad scope database of immune reactivities. Immunome Res 1(1):1
Schlessinger A, Ofran Y, Yachdav G, Rost B (2006) Epitome: database of structure-inferred antigenic epitopes. Nucleic Acids Res 34(Suppl 1):D777–D780
Shannon P, Markiel A, Ozier O, Baliga NS, Wang JT, Ramage D, Amin N, Schwikowski B, Ideker T (2003) Cytoscape: a software environment for integrated models of biomolecular interaction networks. Genome Res 13(11):2498–2504
Smith CM, Wilson NS, Waithman J, Villadangos JA, Carbone FR, Heath WR, Belz GT (2004) Cognate CD4(+) T cell licensing of dendritic cells in CD8(+) T cell immunity. Nat Immunol 5 (11):1143–1148
Sweredoski MJ, Baldi P (2009) COBEpro: a novel system for predicting continuous B-cell epitopes. Protein Eng Des Sel 22(3):113–120
Thirdborough SM, Radcliffe JN, Friedmann PS, Stevenson FK (2002) Vaccination with DNA encoding a single-chain TCR fusion protein induces anticlonotypic immunity and protects against T-cell lymphoma. Cancer Res 62(6):1757–1760
Tomar N, De RK (2010) Immunoinformatics: an integrated scenario. Immunology 131 (2):153–168
Tomar N, De RK (2014) In: De RK, Tomar N (eds) Immunoinformatics: a brief review, Immunoinformatics. Humana Press, New York, pp 23–55
Tong JC, Ren EC (2009) Immunoinformatics: current trends and future directions. Drug Discov Today 14(13):684–689
Wan YY, Flavell RA (2009) How diverse – CD4 effector T cells and their functions. J Mol Cell Biol 1(1):20–36
Wang Y (2004) Immunostaining with dissociable antibody microarrays. Proteomics 4(1):20–26
Wang X, Zhao H, Xu Q, Jin W, Liu C, Zhang H, Huang Z, Zhang X, Zhang Y, Xin D, Simpson AJ, Old LJ, Na Y, Zhao Y, Chen W (2006) HPtaa database-potential target genes for clinical diagnosis and immunotherapy of human carcinoma. Nucleic Acids Res 34(Database issue): D607–D612

Wang SS, Bratti MC, Rodriguez AC et al (2009) Common variants in immune and DNA repair genes and risk for human papillomavirus persistence and progression to cervical cancer. J Infect Dis 199:20–30

Yan Q (2010) Immunoinformatics and systems biology methods for personalized medicine, systems biology in drug discovery and development. Methods Mol Biol 662:203–220

Yates A, Chan CC, Callard RE, George AJ, Stark J (2001) An approach to modelling in immunology. Brief Bioinform 2(3):245–257

Zhang GL, Petrovsky N, Kwoh CK, August JT, Brusic V (2006) PredTAP: a system for prediction of peptide binding to the human transporter associated with antigen processing. Immunome Res 2

Metabolic Pathway Analysis Employing Bioinformatic Software

10

Soma S. Marla, Neelofar Mirza, and K. D. Nadella

Abstract

Metabolomics can be defined as the entire content of metabolites in a system and their roles and interactions in various metabolic pathways reflecting the genetic information encrypted in a genome. Of late, various biochemical and metabolic studies are being applied for monitoring the dynamics of growth and development of model plants. Observed variations in composition of metabolites are used for understanding course of gene expression in stress-affected plants. Critical analysis of metabolic pathways in a system when challenged with stress, for example, comparison of metabolic flux with data obtained from physiological, genetic, genomic studies, is helping plant researchers in identification of candidate genes responsible for regulation of a given trait. In this chapter attempts have been made to describe recent developments in plant metabolomics, and application of bioinformatics and databases in metabolic pathway analysis is reviewed.

Keywords

Plants · Metabolism · Pathways · Bioinformatics · Software · Databases

S. S. Marla (✉) · N. Mirza
Indian Council of Agricultural Research, National Bureau of Plant Genetic Resources, New Delhi, India
e-mail: soma.marla@icar.gov.in

K. D. Nadella
Directorate of Knowledge Management Units (DKMU), ICAR, New Delhi, India

Genetics division, ICAR, IARI, New Delhi, India

G. Wadhwa et al. (eds.), *Current trends in Bioinformatics: An Insight*,
https://doi.org/10.1007/978-981-10-7483-7_10

10.1 Background: Significance of Metabolic Pathway Analysis

Life involves execution of a networked metabolic reactions dictated by the genetic code of the organism where different enzymes, substrates, and their synthesized products are the active players in the game. Analysis of metabolic pathways enables us to understand and investigate the enzyme complex involved in synthesis of various products that represent the expression of a gene for specific trait.

Development of molecular biology techniques and the sequencing of genomes of humans (International Human Genome Sequencing Consortium 2001, 2004), rat, *Drosophila*, *E. coli*, yeast, and rice plant (Sasaki and Burr 2000) are providing huge genetic data that can be analyzed and potentially translated for their improvement. Bioinformatic tools are employed in analyzing this genome sequence data and help in elucidating various regulatory mechanisms underlying expression of genes involved in diseases and various biological functions. For example, biomarkers reflecting presence or absence of a specific enzyme or protein that can be easily scored from body fluids are fast emerging as diagnosis tools for important diseases. Bioinformatic methods integrate available genome information from various databases to analyze, model, and simulate metabolic pathways linked to a disease or any trait of interest. This leads to better understanding the disease, predicting the genes involved in disease process, and thereby discovering new drugs using metabolic pathway analysis.

A great number of biomolecules such as nucleic acids, amino acids, carbohydrates, lipids, and other compounds and metabolites are present in a living cell. Synthesis of these biomolecules and metabolites that support the functioning of a living cell is carried through hundreds of biochemical reactions occurring parallel in a cell. For example, in plants, metabolic pathway analysis reveals content and compositional changes are used to discriminate between different stresses (both for disease and drought). Schauer (Schauer et al. 2008) observed elevated levels of amino acids (proline and beta-glycine), TCA cycle intermediates, sugars, and polyols in drought-stressed tomatoes. Saneoka (Saneoka et al. 1995) observed similar results in water-stressed maize. For characterizing plant germplasm and assessing extent of diversity, Schauer (Schauer et al. 2006) used metabolic profiles generated from GC-MS to characterize differences in wild and cultivated varieties. Visible compositional variations in lysine, methionine, and tocopherol were observed in wild introgression lined in comparison to cultivated varieties of tomato. The biochemical pathway maps of different microorganisms, plants, and animals are available in most biochemistry textbooks and online. KEGG (Kyoto Encyclopedia of Genes and Genomes, http://www.genome.jp/kegg/), EcoCyc (*E. coli* database, http://ecocyc.org/), BioCyc (http://biocyc.org/), and MetaCyc (metabolic pathway database, http://metacyc.org/) for microorganisms and AraCyc (TAIR-Plant Metabolic Network, https://www.arabidopsis.org/biocyc/) and RiceCyc (rice metabolic pathways, http://pathway.gramene.org/gramene/ricecyc.shtml) for many other plants are the major source for obtaining information on metabolism. These biochemical databases combine the molecular information from the genome databases with metabolic networks. These are powerful interactive tools with applications in disease biology and drug development. There are various integrative tools that offer wonderful opportunities to analyze the metabolism of pathogenic microbes to understand the process of pathogenesis in relation with human hosts (Marla 2006).

Metabolic pathway analysis involves a qualitative focus on which enzymes are involved, which pathways are possible in a complex network, and what happens if one enzyme is knocked out or a new one is added to the set. Techniques employed in our laboratory to analyze metabolic pathways are discussed in detail. There is also a qualitative insight involved in analysis of metabolic pathways – underlying synthesis, concentrations of metabolites, substrates, products, and kinetic data.

It should however be noted that there is also a detailed quantitative perspective from which to analyze metabolic pathways – in their conversions and transitions – including metabolic concentrations, substrate and product affinities, and kinetic data on enzymes such as their turnover number which nonetheless will require much more experimental data related to the enzymes in question.

Isotope labeling of metabolites and NMR are modern techniques employed in flux analysis and separation of individual metabolites (Christensen and Nielsen 2000). These questions are only briefly discussed in here. Interested readers can refer to excellent reviews and textbooks on these topics (Hatzimanikatis et al. 1999; Stephanopoulos 1999). Dynamics of metabolites may also be tackled in comprehensive cellular simulations (Womble and Rownd, 1986).

Metabolic pathways present the next level of complexity in understanding working metabolism in a cell. It is remarkable to learn pathway identification and completeness and presence of alternative pathways from fluxes through a given metabolic network and indicate the importance and biological significance in pathway variations. It is important to choose a suitable bioinformatic tool for their analysis and their synthesis.

All enzymatic reactions can be divided into two types. There are biochemical reactions that can proceed from substrate to product and vice versa – these are reversible reactions. There are also reactions that proceed in only one direction – these are irreversible reactions. In most cases, the latter type is achieved as the product is either moved from the cell into outer compartments or by rapid conversion of the reaction product further down the enzymatic pathway. There are over 2000 known biochemical reactions compiled in databases such as Swiss-Prot, enzyme nomenclature database at Swiss Institute of Bioinformatics (http://www.expasy.ch/enzyme/index.html), and The Comprehensive Enzyme Information System – BRENDA database (http://www.brenda.uni-koeln.de).

In a pathway of enzymatic reactions, the flow through system is called the metabolic flux (e.g., the rate of flow in micromoles per minute) of substrates converted into products. Another very general division of enzymes is between energy-producing (exergonic reactions) and energy-consuming (endergonic reactions) ones. Similarly, on the metabolic level, metabolic pathways are also divided into catabolic (metabolite degrading or any type of metabolic "digestion") and anabolic (producing more complex biological metabolites). In contrast, all animals (including man), as well as fungi, parasites, and saprophytic organisms, rely on plants for their energy supply. Few common intermediates are produced from a large number of substrates. An important first end product is often acetyl-CoA and, ultimately, via oxidative pathways as oxidative phosphorylation of all available reduction equivalents (NADH, FAD, etc.) to generate a maximum of ATP molecules. NADPH and ATP serve as major high-energy components in transfer reactions in between.

A metabolic pathway can be understood as a series of enzymatic reactions dedicated to a specific product synthesis. Plentiful examples are curated in databases, viz., KEGG metabolic database (http://www.genome.ad.jp/keeg/pathway/map/map01100.html). Furthermore, these pathways may be branched or integrated with other pathways. Until now, mostly traditional or chemical knowledge has been used to delineate a pathway from the metabolic network. Algorithms are very helpful in gaining rationality and precision. The abundance of enzymes varies with the kind of organism, cell type, nutrient status, and developmental stage. Another aspect, which can be partly revealed by meticulous genome comparison and annotation, is the detection of isozymes – different forms of an enzyme expressed in different tissues or within the same tissue at different developmental states. Different regulatory domains often confer differential regulation on isozymes, and thus tissue-specific or condition-specific metabolites for the same type of reaction can be achieved. An example is lactate dehydrogenase that exits in two different forms for optimal metabolism in two different types of tissues – heart and muscle.

After the protein reading frames, metabolic pathways present the next level of complexity. It is remarkable to learn pathway identification and completeness and identification of alternative pathways and all possible fluxes through a given metabolic network and comprehend the importance and the specific biological significance of pathway variations. Then again, in reality there are always metabolic networks; hence, they are only an abstraction. In that respect, it is important to have tools for their analysis as well as their synthesis.

10.2 Pathway Mapping and Metabolic Networks

Cellular reactions are highly regulated and entangled. In order to get a good overview of the complexity of metabolism, one can use standard literature and databases available on the Internet (e.g., KEGG, EcoCyc, and Boehringer Mannheim Map). However, metabolism is much more elastic than the well-known set of pathways available in textbooks. Using known information about metabolism, computational calculations can help to find new possible routes in the biochemical network that can then be proven by experimental approaches such as isotope labeling of metabolites, use of metabolic inhibitors, and knocking out of enzymes. Through chemical reasoning or tradition, standard pathways such as glycolysis are known for well-known metabolites. However, given the network nature of enzymatic interactions, are these the only possible routes for defining the behavior of an enzyme in a pathway? In a similar way, new genome expression data try to predict roles of enzymes and in this way provide another possible path in its behavior in a pathway.

Three different approaches are generally used to map the pathways in a biochemical network from scratch. They do not require information about enzyme kinetics, affinity, or metabolite concentrations where each of them only state whether a certain metabolite (or a set of enzymes working together) is accessible to the set of enzymes present in the cell.

A general approach is to follow a pathway for synthesis of a known compound, taking clues from it, and comparatively locate presence of homologous enzymes in the pathway in studied organism (Seressiotis and Bailey 1986). To accomplish this, a minimal set of enzymes are chosen in a relative flux that operate at steady state, employing elementary mode. We mean "minimal" by the very necessary enzymes that are required for complete inhibition of other enzymes resulting in termination of this steady state of flux in the system. Elementary modes help in identification of key enzymes and also the minimum required set of enzymes in a flux for operation of a metabolic pathway. The stoichiometry of the network is critical for identifying the elementary flux modes (Clark 1998).

The presence of external and internal metabolites is described. The internal metabolites should fulfill all the requirements to meet the "minimal set" and support steady state in a flux. Internal metabolites are generally described used stoichiometric equations.

In the next step, external and internal metabolites are described. Information relating to calculations of elementary modes is available at http://bms-mudshark.Brookes.ac.uk/algorithm.pdf. The algorithm has been implemented in computer programs in Smalltalk (EMPATH, John Woods, Oxford, Year) and C (METATOOL, Pfeiffer et al. 1999) available at ftp://bmshuxley.brookes.ac.uk/pub/mca/software/ibmpc. This algorithm and accessory tools enable to test whether a given pathway can be supported by a minimum set of "essentially" required set of enzymes. Pathway alignment and comparison of annotated information from different genomes help in defining minimum sets.

- For testing the essentiality of the given set of enzymes for synthesizing a given metabolic product
- For determining only required pathways (e.g., in drug design) and identifying the pathways yielding high and low molar yields (for metabolic engineering)
- For comparative genome analysis and finding similarities and gaps (thereby filling them)
- For studying diseases in pharmacology and identifying deficient enzymes
- For identifying toxic metabolites

This analysis is very beneficial in drug design, testing, and dose optimization. Schuster et al. (2000) studied glycolysis in detail and identified minimum set of enzymes for several reactions. An alternative to elementary mode analysis is to think of all possible metabolic fluxes as a type of space – the metabolic flux space. As in analytical geometry, one may then ask how this space is set up by vectors. These are the so-called basis vectors of flux space (Schilling and Palsson 2000; Simpson et al. 1995; Fell 1993). This however has a problem; there are many possible solutions for the set of vectors defining the flux space. Nevertheless, a good study was done using flux balance studies, for example, investigating the comparative robustness of the *E. coli* metabolic network to supply cell growth over a wide range of different flux conditions (Edwards and Palsson 2000b) or the effect of *E.coli* gene mutants (Edwards and Palsson 2000a; Schilling and Palsson 2000) assessed the metabolic capabilities of *Haemophilus influenzae* Rd. employing flux balance studies. The different methods are compared and discussed in detail in Schilling et al. (1999).

AMDIS (Automated Mass Spectral Deconvolution and Identification System) is open software for metabolic flux analysis. It can find any set of target compounds provided in a GC-MS data file (Figs. 10.1a and 10.1b). It first finds all separated components by deconvoluting the data file and then compares them against a library of target compounds. AMDIS is free software provided by the National Institute of Standards and Technology (NIST), available at http://chemdata.nist.gov/mass-spc/amdis/downloads/.

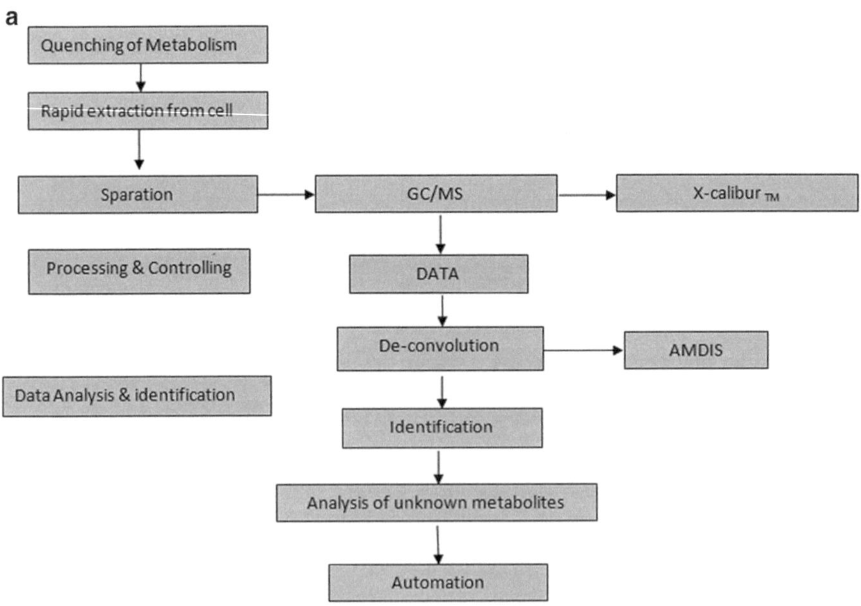

Fig. 10.1a Analysis of metabolic data

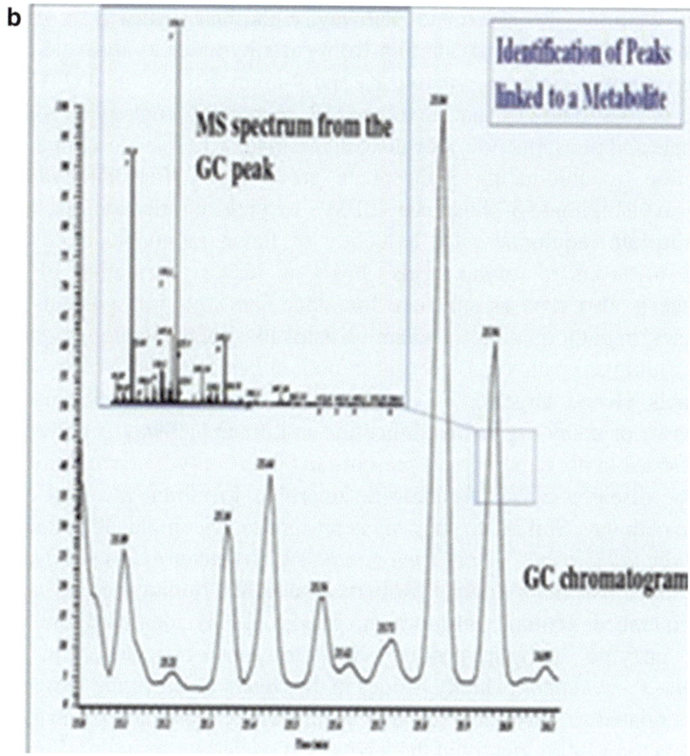

Fig. 10.1b Identification of metabolites

10.3 Comparison of Genomes and Discovery of New Pathways and Enzymes

Pathway comparison yields significant information about evolution, specific pharmacological targets among different genomes, and their biotechnological applications. An important part in today's drug discovery strategy is the alignment of biochemical pathways from diverse species. Pathway alignment involves comparison of a specific pathway from different species and the set of enzymes involved and marking the presence or absence of an enzyme. We will discuss the pathway alignment approach which combines three methods: (1) analysis and comparison of biochemical data, (2) analysis based on the idea of elementary modes, and (3) comparative genome analysis. The genome/sequence information available online from several organisms is of great aid (http://www.tigr.org/tdb). Apart from availability of this, accessibility to high-throughput biochemical data is aiding lofty comparative analysis.

Let us examine the shikimate pathway. Shikimate pathway is essential for synthesis of many compounds starting from carbohydrates to aromatic compounds such as tryptophan, tyrosine, vitamin K, and 4-aminobenzoic acid (PABA). Seven enzymes of the pathway are involved in sequential conversion of erythrose 4-phosphate and phosphoenolpyruvate to chorismate. Chorismate synthase catalyzes this reaction by eliminating 3-phosphate group and C-(6proR) hydrogen from 5-enolpyruvylshikimate-3-phosphate (EPSP) to yield chorismate, as this enzyme has an absolute requirement for reduction of flavin mononucleotide (FMN) for synthesis of aromatic amino acids. FMN is also a derivative of riboflavin. Chorismate is also used as substrate for other pathways for synthesis of folates, ubiquinones, naphthoquinones, and amino acids like phenylalanine, tryptophan, and tyrosine. Shikimate pathway is present in both eukaryotes and prokaryotes, but not in mammals. Hence, targeting chorismate and inhibition of this enzyme define the very survival of many organisms. Since the shikimate pathway is present in pathogenic bacteria but not in humans, it presents an opportunity for development of drugs against the diseases caused by parasitic microbes to inhibit the vital enzymes of shikimate pathway. Shikimate pathway is present in apicomplexan parasites such as *Plasmodium falciparum*, *Toxoplasma gondii*, *Helicobacter pylori*, *Cryptosporidium parvum*, and *Clostridium difficile* (Roberts et al. 2002; Sunita and Ramadevi 2004). Using sequenced genome information, Marla (2006) identified the chorismate synthase enzyme in gram-positive obligate anaerobic bacterium *Prevotella ruminicola*. *P. ruminicola* chiefly resides in the rumen of cattle and humans, causing infections related to bowel and gastroenteritis, where shikimate pathway is not yet reported. Marla (2006) identified the enzyme, modeled its structure (Fig. 10.2), and detected active binding sites (Fig. 10.2) using structure template from *H. pylori*. Of

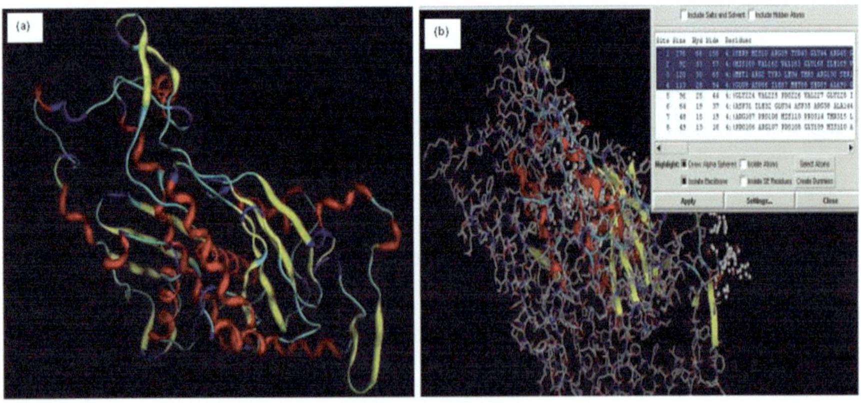

Fig. 10.2 (**a**) Structural variation in enzyme chorismate synthase (*shikimate pathway*) from *P. ruminicola* as compared to *Helicobacter pylori*. Different colors indicate similar and dissimilar regions (*procheck output*). (**b**) Location of four active binding sites for cofactor FMN in chorismate synthase (MOE V.11.0 output) (Marla 2006)

late, alignment of metabolic pathways from different organisms in conjunction with sequenced genome information is widely being used for identification of enzymes from parasite-specific pathway variants that may potentially block pharmacologically and serve as drug targets.

Of late, these databases on the World Wide Web provide biochemical and genomic data and facilitate to establish presence or absence of enzymes prior to pathway alignment (KEGG, http://www.genme.ad.jp/kegg/kegg2.html; WIT, http://wit.mcs.anl.gov/WIT2/). These systems are constantly updated (e.g., the ortholog group tables for KEGG; (Bono et al. 1998)). Another useful tool for pathway visualization and exploration is differential metabolic display which is based on Petri nets to generate all pathways fulfilling the given set of constraints ((Küffner et al. 2000); http://cartan.gmd.de/ToPL/ign.html). For an accurate analysis, however, these automated predictions need to be further refined since databases are derived from automated processes, viz., preformed pathway charts, for an accurate analysis.

At a given time, several hundreds of reactions are going on in parallel in a living cell, and, hence, investigating the full metabolic network of the cell is still an open research area, and fortunately there are a number of very useful bioinformatic tools to analyze metabolic pathways.

10.4 Metabolomics in Crop Nutritional Improvement

One of the major problems the world is facing today is malnutrition or hidden hunger caused by a lack of essential minerals and vitamins in food and is affecting an estimated 40–50% of the world's population. Only marginal success has been recorded with conventional breeding-based biofortification programs. Cereals are in general poor in vitamins and minerals as key enzymes linked to their synthesis or transport, respectively, are either absent, truncated, or inhibited in the grain endosperm (Naqvi et al. 2009). Numerous examples of nutritional improvement in crops through metabolic engineering are reported such as carboxylic acid content in tomato (Morgan et al. 2013), yield in tomato (Schauer et al. 2008), fruit quality in tomato (Genard et al. 2007) for enhanced protein quality (essential amino acids Met, Thru, Lys, Leu, etc.) in sweet potato (Prakash and Egnin 1997) and folate content (DellaPenna 2007), iron and vitamin A in rice, and tocopherol and carotenoids in tomato (DellaPenna and Pogson 2006; Fernie et al. 2006) and sesame (Marla 2014, personal communication). Further, several agronomically important and quality-related traits, e.g., biomass, stress tolerance, fruit quality, starch content, etc., have been improved by using metabolic approaches as well, viz., biomass accumulation in *Arabidopsis thaliana* (Lisec et al. 2008); drought tolerance in maize (Larkin and Harrigan 2007); glucosinolates in *Brassica napus* (Magrath et al. 1993); alcohol acyl transferase (MpAAT1) for flavor in apple (Souleyre et al. 2005); linalool for flavor enhancement in flowers, kiwi, and apple (Friel et al. 2006); starch quality in rice

(Smytha 1998) and potato (Schwall et al. 2000); and flavor and pigmentation traits in potato (Beckmann et al. 2007).

10.5 Plant Metabolomic Databases and Tools

The numerous metabolic databases available online in public domain can broadly be classified into:

1. Universal databases such as KEGG, PlantCyc, and MetaCyc
2. Species-/organism-specific databases like RiceCyc, AraCyc, PoplarCyc, etc.

PlantCyc is a comprehensive plant metabolic pathway database created to house a full spectrum of plant pathways and enzymes. PlantCyc includes plant-specific pathways from MetaCyc (Caspi and Karp 2007), AraCyc (Mueller et al. 2003; Zhang 2005), RiceCyc, and MedicCyc (Urbanczyk-Wochniak and Sumner 2007) that have been manually validated and curated and several other annotated pathways mined from the literature. PlantCyc contains 714 pathways, with 375 primary and 339 secondary metabolites and over 300 plant species linked to one or more of these pathways. While PlantCyc provides a large set of general metabolic pathways found in many plants, it also has a large number of pathways for the biosynthesis of rare but valuable compounds and serves as a central resource to access them.

Major organism-specific metabolic databases (Table 1) include AraCyc (*Arabidopsis thaliana*), RiceCyc (*Oryza sativa*), and PoplarCyc (*Populus trichocarpa*). The pathway section of Gramene database (a database for grasses and a resource for comparative grass genomics, http://www.gramene.org/; (Zhao et al. 2008)) contains known and predicted biochemical pathways of rice (RiceCyc) and sorghum (SorghumCyc) both of which are curated. The website also mirrors known and predicted biochemical pathways from SolCyc, AraCyc, EcoCyc, MetaCyc reference databases, and others. MetaCyc is a major database of nonredundant, experimentally elucidated metabolic pathways and contains more than 1400 pathways from more than 1800 different organisms.

Several databases for Solanaceae species are also available (Sol Genomics Network (SGN, http://solgenomics.net)) for species in the Solanaceae and their close relatives. The Metabolome of Tomato Database (MoTo DB) was developed as an LC-MS-based metabolome database (http://appliedbioinformatics.wur.nl/moto/, (Moco et al. 2006)). The KOMICS (Kazusa-Omics) database collects annotations of metabolite peaks detected by LC-FT-ICR-MS and contains a representative metabolome data set for the miniature dwarf model tomato cultivar “Micro-Tom” (Iijima et al. 2008). The Armec Repository Project on potato serves as a storehouse for metabolome data detected by ESI-MS (http://www.armec.org/MetaboliteLibrary/index.jsp). The Golm Metabolome Database (GMD) provides public access to customized mass spectral libraries and metabolite profiling experimental data and related tools (http://csbdb.mpimp-golm.mpg.de/csbdb/gmd/gmd.html; (Kopka et al. 2005)). The MS/MS spectral tag (MS2T) libraries at the Platform for RIKEN Metabolomics (PRIMe) provide access to libraries of phytochemical LC-MS2 spectra obtained from LC-ESI-Q-TOF/MS experiments for plant species (http://prime.psc.riken.jp/lcms/ms2tview/ms2tview.html; (Matsuda et al. 2009)). These databases are important information resources and repositories of large-scale data sets and serve as tools for further integration of metabolic profiles with data from other omics experiments (Akiyama et al. 2008).

In addition to the above listed databases, crop gene co-expression network databases are rich sources to search and locate orthologous genes and enzymes of interest. Often, key enzymes from essential reactions of pathways, active sites for drugs, and pathogens are evolutionarily conserved and are important in context of identifying, for example, disease-resistant genes. For locating metabolite orthologs in crops, gene co-expression databases, developed from public microarray databases in rice (RiceArrayNet (Lee et al. 2009)); PLEXdb database including expression related and other data for rice, barley, poplar, and several other crops (Edwards et al. 2007); and CoP database (Ogata et al. 2010) that has gene modules for *Arabidopsis*, rice, wheat, barley, poplar, etc. are excellent sources. Various biochemical databases employed in crop improvement are described in detail by Nadella et al. (2012).

10.5.1 Softwares for Modeling and Simulation and Integrating Metabolism with Growth

For modeling plant growth and development, modeling software tools use metabolic signatures and their compositional variations (http://csbe.bio.ed.ac.uk/software.php).

Plant Systems Biology Modeling (PlaSMo) and Bio-PEPA (Ciocchetta and Hillston 2008) are plant-specific model-generating tools that aid to develop plant-based models for applications in plant studies. Other important tools for modeling and simulation of metabolic pathways include MMT (metabolic modeling tool (j.hurlebaus@fz-juelich.de)), CellDesigner, and SYCAMORE that allows metabolite/enzyme parameter input, simulation, and modeling specific to user organism.

Of late tools started appearing for metabolic analysis of whole-genome sequenced data sets, for example, Altered Pathway Analyzer (APA, http://bioinfo.icgeb.res.in/APA) that analyzes microarray data sets and identifies and prioritizes altered pathways which may be differentially regulated during organism's development. Metabolic changes occurring in stress-challenged rice plants were analyzed using transcriptomics (Mohanty et al. 2015). Various parameters including listing key enzymes affected, Cis regulatory elements, and transcription factors were analyzed to verify the observed spatiotemporal gene expression of drought-affected rice plants.

10.6 Identification and Annotation of Candidate Enzymes from Unsequenced Crop Genomes

Researchers studying plant species that are not fully sequenced may want to identify enzymes that are likely to participate in a particular metabolic process. The availability of experimentally verified enzyme data from a variety of species distributed throughout the plant kingdom makes PlantCyc a useful starting point for identification of candidate enzymes involved in a wide range of biochemical pathways in other crops and under studied species. RESD is a tool from PlantCyc to predict enzymes from newly sequenced genomes. The RESD was built by extracting sequences that are annotated based on specific enzymatic activities from curated databases such as BRENDA (Chang et al. 2009), UniProt (The UniProt Consortium 2011), MetaCyc (Caspi et al. 2006, 2014; Caspi and Karp 2007), and The Arabidopsis Information Resource (TAIR; (Swarbreck et al. 2008)). RESD (v. 1.0) contains 14,187 literature-supported enzymes across all kingdoms. RESD module illustrates functions of all enzymes found in PlantCyc and contains available protein sequence information. The automatic function analysis pipeline from PMN (Plant Metabolic Network; http://www.plantcyc.org/) can be combined with RESD to perform genome-wide annotation of groups of enzymes from the given metabolic pathway (http://plantcyc.org/downloads/data_downloads.faces).

PMN BLAST can also be employed to locate putative homologs from sequence comparison. It provides partial information revealing functions in metabolic context of the identified homologs. For example, with the available EST library from a yet unsequenced organism such as cowpea, user can search against information available from sequenced organism such as *Medicago truncatula* for easy location of candidate enzymes by employing TBLASTX. A researcher involved in plant diversity studies and possessing transcript data of wild type and the mutants can locate potential quantitative changes occurring in gene expression across many metabolic domains instantly, can infer the effects of mutated gene on metabolism, and can develop a hypothesis explaining the observed metabolism in the mutant appearing under the influence of stress conditions.

10.7 Integration of Metabolomics and Other *Omics* Data

Metabolomic approaches also support the understanding of global relationships among cellular metabolic systems in combination with other omics instances such as profiles of the transcriptome and proteome and also genetic variation support systems approaches. So far, these combinatorial approaches have been successfully demonstrated in potato, in tomato, and in *Arabidopsis* by taking advantage of the many other available omics resources, using data from whole-genome sequencing, large-scale transcriptome data sets and related co-expression data, and bio-resources including collections of mutants and full-length cDNA clones. An integrated approach that comprises metabolome and transcriptome analysis was conducted for investigation of an activation-tagged mutant and MYB TF overexpressers of

PAP1 gene in order to identify genes involved in anthocyanin biosynthesis in *Arabidopsis* (Tohge et al. 2005). Co-expression data of the *Arabidopsis* transcriptome provided by the ATTED-II database (Obayashi et al. 2009) was applied to detect key genes involved in specific metabolic pathways using mutant lines of the targeted genes by several workers. For example, ATTED-II database was successfully employed to identify novel genes involved in lipid metabolism, leading to identification of a novel gene, UDP-glucose pyrophosphorylase3 (UGP3) from sulfolipid biosynthesis (Okazaki et al. 2009). Yonekura-Sakakibara et al. (2008) employed flavone profiling and transcriptome co-expression analysis to identify all of the genes related to flavonoid biosynthesis, which led to further detailed analysis of two flavonoid pathway genes *UGT78D3* and *RHM1* in *Arabidopsis*.

Approaches that integrate metabolome and transcriptome data also aid to understand regulatory networks that act in response to environmental stresses in plants. The metabolic pathways that act in response to cold and dehydration conditions in *Arabidopsis* were investigated recently using various experiments combining MS (mass spectroscopy) with microarray analysis of overexpressed genes encoding two transcription factors – DREB1A/CBF3 and DREB2A (Maruyama et al. 2009). Metabolomic profiling was also used to investigate chemical phenotypic changes between wild-type *Arabidopsis* and a knockout mutant of the *NCED3* gene under dehydration stress conditions. The metabolic data was later integrated with transcriptome data to reveal ABA-dependent regulatory networks (Urano et al. 2009).

MapMan is a graphical tool to project omics data sets including gene expression data onto diagrams of metabolic pathways or other processes (http://mapman.gabipd.org/web/guest, (Thimm et al. 2004)). KaPPA-View is another Web-based analysis tool that can be used to superimpose transcriptome and metabolome data onto plant metabolic pathway maps (http://kpv.kazusa.or.jp/kappa-view/; (Tokimatsu et al. 2005)). PRIMe is a Web-based service that provides data sets of metabolites measured by multidimensional NMR spectroscopy, GC-MS, LC-MS, and CE-MS together with analytical tools that use metabolome and transcriptome data sets to promote data integration (http://prime.psc.riken.jp/; (Akiyama et al. 2008)).

10.8 Metabolic Data Analysis

Metabolomics produces high-dimensional data sets like other functional genomics technologies and has many features common with transcriptomics and proteomics. However, it has a number of unique problems associated with most features specific to it (Mendes 2002). Any multivariate analysis can be applicable to analysis of metabolic profiles and statistical clustering, and also machine learning algorithms such as neural networks (Marla et al. 2010), Kohonen's self-organizing maps, and support vector machines are beginning to be employed to analyze metabolic profiles and deduce information that can potentially be of importance for crop improvement. Statistical analyses using multivariate analysis, such as principal component analysis

(PCA), hierarchical clustering analysis (HCA), and self-organization mapping (SOM), are typically used to classify samples and/or metabolites (Kose et al. 2001; Hirai et al. 2004; Jonsson et al. 2004; Matsuda et al. 2009).

Bioinformatic analysis enables classification and assignment of genes into certain functional classes based on metabolic profiles of mutants and is broadly conducted with the following objectives:

1. To classify profiles and detect key molecules or enzyme(s) that can be functionally linked with genes
2. To perform subsequent gene functional analysis

Nicholson et al. (1999) successfully applied multivariant analysis of metabolic profile data generated from NMR in fingerprinting and identified crop diseases. In metabolomics, primary objective is development of patterns of metabolites or enzymes that are characteristic to the given plant sample and thereby identification of functional gene(s) associated with them. Bioinformatic analysis of profiles and generation of fingerprint signatures of a given crop species involve the following steps:

1. Deconvolution of fingerprints (spectrum, chromatogram, etc.) into component peaks
2. Identification of biochemical species by lookup functions in databases containing relevant physicochemical properties

After metabolite identification, the inference of biochemical networks can be performed using metabolite profiles by methodologies such as time series collection of metabolite profiles followed by temporal analysis of the dynamics (Mendes 2002). Various bioinformatic algorithms employed in data analysis significantly contributed to analytical chemical experimentation in elucidating knowledge about new classes of functional genes associated with plant phenotypes. Scott et al. (2010) successfully employed machine learning classification techniques for metabolite fingerprinting of *Arabidopsis thaliana* mutants and found the machine learning techniques, viz., support vector machine (SVM) and random forest (RF) classifiers, were unsurpassed for phenotype discrimination compared to standard PCA. Software packages XCalibur and MetAlign are designed for analysis of metabolic profiles generated from large-scale metabolic profiling techniques such as GC-MS. By aligning the full mass spectrometry (MS) metabolic profiles using MetAlign software, conducting multivariate comparative analysis of the metabolic phenotypes at individual molecular fragment level and by multivariate mass spectral reconstruction that enables faster metabolite discrimination, recognition, and identification, Tikunov et al. (2005) successfully discriminated 94 tomato genotypes (*Lycopersicon esculentum* Mill.) based on intensity patterns of more than 20,000 individual molecular fragments using 198 GC-MS data sets. AMDIS is another user-friendly software package for analysis of mass spectrometry generated flux sample data by platforms like GC-MS (http://chemdata.nist.gov/mass-spc/amdis/downloads/), which helps in automatic mass spectral deconvolution, locating separated compounds, their comparison (against a library of compounds), and finally identification of target compounds. AMDIS also aids to compare the detected metabolic profiles and draw relationships between pool and flux. Thus identified metabolic signatures have wide applications in genetic diversity analysis, cultivar identification, and new gene discovery in crop improvement.

Visualization of metabolic profiles on metabolic pathway maps is often used with other omics methods, including expression profiles of genes encoding enzymes involved in particular pathways (Thimm et al. 2004; Tokimatsu et al. 2005). Metabolite QTL (mQTL) analysis using segregated populations is described above in detail for several crops. Furthermore, recent availability of data sets of genome-wide variation acquired from high-throughput genotyping including re-sequencing and parallel sequencing has led to discovery of the genetic association between nucleotide variation and phenotyping in identification of key genes. Laurentin et al. (2008) and Mochida et al. (2009) successfully derived correlative patterns between metabolism and genomic diversity in sesame and rice using seed stocks of natural variations.

10.9 Conclusion and Future Perspectives

Metabolomics has started gaining importance in modern plant research in both basic and applied context. A few studies have already shown how detailed insights gained from chemical composition can help us to understand the various physiological and biochemical changes occurring in the plants and their influence on the phenotype. A better understanding of pathways responsible for biosynthesis of compounds related to nutrition therapeutics and product quality will be key to future development of agricultural metabolomics. Computationally robust bioinformatic algorithms and techniques assisting to link the pathway information and knowledge of genetics, through metabolic quantitative trait loci (mQTLs), modeling, and simulation, are contributing greatly toward designing breeding programs aimed at improvement of important traits. Recent advances witnessed in instrumentation, technology, genomics, and bioinformatics appear to be promising in turning metabolomics a potential tool for improvement of product quality in agriculture in years to come.

References

Akiyama K et al (2008) PRIMe: a web site that assembles tools for metabolomics and transcriptomics. In Silico Biol 8:339–345

Beckmann M et al (2007) Representation, comparison, and interpretation of metabolome fingerprint data for total composition analysis and quality trait investigation in potato cultivars. J Agric Food Chem 55:3444–3451

Bono H et al (1998) Reconstruction of amino acid biosynthesis pathways from the complete genome sequence. Genome Res 8:203–210

Caspi R, Karp PD (2007) Using the MetaCyc pathway database and the BioCyc database collection. Curr Protoc Bioinform. https://doi.org/10.1002/0471250953.bi0117s20

Caspi R et al (2006) MetaCyc: a multiorganism database of metabolic pathways and enzymes. Nucleic Acids Res 34:D511–D516

Caspi R et al (2014) The MetaCyc database of metabolic pathways and enzymes and the BioCyc collection of pathway/genome databases. Nucleic Acids Res 44(D1):D471–D480

Chang A et al (2009) BRENDA, AMENDA and FRENDA the enzyme information system: new content and tools in 2009. Nucleic Acids Res. https://doi.org/10.1093/nar/gkn820

Christensen B, Nielsen J (2000) Metabolic network analysis of Penicillium chrysogenum using (13) C-labeled glucose. Biotechnol Bioeng 68:652–659

Ciocchetta F, Hillston J (2008) Bio-PEPA: an extension of the process algebra PEPA for biochemical networks. Electron Notes Theor Comput Sci 194:103–117

Clark BL (1998) Stability of complex reaction network analysis. Cell Biophys 12:237–253

Consortium, I. H. G. S (2001) Initial sequencing and analysis of the human genome. Nature 412:860–921

DellaPenna D (2007) Biofortification of plant-based food: enhancing folate levels by metabolic engineering. Proc Natl Acad Sci U S A 104:3675

DellaPenna D, Pogson BJ (2006) Vitamin synthesis in plants: tocopherols and carotenoids. Annu Rev Plant Biol 57:711–738

Edwards JS, Palsson BO (2000a) Metabolic flux balance analysis and the in silico analysis of Escherichia coli K-12 gene deletions. BMC Bioinforma 1:1

Edwards JS, Palsson BO (2000b) The Escherichia coli MG1655 in silico metabolic genotype: its definition, characteristics, and capabilities. Proc Natl Acad Sci U S A 97:5528–5533

Edwards D et al (2007) BarleyBase/PLEXdb a unified expression profiling database for plants and plant pathogens. Plant Bioinforma 406:347–363

Fell DA (1993) In: Chuster S, Rigoulet M, Ouhabi R, Mazat JP (eds) Modern trends in Biothermakinetics. Plenum Press, New York, pp 97–101

Fernie AR, Tadmor Y, Zamir D (2006) Natural genetic variation for improving crop quality. Curr Opin Plant Biol 9:196–202

Friel E, Green S, Matich A, Beuning L, Yauk YK, Wang M, MacRae E (2006) Pathway analysis in horticultural crops: linalool as an example. Dev Food Sci 43:93–96

Génard M et al (2007) Towards a virtual fruit focusing on quality: modelling features and potential uses. J Exp Bot 58:917–928

Hatzimanikatis V, Lee KH, Bailey JE (1999) A mathematical description of regulation of the G1-S transition of the mammalian cell cycle. Biotechnol Bioeng 65:631–637

Hirai MY et al (2004) Integration of transcriptomics and metabolomics for understanding of global responses to nutritional stresses in Arabidopsis thaliana. Proc Natl Acad Sci U S A 101:10205–10210

Iijima Y et al (2008) Metabolite annotations based on the integration of mass spectral information. Plant J 54:949–962

International Human Genome Sequencing Consortium (2004) International human genome sequencing consortium. Finishing the euchromatic sequence of the human genome. Nature 431:931–945

Jonsson P et al (2004) A strategy for identifying differences in large series of Metabolomic samples analyzed by GC/MS. Anal Chem 76:1738–1745

Kopka J et al (2005) GMD@CSB.DB: the Golm metabolome database. Bioinformatics 21:1635–1638

Kose F et al (2001) Visualizing plant metabolomic correlation networks using clique-metabolite matrices. Bioinformatics 17:1198–1208

Küffner R, Zimmer R, Lengauer T (2000) Pathway analysis in metabolic databases via differential metabolic display (DMD). Bioinformatics 16:825–836

Larkin P, Harrigan GG (2007) Opportunities and surprises in crops modified by transgenic technology: metabolic engineering of benzylisoquinoline alkaloid, gossypol and lysine biosynthetic pathways. Metabolomics 3:371–382

Laurentin H, Ratzinger A, Karlovsky P (2008) Relationship between metabolic and genomic diversity in sesame (*Sesamum indicum* L.) BMC Genomics 9:250

Lee TH et al (2009) RiceArrayNet: a database for correlating gene expression from transcriptome profiling, and its application to the analysis of coexpressed genes in rice. Plant Physiol 151:16–33

Lisec J et al (2008) Identification of metabolic and biomass QTL in Arabidopsis thaliana in a parallel analysis of RIL and IL populations. Plant J 53:960–972

Magrath R et al (1993) The inheritance of aliphatic glucosinolates in Brassica napus. Plant Breed 111:55–72

Marla S (2006) Comparative structure analysis of chorismate synthase comparative structure analysis of Chorismate synthase. Online J Bioinforma 7(1):35–45

Marla S (2014) Annotation of genome of *Sessamun radiatum*, ICAR.NBPGR, New Delhi, Unpublished, Personal communication

Marla S et al (2010) Classification of rice seed storage proteins using neural networks. J Plant Biochem Biotechnol, 19:123–126

Maruyama K et al (2009) Metabolic pathways involved in cold acclimation identified by integrated analysis of metabolites and transcripts regulated by DREB1A and DREB2A. Plant Physiol 150:1972–1980

Matsuda F et al (2009) MS/MS spectral tag-based annotation of non-targeted profile of plant secondary metabolites. Plant J 57:5550–5577

Mendes P (2002) Emerging bioinformatics for the metabolome. Brief Bioinform 3:134–145

Mochida K et al (2009) Correlation exploration of metabolic and genomic diversity in rice. BMC Genomics 10:568

Moco S et al (2006) A liquid chromatography-mass spectrometry-based metabolome database for tomato. Plant Physiol 141:1205–1218

Mohanty B et al (2015) Identification of candidate network hubs involved in metabolic adjustments of rice under drought stress by integrating transcriptome data and genome-scale metabolic network. Plant Sci 241:224–239

Morgan MJ et al (2013) Metabolic engineering of tomato fruit organic acid content guided by biochemical analysis of an introgression line. Plant Physiol 161:397–407

Mueller LA, Zhang P, Rhee SY (2003) AraCyc: a biochemical pathway database for Arabidopsis. Plant Physiol 132:453–460

Nadella KD, Marla SS, Kumar PA (2012) Metabolomics in agriculture. OMICS 16:149–159

Naqvi S et al (2009) Transgenic multivitamin corn through biofortification of endosperm with three vitamins representing three distinct metabolic pathways. Proc Natl Acad Sci U S A 106:7762–7767

Nicholson JK, Lindon JC, Holmes E (1999) "Metabonomics": understanding the metabolic responses of living systems to pathophysiological stimuli via multivariate statistical analysis of biological NMR spectroscopic data. Xenobiotica 29:1181–1189

Obayashi T et al (2009) ATTED-II provides coexpressed gene networks for Arabidopsis. Nucleic Acids Res. https://doi.org/10.1093/nar/gkn807

Ogata Y et al (2010) CoP: a database for characterizing co-expressed gene modules with biological information in plants. Bioinformatics 26:1267–1268

Okazaki Y et al (2009) A chloroplastic UDP-glucose pyrophosphorylase from *Arabidopsis* is the committed enzyme for the first step of sulfolipid biosynthesis. Plant Cell 21:892–909

Pfeiffer T et al (1999) METATOOL: for studying metabolic networks. Bioinformatics:251–257

Prakash CS, Egnin M (1997) Engineered Sweet potato (*Ipomea batatas*) plants with a synthetic protein storage gene show high protein and essential amino acid levels. Concurrent session 35. In: Dean JFD (ed) 5th International Congress of Plant Molecular Biology, 21–27 Sept, Singapore, Kluwer Academic Publishers

Roberts CW et al (2002) The shikimate pathway and its branches in apicomplexan parasites. J Infect Dis 185(Suppl):S25–S36

Saneoka H et al (1995) Salt tolerance of glycinebetaine-deficient and -containing maize lines. Plant Physiol 107:631–638

Sasaki T, Burr B (2000) International rice genome sequencing project: the effort to completely sequence the rice genome. Curr Opin Plant Biol 3:138–141

Schauer N et al (2006) Comprehensive metabolic profiling and phenotyping of interspecific introgression lines for tomato improvement. Nat Biotechnol 24:447–454

Schauer N et al (2008) Mode of inheritance of primary metabolic traits in tomato. Plant Cell 20:509–523

Schilling CH, Palsson BO (2000) Assessment of the metabolic capabilities of Haemophilus influenzae Rd through a genome-scale pathway analysis. J Theor Biol 203:249–283

Schilling CH, Edwards JS, Palsson BO (1999) Toward metabolic phenomics: analysis of genomic data using flux balances. Biotechnol Prog 15:288–295

Schuster S, Fell DA, Dandekar T (2000) A general definition of metabolic pathways useful for systematic organization and analysis of complex metabolic networks. Nat Biotechnol 18:326–332

Schwall GP et al (2000) Production of very-high-amylose potato starch by inhibition of SBE A and B. Nat Biotechnol 18:551–554

Scott IM et al (2010) Enhancement of plant metabolite fingerprinting by machine learning. Plant Physiol 153:1506–1520

Seressiotis A, Bailey JE (1986) MPS: an algorithm and data base for metabolic pathway synthesis. Biotechnol Lett 8:837–842

Simpson TW, Colon GE, Stephanopoulos G (1995) Two paradigms of metabolic engineering applied to amino acid synthesis. Biochem Soc Trans 23:381–387

Smytha DA (1998) Some properties of starch branching enzyme from indica rice endosperm (*Oryza sativa* L.) Plant Sci 57:1–8

Souleyre EJF et al (2005) An alcohol acyl transferase from apple (cv. Royal Gala), MpAAT1, produces esters involved in apple fruit flavor. FEBS J 272:3132–3144

Stephanopoulos G (1999) Metabolic fluxes and metabolic engineering. Metab Eng 1:1–11

Sunita T, Ramadevi S (2004) In silico gene identification and homology modeling of chorismate synthase in clostridium difficile. Online J Bioinforma 5:129–131. 1443–50

Swarbreck D et al (2008) The Arabidopsis information resource (TAIR): gene structure and function annotation. Nucleic Acids Res. https://doi.org/10.1093/nar/gkm965

The UniProt Consortium (2011) Ongoing and future developments at the universal protein resource. Nucleic Acids Res 39:D214–D219

Thimm O et al (2004) MAPMAN: a user-driven tool to display genomics data sets onto diagrams of metabolic pathways and other biological processes. Plant J 37:914–939

Tikunov Y et al (2005) A novel approach for nontargeted data analysis for metabolomics. Large-scale profiling of tomato fruit volatiles. Plant Physiol 139:1125–1137

Tohge T et al (2005) Functional genomics by integrated analysis of metabolome and transcriptome of \emph{Arabidopsis} plants over-expressing an MYB transcription factor. Plant J 42:218–235

Tokimatsu T et al (2005) KaPPA-view: a web-based analysis tool for integration of transcript and metabolite data on plant metabolic pathway maps. Plant Physiol 138:1289–1300

Urano K et al (2009) Characterization of the ABA-regulated global responses to dehydration in Arabidopsis by metabolomics. Plant J 57:1065–1078

Urbanczyk-Wochniak E, Sumner LW (2007) MedicCyc: a biochemical pathway database for Medicago truncatula. Bioinformatics 23:1418–1423

Womble DD, Rownd RH (1986) Regulation of IncFII plasmid DNA replication. A quantitative model for control of plasmid NR1 replication in the bacterial cell division cycle. J Mol Biol 192:529–547

Yonekura-Sakakibara K et al (2008) Comprehensive flavonol profiling and transcriptome coexpression analysis leading to decoding gene-metabolite correlations in Arabidopsis. Plant Cell 20:2160–2176

Zhang P (2005) MetaCyc and AraCyc. Metabolic pathway databases for plant research. Plant Physiol 138:27–37

Zhao X et al (2008) Extracellular Ca2+ regulating stomatal movement and plasma membrane K+ channels in guard cells of Vicia faba under salt stress. Acta Agron Sin 34:1970–1976

The Interactomics of the RNA-Induced Silencing Complex

11

Abhijit Datta and Sayak Ganguli

Abstract

The posttranscriptional gene silencing mechanism initially mistaken as co-suppression but later identified as RNA interference mediated and regulated by small interfering RNAs (siRNAs), and microRNAs (miRNAs) has gradually emerged as a landmark discovery of the last decade. siRNA-based therapeutics is currently being investigated as an emerging opportunity for healthcare. The lack of structural data of the RNA-induced silencing complex and its key players has hindered the progress of utilization of this unique mechanism in the betterment of humanity. Crystallographic information regarding the Argonaute (Ago)-DNA-RNA complexes have helped in understanding the chemistry of the complex, but other valuable structural details still remain elusive. It is an immediate requirement to understand the exact mechanisms of interactions that occur between the key players of the microprocessor complex or for that matter the holo-RISC. Unless these interaction maps are obtained, complete effective usage and manipulation of this natural phenomenon shall remain uphill tasks for researchers worldwide. To harness the complete potential of siRNAs as therapeutic agents, various chemical modifications need to be performed to prevent nuclease attack, immune activation, increase the specificity of the interaction, and improve pharmacodynamics of the interacting components. Computational molecular dynamics simulations provide a probabilistic alternative for studying such complex structures. Both flexible and rigid docking processes can be utilized to understand the specificities of interactions as both protein-protein and protein nucleic acid docking algorithms have been implemented in free and licensed softwares, complexes obtained from which can then be subjected to analyses using the

A. Datta
Department of Botany, Jhargram Raj College, Medinipur, West Bengal, India

S. Ganguli (✉)
Theoretical and Computational Biology Division, Amplicon Institute of Interdisciplinary Science and Technology, Palta, West Bengal, India

G. Wadhwa et al. (eds.), *Current trends in Bioinformatics: An Insight*,
https://doi.org/10.1007/978-981-10-7483-7_11

various interaction mapping tools to formulate a classical map of the RNAi interactome. In this chapter we attempt to elucidate the interactions of the key members of the holo-RISC complex using molecular docking and simulation. Specific interacting residues having the potential to serve as interacting hotspots were identified.

Keywords
Argonaute · Dicer · Interactome · PAZ domain · Oligonucleotide · Therapeutics

11.1 Introduction

The work of Fire et al. (1998) is a citation classic brought into the forefront the identification of a mechanism of sequence-specific gene silencing. In the model nematode *C. elegans*, they identified the above mechanism where minuscule amounts of double-stranded RNA (dsRNA) having gene specificity were enough to suppress the expression of a particular gene by mRNA elimination. This new mechanism of silencing was experimentally proven to be more effective than the established methods of antisense reagents (Amarzguioui et al. 2003), most single stranded as practiced by scientists worldwide. It was observed that a few dsRNA molecules were enough per cell to silence its target gene, which lead to the prediction that darn-mediated gene silencing was a catalytic or enzymatic process rather than a stoichiometric binding one (Chiu and Rana 2003; Nykanen et al. 2001).

Over the years we have been apprised of the facts that this mechanism and its components if not in diversity but surely in function are conserved across the major domains of organisms – plants, worms, and humans. Even viruses have been reported to encode and induce small RNAs. To put it categorically, a collection of cellular proteins exists that recognize and interact with dsRNAs in order to process and compartmentalize them into the effector enzyme complex which in turn bind and cleave complementary or target mRNAs. Drosha and Dicer are two dedicated ribonucleases (RNAse III) which mediate the production of small interfering RNAs (siRNAs) from their precursor molecules (Das et al. 2011; Elbashir et al. 2001a, b, c). siRNAs on an average are 21 nucleotides in length and contain 3′ overhangs. The double strand possesses the guide and passenger strands (Tuschl et al. 1999; Parrish et al. 2000). The guide strand is the one complementary to the target mRNA, while the passenger strand is involved in interaction. RNA-induced silencing complex (RISC) incorporates the guide strand as a part of its macromolecular assembly (Elbashir et al. 2001a, b, c; Ganguli et al. 2011; Hamilton and Baulcombe 1999). A key component of the RISC is the endonuclease Argonaute 2 (AGO2), which plays the role of cleaving the bound mRNA only when the latter is perfectly complementary to the guide strand. This cleaved mRNA is rapidly destroyed (Orban and Izaurralde 2005; Newman and Scott 2010), resulting in the renewal of the RISC-guide strand complex for additional cycles of mRNA binding and cleavage (Fig. 11.1). This process causes the significant suppression of cognate gene expression.

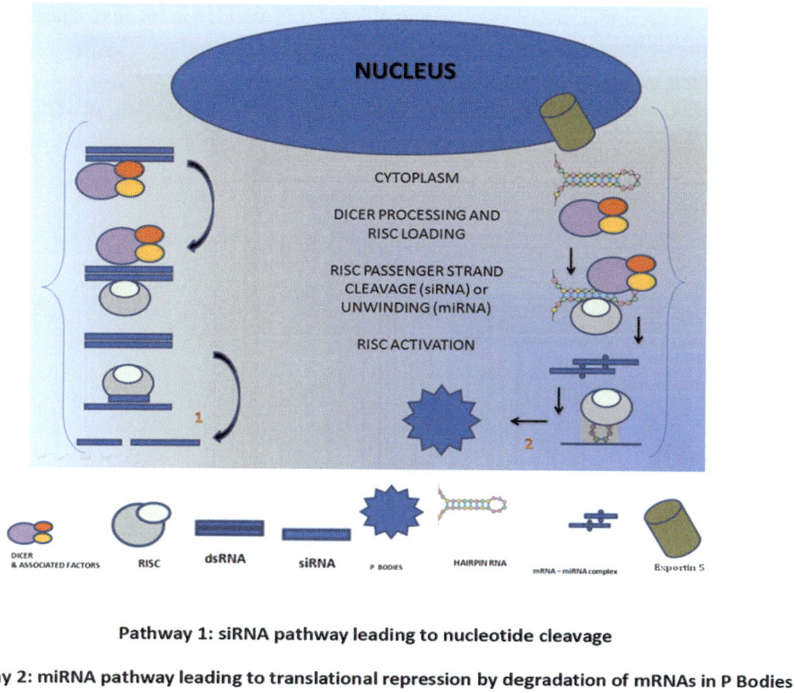

Fig. 11.1 The generation and action of siRNA and miRNA in RNA interference

11.2 RNA-Induced Silencing Complex (RISC)

Reports suggest that the link between the early and late phases of RNAi involves the sequential sequestration of the siRNAs into the RNA-induced silencing complex (Martinez et al. 2002). The essential prerequisite is the formation of the R1 initiation complex which comprises of cellular factors Dicer 2 (Dcr-2) and a Dcr-2-associated factors – R2D2. This factor in *Drosophila* plays a major role as a connector between the initiation and effector phases of RNAi (Liu et al. 2003; Tomari et al. 2004a, b). This initial R1 complex is followed by the formation of the intermediate complex R2 by automatic addition of some unidentified factors. This addition is ATP-independent. The holo-RISC (formerly R3) a ~80S complex is formed once the R2 housed siRNA is unwound, and this acts as the key effector for RNAi in *Drosophila*.

A similar pathway has been proposed by Zamore and co-workers (2000). According to them the process is initiated with the siRNAs binding to a complex referred to as B. This is thought to be a direct precursor to the RISC-loading complex and lacks both Dcr-2 and R2D2. The two siRNA ends are thermodynamically stable which in turn stabilizes the orientation and cross-linking of the Dcr-2-R2D2 complex with the ends of the siRNA duplex (Liu et al. 2003). R2D2 prefers binding to stable base pairs which is exactly opposite to the affinity of Dcr-2 for less stable pairing. The sensing of the asymmetry in the siRNA strand which enters RISC is attributed to R2D2 which enables the proper strand to be degraded.

Though you can find similarities between the proposed models of RISC assembly, gray areas are still persistent. For example, the assembly scheme comprising of R1 → R2 → holo-RISC steps does not allow the fitting in of complex B and RLC into it (Fig. 11.2).

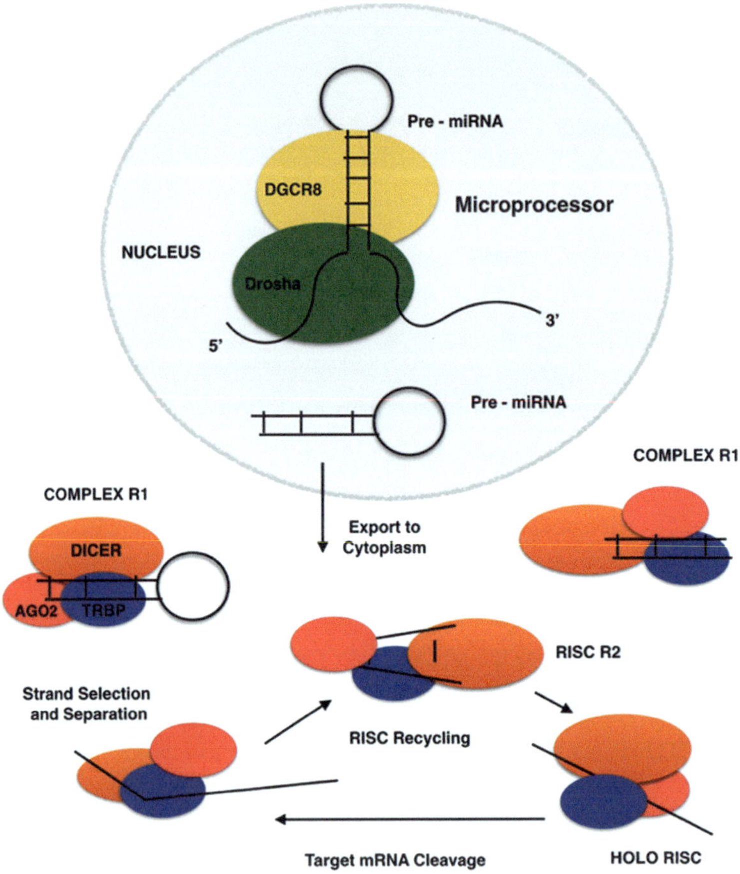

Fig. 11.2 Steps and stages of RNA-induced silencing complex formation and recycle

11.3 Dicer

Entry into the cytoplasm brings pre-miRNAs into contact with Dicer, a predominantly cytoplasmic enzyme. Mature ~22mer miRNAs are produced by Dicer following cleavage of the precursor molecule; it can also cleave dsRNA into ~22mer siRNAs. Two RNase III domains form the DICER core (Fig. 11.3). Apart from these two domains, almost all Dicer enzymes are composed of aDExH/DEAHbox RNA helicase domain and a PAZ domain as well as a domain of unknown function (DUF283) along with a double-stranded RNA-binding motif. The helicase domain of dicer is most phylogenetically conserved. Spindle-E (homeless) domain is required for RNAi activation during oocyte maturation in *Drosophila* (Hammond et al. 2000; Hutvagner and Zamore 2002; Pham et al. 2004). Dicer and Argonautes share the PAZ domain which is conserved in both. Recent crystallographic structures of PAZ have revealed the presence of a deviant OB-fold structure with consistent nucleic acid affinity though weak in comparison with other proteins. Single-stranded RNA ends particularly attract the PAZ domain (Okamura et al. 2004), and special attention is received by the two nucleotide overhang structure which is generated as a result of RNase III processing event. This affinity and specificity of interaction probably allows for the selection of the pre-miRNAs along with substrates preprocessed by Drosha.

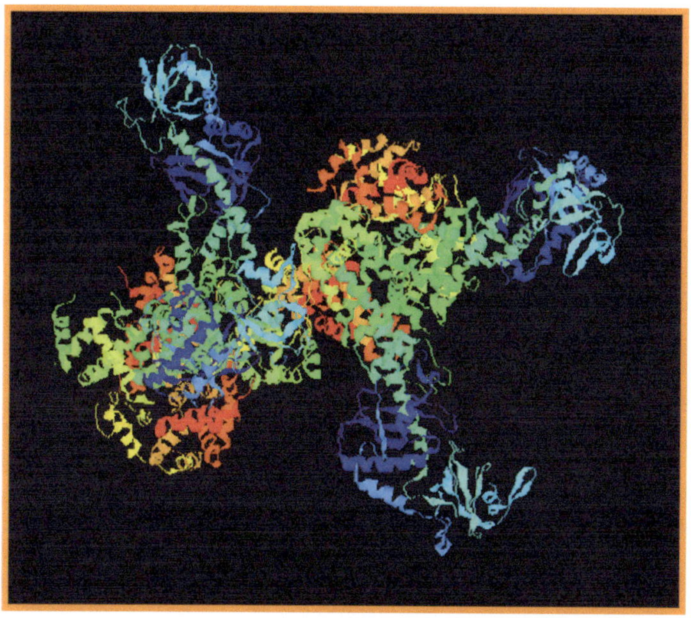

Fig. 11.3 The structure of DICER showing the RNA-binding domains (blue) and the Argonaute-interacting areas (red and orange)

11.4 Argonautes

Argonautes get their name from the ancient mythological Greek warriors – The Argonautes. They were first described as developmental important proteins in plants and in stem cell division of the gremlin in *Drosophila melanogaster*. However, over the years it has been understood that they have a huge significance in the proper execution of the RNA interference pathways (Ganguli et al. 2011; Parker et al. 2005). Established works suggests that Argonaute proteins are involved in maintaining genome integrity, in controlling protein synthesis and RNA stability, and in the production of a specific set of small noncoding RNAs. The N-terminal, PAZ, Mid, and PIWI domains make up the four distinct domains of these proteins. These domains are consistent in the eukaryotic Argonaute proteins that play roles in gene regulation and small RNA-mediated pathways (Rand et al. 2004; Schwarz et al. 2003).

11.5 The PAZ Domain

As mentioned earlier, the common domain of Dicer and Argonaute proteins is the PAZ domain. This can be divided into two subdomains – the first with a distinct oligonucleotide/oligosaccharide-binding (OB) fold which enables the interaction of PAZ with single-stranded nucleic acids. Combinatorial approaches with crystallography and biochemical analyses have showed that there is low-affinity binding of PAZ with single-stranded nucleic acids independent of sequence specificity. The ability of the PAZ domain to recognize the 3′ sRNA ends makes it remarkable as these sites possess overhangs which are aftermaths of the sequential processing of RNAse III enzymes (animal Drosha/Dicer and only Dicer in plants/yeast). Thus this feature enables the proper delimitation of other small RNAs which are produced as a result of degradation events or from other pathways (Song et al. 2004).

11.6 The PIWI Domain

The PIWI domain has a RNase H-like fold, which was elucidated by comparing the crystal structures of the archaeal and eubacterial Argonaute proteins and the piwi-like protein of *Archaeoglobus fulgidus* another member of the archaea. The second protein lacked the N-terminal domain and the PAZ domain. Interestingly RNAse H-like enzymes are responsible for cleavage of RNA in a DNA-dependent manner, and this reaction requires the conserved Asp-Asp-Glu/Asp motif to be part of the catalytic core. Divalent metal ion binding to the core is also an important prerequisite. This pattern of interaction is slightly dissimilar to that of Argonaute proteins where the conserved motif is slightly more degenerate Asp-Asp-Asp/Glu/His/Lys and the activity is controlled by the binding of a divalent cation. 3′ -OH and phosphate are present in the RNAse H cleavage product, which is also a special signature of post-processing products of RNAse H.

Substrate recognition by terminal or semi-terminal nucleotides generally do not play important roles in substrate recognition which has been substantiated by several reports and structural studies where the 5′ end of the guide strand of a small RNA remains separated from the targeted RNA (Ma et al. 2005). The mechanism of RNA-mediated cleavage always tends to occur at a fixed place as the ribonuclease catalytic motif gets placed in front of the scissile phosphate on the target RNA in-between the guide strand tenth and eleventh nucleotides if you count from the anchored 5′ end. Recombinant protein-based reconstitution studies of minimal RNA-induced silencing complex (RISC) revealed that the essential components are an Argonaute PIWI domain and a bound small RNA. This justifies the observations that endonuclease activity of the Argonaute-like and Piwi-like proteins in fission yeast, fungi, plants, flies, and mammals is an essential prerequisite for proper RISC functioning (Parker et al. 2004, 2005).

Cleavage of target RNAs in a sequence-specific manner is initiated only when the non-active siRNA strand is removed following maturation of the siRNAs. This process is influenced by Argonaute proteins as well. Piwi-like proteins impart their cleavage activity during the process of maturation of the repeat-associated small interfering (rasi) RNAs and piRNAs in mammals. However, it should be noted that correlating cleavage activity of argonautes with the presence of intact PIWI domain is just one half of the story, for example, despite having a canonical active site, human argonaute 3 (AGO3) does not bind miRNAs. This implies the role of additional factors which may play critical roles in the process (Fig. 11.4).

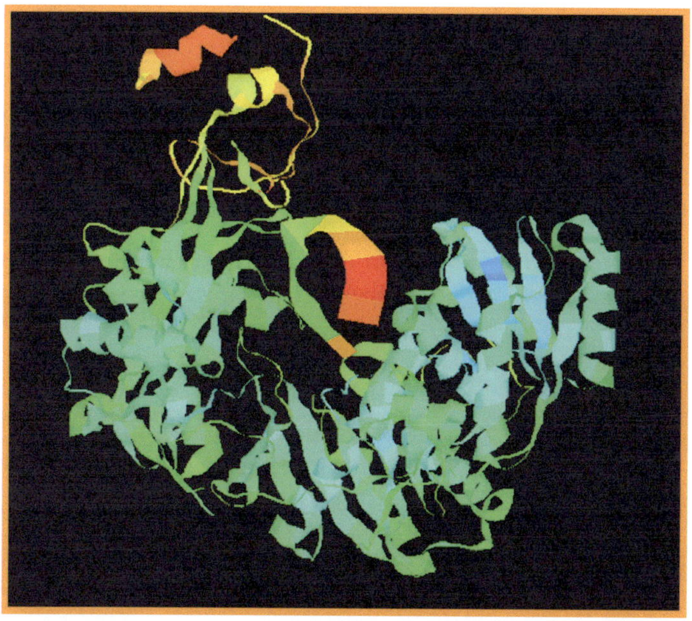

Fig. 11.4 The structure of Argonaute protein with miRNA (red) and target RNA (orange) (Molecular models viewed using RasMol)

11.7 Dissecting the Interactomics of the Members of the RISC

The proper elucidation of the interactomics of the different partners of the RISC was performed using the following approach (Fig. 11.5):

Threading involved the following steps:

- 3D coordinates of known template structures were noted.
- The annotated backbone was then combined with the coordinates which provide potential positions for the residues of the query sequence.
- Threading is then performed to align the query on the template structure, while loops and coiled regions are modeled separately.
- The above mapping of the target query sequence against the template results in several candidate threadings (alignments).
- The decoys are eliminated using a score function, and the best template is retained (Parisien and Major 2008).

Molecular dynamic simulations were performed at four different levels (Figs. 11.6 and 11.7):

- Nucleic acid-nucleic acid interaction
- Nucleic acid-protein interaction
- Multiple nucleic acid-protein interaction
- Multiple nucleic acid-multiple protein interaction

For the first level of the study, it was found that noncanonical Hoogsteen pairing was the predominant base pairing pattern in the miRNA-target mRNA complex. Though Watson and Crick base pairs were also detected, the seed sequence-target interactions were mostly noncanonical Hoogsteen in almost 90% of the cases. Nucleic acid protein interactions were performed separately using miRNA + Argonaute and miRNA + Dicer. In the first complex, it was found that the Argonaute interacting surface was devoid of any neutral but slightly polar amino acids like

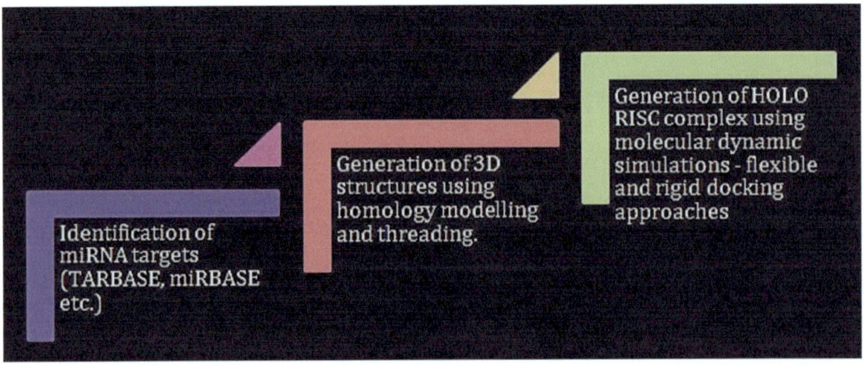

Fig. 11.5 Workflow for analysis

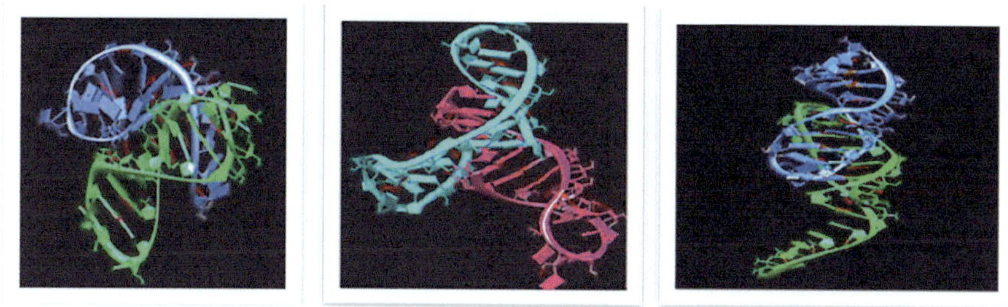

Fig. 11.6 MiRNA-target mRNA complex

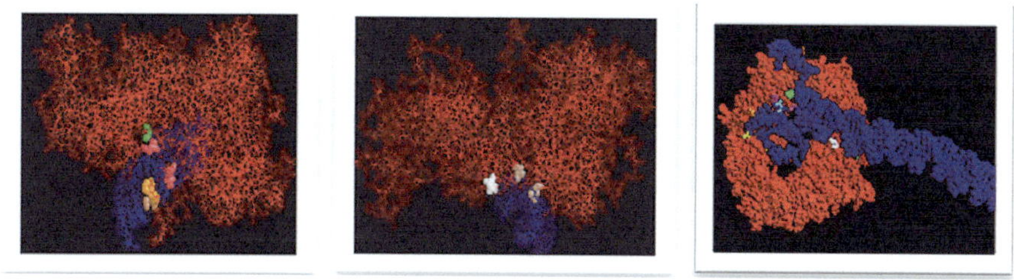

Fig. 11.7 Interactions of DICER + miRNA and ARGONAUTE + miRNA

cysteine and tryptophan, while the grooves of the proteins did not have any asparagine residue which is polar and neutral. Neutral and nonpolar amino acids like proline, methionine, and phenylalanine were found to be the most abundant and conserved interacting residues. In case of the latter, complex-specific residues were found to be present in most of the interactions that were studied – arginine (Arg) at 708, asparagine (Asn) at 721, glutamic acid (Glu) at 726, histidine (His) at 802, methionine (Met) at 722, phenylalanine (Phe) at 182, proline (Pro) at 180, tryptophan (Trp) at 725, glutamine (Gln) binds specifically at 186, and serine (Ser) at 184.

In the third set of interactomics study, the miRNA-target complex was made to interact with the ARGONAUTE and DICER proteins separately to test whether the stereochemistry of the nucleic acid complex hindered with the actual binding sites. But it was found that the changes induced by the binding of the target mRNA with the miRNA did not have any negative influence over the binding affinity of the argonaute protein or dicer as a matter of fact. This result leads us to conclude that the binding of proteins and nucleic acids to the proteins in the RISC is specific event which is not dependent on the intramolecular affinity.

The fourth phase of the study was the attempt to construct the assembly of the HOLO-RISC in the sequential steps (Figs. 11.8 and 11.9). It was found that the Argonaute and Dicer interacting sites were almost the same in most of the case, and

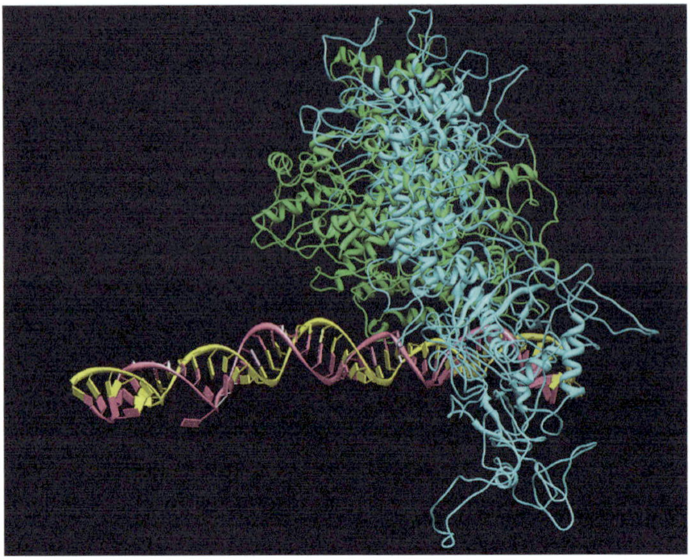

Fig. 11.8 The Holo-RISC sans the associated factors – [green, AGO; cyan, Dicer; pink, target mRNA region; yellow, miRNA]

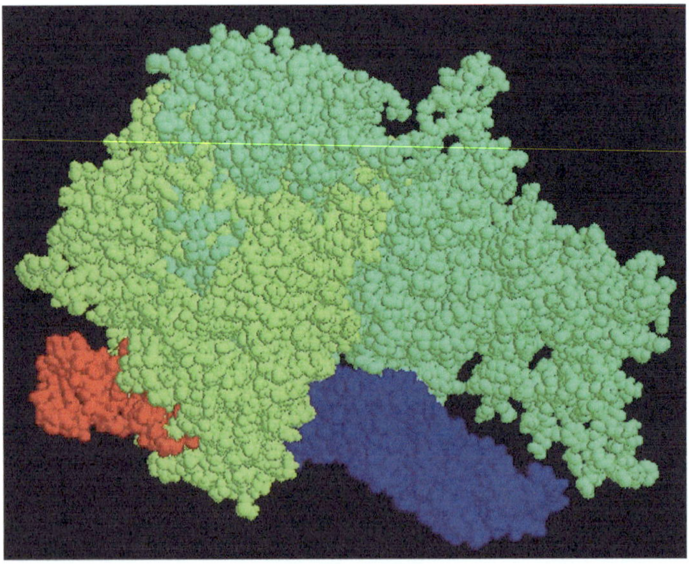

Fig. 11.9 The allosteric property of Argonaute proteins in interaction with Dicer and miRNA-target complex. Red molecule is another miRNA, the binding of which indicates that the Argonaute protein is capable of binding another miRNA and its target while it is in the process of dealing with the first miRNA

the bias toward interaction was the predominance of arginine residues. The total complex was found to be highly stable as the average relative free energy was found to be $= -1131.65$.

To check whether the RISC recycling was brought about by the ability of the Argonaute proteins to bind multiple miRNAs, we introduced another miRNA, into the system, while the HOLO-RISC was still operational, and it was found that indeed the miRNA bound to a different site on the ARGONAUTE protein indicating the allostericity of the protein and possible insight on the recycling of RISC.

11.8 Interaction During Regulation of the MicroRNA Biogenesis

The key understanding of the phenomenon of RNA interference is the generation of the mature noncoding RNAs. The biogenesis of the mature molecules has been reported to be regulated by corepressors and activators (Ganguli et al. 2011). Many reports have suggested the presence of dedicated promoters in intergenic miRNAs, which have been exhibited to share common features with transcription (Pol II-mediated) conserved transcription factor binding sites and other genomic landmarks such as CpG islands etc. Clustered miRNAs on the other hand have a common promoter and are naturally under coregulation as parts of long pri-miRNAs. The presence of these specific promoters and the need for functional transcription factors become necessary which could then serve as regulatory molecules (Fig. 11.10). For example, the classic case of p53 emerges which has been extensively studied and has been established to upregulate transcription of miR-34 family transcription members, which are involved in the repression of important factors for cell proliferation and survival.

Several RNA-binding proteins have also been implicated in the microprocessor regulation which probably plays a dual role in the cell, their primary activity being regulation of splicing in the cell. One of the most prominent of these members is the KH-type splicing regulatory protein (KHSRP) which has been reported to bind to conserved terminal loop regions of a set of miRNAs, and free energy values suggest that they have very strong binding affinity. When workers investigated the binding site of this particular protein, a very conserved binding motif was identified which was similar to binding sites in mRNA and consisted of a single-stranded GGG triplet (Ganguli and Datta 2014).

Apart from the regulatory interactions that can govern the production of microRNAs from its genes, a second level of control can be achieved at the microprocessor which allows for the integration of signaling events from many pathways and regulates the expression of members of many divergent microRNA families. All such events have been established to be mediated by the helicases p68 and p72. Questions have been raised about the specificity of such pathways since varied microRNA families are under regulation. The answer probably lies in the interaction with the primary RNA. The SMAD interacting protein SNIP 1 has been implicated in regulating the interactions that govern the processing of mir21 processing by Smad proteins. The homolog of this protein in *Arabidopsis* has been

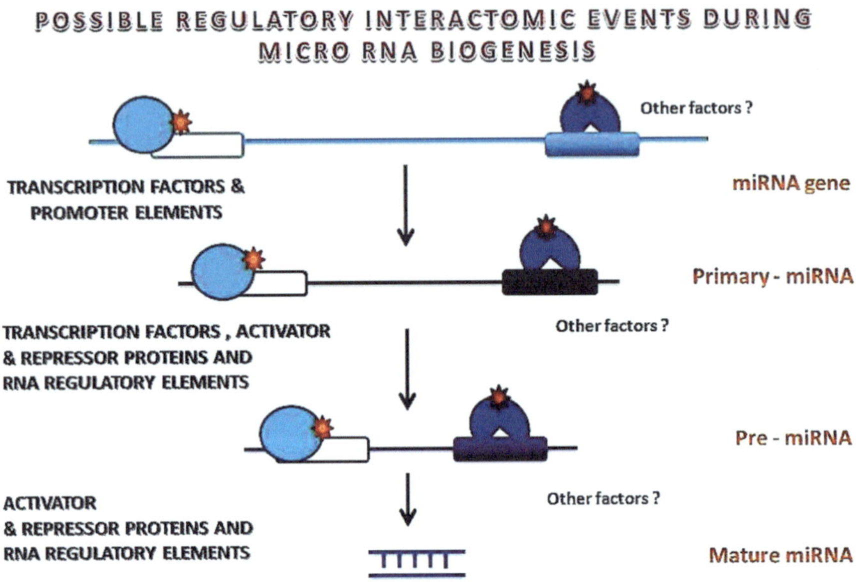

Fig. 11.10 Possible regulatory events governing biogenesis of miRNAs

shown to exhibit RNA-binding activity, and this protein in mammals possibly acts as coactivator. Of course, many unidentified proteins which are expected to be part of the microprocessor possibly are members of a transient complex which may get assembled after signal reception. Defined activity assays need to be established to understand and characterize the interactions such that the complex can be completely elucidated.

Sketchy information regarding other regulating events have also emerged, for example, the metabolite heme has been reported to modulate microprocessor function, although the exact biological role of this effect is still not completely understood. ARS2 which is an established component of the nuclear-cap binding complex has been envisaged to be an important regulator of microprocessor function (Denli et al. 2004). Through this vast expanse of regulatory phenomenon, we are still at vary nascent stage, and understanding of the complete picture would be the challenge for future to come.

References

Amarzguioui M et al (2003) Tolerance for mutations and chemical modifications in a siRNA. Nucleic Acids Res 31:589–595

Chiu YL, Rana TM (2003) siRNA function in RNAi: a chemical modification analysis. RNA 9:1034–1048

Das AK et al (2011) Secondary structural analysis of MicroRNA and their precursors in plants. Int J Agric Sci 3(1):62–64

Denli AM et al (2004) Processing of primary microRNAs by the microprocessor complex. Nature 432(7014):231–235

Elbashir SM, Lendeckel W, Tuschl T (2001a) RNA interference is mediated by 21 and 22 nt RNAs. Genes Dev 15:188–200

Elbashir SM et al (2001b) Duplexes of 21-nucleotide RNAs mediate RNA interference in cultured mammalian cells. Nature 411(6836):494–498

Elbashir SM et al (2001c) Functional anatomy of siRNAs for mediating efficient RNAi in *Drosophila melanogaster* embryo lysate. EMBO J 20(23):6877–6888

Fire A et al (1998) Potent and specific genetic interference by double-stranded RNA in *Caenorhabditis elegans*. Nature 391(6669):806–811

Ganguli S, Datta A (2014) In silico mutagenesis reveals specific binding residues that regulate KSRP – microRNA precursor interactions in human. Ann Res Rev Biol 4(1):143–153

Ganguli S, De M, Datta A (2011) Analyses of argonaute– microRNA interactions in *Zea mays*. Int J Comput Biol 2(1):32–34

Hamilton AJ, Baulcombe DC (1999) A species ofsmall antisense RNA in posttranscriptional gene silencing in plants. Science 286:950–952

Hammond SM et al (2000) An RNA-directed nuclease mediates post-transcriptional gene silencing in Drosophila cells. Nature 404(6775):293–296

Hutvagner G, Zamore PD (2002) A microRNA in a multiple-turnover RNAi enzyme complex. Science 297:2056–2060

Liu Q et al (2003) R2D2, a bridge between the initiation and effector steps of the Drosophila RNAi pathway. Science 301:1921–1925

Ma JB et al (2005) Structural basis for 5′-end-specific recognition of guide RNA by the A. fulgidus Piwi protein. Nature 434(7033):666–670

Newman MA, Scott M (2010) Hammond emerging paradigms of regulated microRNA processing. Genes Dev 24:1086–1092

Martinez J et al (2002) Single-stranded antisense siRNAs guide target RNA cleavage in RNAi. Cell 110:563–574

Nykänen A, Haley B, Zamore PD (2001) ATP requirements and small interfering RNA structure in the RNA interference pathway. Cell 107:309–321

Okamura K et al (2004) Distinct roles for Argonaute proteins in small RNA-directed RNA cleavage pathways. Genes Dev 18:1655–1666

Orban TI, Izaurralde E (2005) Decay of mRNAs targeted by RISC requires XRN1, the Ski complex, and the exosome. RNA 11:459–469

Parisien M, Major F (2008) The MC-Fold and MC-Sym pipeline infers RNA structure from sequence data. Nature 452:51–55

Parker JS, Roe SM, Barford D (2004) Crystal structure of a PIWI protein suggests mechanisms for siRNA recognition and slicer activity. EMBO J 23:4727–4737

Parker JS, Roe SM, Barford D (2005) Structural insights into mRNA recognition from a PIWI domain-siRNA guide complex. Nature 434:663–666

Parrish S et al (2000) Functional anatomy of a dsRNA trigger: differential requirement for the two trigger strands in RNA interference. Mol Cell 6:1077–1087
Pham JW et al (2004) A Dicer-2-dependent 80s complex cleaves targeted mRNAs during RNAi in Drosophila. Cell 117:83–94
Rand TA et al (2004) Biochemical identification of Argonaute 2 as the sole protein required for RNA-induced silencing complex activity. Proc Natl Acad Sci U S A 101:14385–14389
Schwarz DS et al (2003) Asymmetry in the assembly of the RNAi enzyme complex. Cell 115:199–208
Song JJ et al (2004) Crystal structure of Argonaute and its implications for RISC slicer activity. Science 305:1434–1437
Tomari Y et al (2004a) RISC assembly defects in the Drosophila RNAi mutant armitage. Cell 116:831–841
Tomari Y et al (2004b) A protein sensor for siRNA asymmetry. Science 306:1377–1380
Tuschl T et al (1999) Targeted mRNA degradation by doublestranded RNA in vitro. Genes Dev 13:3191–3197
Zamore PD et al (2000) RNAi: double-stranded RNA directs the ATP-dependent cleavage of mRNA at 21 to 23 nucleotide intervals. Cell 101:25–33

Computational Tools: RNA Interference in Fungal Therapeutics

12

Chakresh Kumar Jain and Gulshan Wadhwa

Abstract

There is steady rise in the number of immunocompromised population due to increased use of potent immunosuppression therapies. This is associated with increased risk of acquiring fungal opportunistic infections in immunocompromised patients which account for high morbidity and mortality rates, if left untreated. The conventional antifungal drugs to treat fungal diseases (mycoses) are increasingly becoming inadequate due to observed varied susceptibility of fungi and their recurrent resistance. RNA interference (RNAi), sequence-specific gene silencing, is emerging as a promising new therapeutic approach. This chapter discusses various aspects of RNAi, viz., the fundamental RNAi machinery present in fungi, in silico siRNA features, designing guidelines and tools, siRNA delivery, and validation of gene knockdown for therapeutics against mycoses. Target gene identification is a crucial step in designing of gene-specific siRNA in addition to efficient delivery strategies to bring about effective inhibition of fungi. Subsequently, designed siRNA can be delivered effectively in vitro either by soaking fungi with siRNA or by transforming inverted repeat transgene containing plasmid into fungi, which ultimately generates siRNA(s). Finally, fungal inhibition can be verified at the RNA and protein levels by blotting techniques, fluorescence imaging, and biochemical assays. Despite challenges, several such in vitro studies have spawned optimism around RNAi as a revolutionary new class of therapeutics against mycoses. But, pharmacokinetic parameters need to be evaluated from in vivo studies and clinical trials to recognize RNAi as a novel treatment approach for mycoses.

C. K. Jain (✉)
Department of Biotechnology, Jaypee Institute of Information Technology, Noida, Uttar Pradesh, India

G. Wadhwa
Department of Biotechnology, Apex Bioinformatics Centre, Ministry of Science & Technology, New Delhi, India

G. Wadhwa et al. (eds.), *Current trends in Bioinformatics: An Insight*, https://doi.org/10.1007/978-981-10-7483-7_12

Keywords
RNA interference · Short interfering RNA · Mycoses · Immunosuppression · Opportunistic infection · siRNA design · Target gene · Soaking · Inverted repeat transgene · Therapeutics

12.1 Introduction

The last two decades have seen a rising trend in the immunocompromised patients (Richardson 1991; Krcmery 1996; Warnock and Richardson 1991; Ribes et al. 2000), who are at a heightened risk for "opportunistic" fungal infections (mycoses). The fungal infections can affect any part of the body and cause superficial mycoses (affecting only the outermost layers of the skin and hair), cutaneous mycoses (extending deeper into the layers of the skin, hair, and nail), subcutaneous mycoses (involve the dermis, subcutaneous tissues, muscle, and fascia), or systemic mycoses (dissemination in the body). The fungal diseases (listed in Table 12.1) are classified according to the three important phyla of the fungal taxonomy – Ascomycetes, Basidiomycetes, and Zygomycetes. These diseases turn out to be severe if disseminated, which is common in the immunosuppressed. Candidiasis is one of the highly frequent fungal infections in 90% of untreated, advanced HIV cases suffering from oropharyngeal candidiasis according to World Health Organization (WHO) 2010 statistics. Another study reported overall mortality rate of approximately 80% for zygomycosis (Ribes et al. 2000) and 80–90% in invasive aspergillosis of high-risk leukemia patients and allogeneic bone marrow transplant patients (Chamilos and Kontoyiannis 2005).

The existing fungal treatments require parallel approaches involving surgical intervention and antifungal therapy (Ribes et al. 2000). Surgical intervention has shown improving survival rates for many patients, but surgery is not the only solution in treating fungal diseases. The currently available antifungal drugs comprise polyenes, macrolides, azole drugs, and echinocandins (Mukherjee et al. 2005). Among them, the polyene amphotericin B is the first-line drug of choice for invasive mycoses (Ribes et al. 2000). These drugs have drawbacks such as nephrotoxicity and hepatotoxicity as exhibited by liposomal amphotericin B, narrow spectrum of activity as shown by fluconazole, and itraconazole causes problems with absorption (Pauw and Picazo 2008). Moreover, the fungi are acquiring resistance recurrently against novel antifungal agents (Rogers 2008) either due to defects in drug import and efflux of drugs, variations in intracellular drug processing, alterations in the enzymes of a target-specific biosynthetic pathway, or its competitive inhibition (Chamilos and Kontoyiannis 2005; Bhanderi et al. 2009).

The existing treatments are not effective due to varied susceptibility of fungi toward the drugs; hence, newer safe and effective therapeutic interventions are desired. There have been reports of the use of gene therapy, antisense oligonucleotides, ribozymes, DNAzymes, and RNAi therapy. RNAi scores several

Table 12.1 Fungal diseases associated with immunocompromised patients

Disease	Causal fungal agent	Symptoms	Treatment
Ascomycetes			
Candidiasis	*Candida albicans, C. tropicalis*	Localized infection of the skin or mucosal membranes like the vagina: itching, burning, soreness, irritation, and a whitish discharge	Clotrimazole, nystatin, amphotericin B, fluconazole, caspofungins
Aspergillosis	*Aspergillus fumigatus* (causes more than 95% of cases), *A. flavus*	Fungus balls in lungs, coughing up blood; fever, chest pain, difficulty breathing; aspergillosis of the ear canal causes itching and occasionally pain and fluid draining	Amphotericin B, voriconazole, caspofungin
Histoplasmosis	*Histoplasma capsulatum, H.duboisii* (causes disease restricted to Africa and Madagascar)	Acute respiratory infection leads to respiratory symptoms, malaise, fever, chest pains, dry cough, flu-like illness, fever, cough, headache	Fluconazole, itraconazole, amphotericin B
Coccidioidomycosis	*Coccidioides immitis*	Only 40% cases symptomatic, flu-like symptoms, fever, cough, headache, rash, myalgias, chest pain, fatigue, arthralgia	Fluconazole, amphotericin B, voriconazole, posaconazole
Blastomycosis	*Blastomyces dermatitidis*	Fever, chills, myalgia, arthralgia, pleuritic chest pain, cough, difficulty breathing can become disseminated to mostly skin, bones, and genitourinary tract	Amphotericin B, itraconazole, voriconazole
Penicilliosis	*Penicillium marneffei*	Fever, weight loss, anemia, pneumonitis, skin lesions, pharyngeal, and palatal lesions	Amphotericin B, itraconazole
Basidiomycetes			
Cryptococcosis	*Cryptococcus neoformans, C.grubii*	Headache, fever, visual loss, osteolytic bony lesions, meningitis, calcification in lungs	Fluconazole, amphotericin B

(continued)

Table 12.1 (continued)

Disease	Causal fungal agent	Symptoms	Treatment
Zygomycetes			
Zygomycosis	*Rhizopus oryzae, R. rhizopodiformis, Rhizomucor pusillus, Absidia corymbifera, Apophysomyces elegans, Mucor circinelloides*	Facial pain, headache, nausea, fever, blood or pus draining from nose, lethargy, impaired vision, bulging eyes, convulsions, ulcers in roof of mouth, may become disseminated	Amphotericin B, posaconazole

benefits over other techniques, for instance, its effective delivery into various organs at low concentrations, thereby increasing safety, its cost-effectiveness as compared to protein/enzymes, and its non-immunogenicity bypassing the interferon pathway (Ruddon 2007). Subsequently, it was marked the "Breakthrough of the Year" in 2002 by *Science magazine*. One of the recent developments in this field has been the FDA approval of siRNA (short interfering RNA) drug DGFi in 2008, developed by Quark Pharmaceuticals, Inc., for use in kidney transplantation. Another siRNA-based therapy, Sirna-027 originally developed by Sirna Therapeutics for the treatment of acute macular degeneration, is moving forward into phase II clinical trials. Hence, the vital relevance of innovative treatments like the siRNA therapy can be envisaged for antifungal infections. RNAi application has also been identified in functional genomics and therapeutics, viz., cancer, neurodegenerative diseases, and multiple sclerosis besides fungal infection. This chapter presents the prospects of RNAi technology in fungal infection.

12.2 RNA Interference in Fungi

Gene cosuppression (RNAi-type phenomenon), which is more commonly known as "quelling" in fungi, was first discovered in the filamentous fungus *Neurospora crassa* by Romano and Macino (1992). The significant aspect of RNAi was established in 1998 when Fire and Mello found out that dsRNA is 10–100 times more effective in triggering silencing as compared to ssRNA in *Caenorhabditis elegans* (Fire et al. 1998) for which they shared the Nobel Prize in Medicine and Physiology in 2006.

RNAi naturally occurring "sequence-specific gene silencing" phenomenon and evolutionarily highly conserved mechanism are well developed in eukaryotic organisms, which is induced by double-stranded small noncoding RNA (dsRNA). These RNA-mediated gene silencing pathways have been comparatively thoroughly known and established in plant and animal system and are widely investigated in a variety of fungi, few of them are reported on opportunistic pathogenic fungi in human. The RNAi machinery in fungi shares similar mechanistic principles with other organisms where RNAi has been discovered which consists of dsRNA precursors, Dicer enzyme, RISC (RNA-induced silencing complex), and the target mRNA. The dsRNA is processed into short interfering RNA (siRNA) by two steps – the initiator step and the effector step. In the *initiator step*, the dsRNA is cleaved by the enzyme Dicer of the RNase III family (Jinek and Doudna 2009) into specific lengths of 21–25 nucleotides. In the *effector step*, the siRNA must segregate into "competent" single strands (the guide strands) which are guided through a ribonucleotide protein complex called the RISC (Siomi and Siomi 2009), a member of the Argonaute family of proteins (Song et al. 2004a, b; Leuschner et al. 2006; Fulci and Macino 2007). Both Dicer and RISC complex require ATP for energy. The Argonaute has a N-terminal domain, PAZ domain, a middle domain, and a PIWI domain. It is the "guide strand" which is incorporated into RISC; the non-incorporated strand known as the "passenger strand" is also crucial for proper target mRNA cleavage. Until the passenger strand is not cleaved at ten nucleotides from the 5′ phosphate of the guide strand, target mRNA cleavage is severely impaired (Leuschner et al. 2006). The cleaved target mRNA is utilized by RNA-dependent RNA polymerase (RdRp) generating more dsRNA that is further recognized and cleaved by Dicer to increase the number of siRNA molecules (Schepers 2005). Genes responsible for RNAi have been discovered in fungi such as *qde-1* (*quelling deficient-1*) which was the first RNAi gene discovered in *N. crassa* which encodes a cellular RdRp. Simultaneously, *qde-2* was cloned and found out to encode Argonaute protein. Further, partially redundant Dicer proteins DCL-1 (Dicer-like-1) and DCL-2 have been characterized from *N. crassa* by reverse genetics (Li et al. 2010).

So far, RNAi mechanism has not been observed in *Saccharomyces cerevisiae*, *Candida guilliermondii*, *C. lusitaniae*, *C. tropicalis*, and *Ustilago maydis* (Nakayashiki 2005; Münsterkötter and Mannhaupt 2008). In *S.cerevisiae*, it has been attributed to the absence of conserved components like Dicer-like RNases, Argonaute or PIWI-like components, and RNA-dependent RNA polymerases. However, RNAi was very recently discovered in *S. castellii*, which is closely related to *S.cerevisiae*, and also in *C. albicans*, a common human pathogen, by David P. Bartel's lab [22] (Drinnenberg et al. 2009). They found out that these species with noncanonical Dicer activity have RNase III domain containing gene to generate siRNA(s). This gene is orthologous to RNase III domains of other Argonaute-containing budding yeasts and is mostly enriched in long inverted repeats and transposable elements.

12.3 siRNA Design and Computational Tools

A significant facet of siRNA design involves identification and characterization of gene target which plays a key role in the survival of the organism such that suppression of the gene by inhibiting translation should limit the growth of the fungus. A particular fungal disease is often caused by more than one fungus, for instance, zygomycosis is caused by *Rhizopus*, *Absidia*, *Mucor*, and *Rhizomucor*. siRNA can be designed taking into consideration a conserved antifungal drug target present ubiquitously in the species, so that the drug discovery process becomes less intricate; but a crucial point to be taken care of is that the protein and its corresponding gene should not have similarities with the human genome, such that the human processes are not affected. For instance, cell wall is a very promising target in disease-causing fungi. The pathway for formation of cell wall is readily accessible (Moussian 2008), and the KEGG database (http://www.genome.jp/kegg/) can be referred. Silencing a gene coding for a significant component of the cell wall biosynthesis pathway can delimit the formation of cell wall and hence restrain the growth of the pathogenic fungus. For instance, a study reported loss of a cell wall polysaccharide, α-(1,3) glucan synthase from the cell walls of *Histoplasma capsulatum* which led to decreased virulence and pathogenesis of the fungi (Rappleye et al. 2004). Apart from the cell wall, targeting can be carried out against lipid biosynthesis pathways or at the translational or posttranslational levels. Previous studies have led to the confirmation of N-myristoyl transferase, lanosterol 14-α demethylase, and geranylgeranyl transferase I as potential drug targets against which antifungal drugs benzofurans, azoles, and azaphilones have been designed, respectively (Kawasaki et al. 2003; Song et al. 2004; Singh et al. 2005). Furthermore, orthologous and paralogous studies can be performed to check the presence of the target gene in other related organisms, which can highlight the evolutionary history of the gene, and information about gene conservation can acquaint us with the vital relevance of the gene.

siRNA should be a perfect complementary match to its target mRNA, and hence, it needs to be cautiously designed. Various computational tools are freely available online for designing siRNA. For instance, the Ambion's siRNA Target Finder, Eurofins MWG Operon's free online siMAX™ Design Tool, the BLOCK-iT™ RNAi Designer from Invitrogen, the SVM RNAi 3.6, and the siDESIGN Center by Dharmacon can be used for siRNA designing against fungal genes. The designing of siRNA molecule is purely dependent on suitable selection of various siRNA features, viz., sequence, motif, GC content, and thermodynamic features, etc. Moreover, various studies have documented the comparison among few siRNA design tools (Yiu et al. 2005; Matveeva et al. 2007). The tools mainly follow the Reynolds et al. (2004) or Tuschl et al. (1999) guidelines for rational siRNA design of usually 21 nucleotides in length. Few guidelines are described in onward section. According to these guidelines, regions within 50–100 bp of the start codon and the termination codon; intronic regions, stretches of 4 or more bases such as AAAA and CCCC; and regions with GC content less than 30% or more than 60%, single-nucleotide polymorphisms, repeats, and low complex sequences should be avoided.

12.3.1 General Guidelines for siRNA Designing

Many guidelines to design siRNA were proposed by different groups which are mentioned below.

12.3.1.1 MPI (Max-Planck-Institute) Rule Set

Tuschl et al. (1999) have provided a set of guidelines (commonly known as the MPI principles) on how to design effective siRNA.

Initially an empirical rules (based on GC content and symmetric 3′ TT overhangs) for effective siRNA designing were established by Tuschl et al. (1999), which have been found to show significant proportion of ineffective siRNAs as described by Yiu et al. (2005) and identify the need of understanding the structural features of sequences (Yiu et al. 2005). Therefore, the new rules were suggested with advancement of technology for effective designing of siRNA, (1) where select targeted region from a given cDNA sequence beginning 50–100 nt. downstream of start codon (2) afterward finds 23-nt sequence motif AA (N<sub>19). If no suitable sequence is found, then find for 23-nt sequence motif NA(N<sub>21) and convert the 3′ end of the sense siRNA to TT or search for [N (any nucleotide base pair)A (adenine) R (adenine or guanine (purines)) (N<sub>17)Y(thymine or cytosine (pyrimidines)) NN], and the GC content must be around 50% in target sequence.

12.3.1.2 Rational siRNA Design

In the designing of efficient siRNA molecule, a rational rule sets (total of eight criteria with weight values) were developed after the experiential analysis of the silencing efficiency of 180 siRNAs targeting the mRNA of two genes, by Reynolds et al. (2004) at Dharmacon, Inc.; these rules/criteria cover the different compositional properties such as sequential features of individual siRNAs. These characteristics are used by rational siRNA design algorithm to evaluate potential targeted sequences and assign scores to them (Reynolds et al. 2004). The higher value or score of these sequences reflects the higher chance of success of RNAi experiment.

The tools offered by Dharmacon, Inc. deploy the eight criteria-based (rational siRNA design rules) guideline, while Maurice Ho et al. have developed the Excel-based template on similar guidelines and available at http://boz094.ust.hk/RNAi/siRNA.

It is clear from Fig. 12.1 that different base pair shows differential stability with respect to their position (H = A, C or U).

12.3.1.3 Challenges

Despite several in silico siRNA design cases being reported in fungal systems, there are challenges that need to be addressed. These include optimal siRNA designing, off targeting, and efficient delivery systems apart from the genomic architecture and levels of organizational complexity. For example, whole-genome duplication events are commonly observed in fungi (Wapinski et al. 2007; Dujon et al. 2004; Ibrahim et al. 2009) such as in *Rhizopus oryzae*, giving rise to paralogous genes might not have 100% nucleotide sequence similarity with each other, which are actively expressed to code for an enzyme. Such constraints are often envisaged while designing effective siRNA against a fungal enzyme, as targeting only one out of many genes for a given enzyme target doesn't result in a very high efficacy for

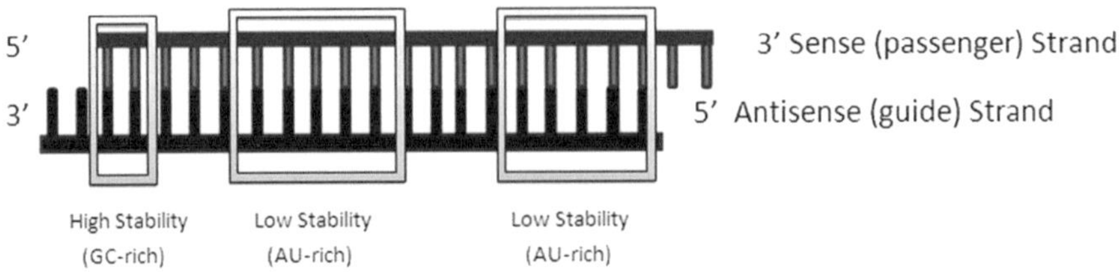

Fig. 12.1 Relative stability of both ends of the siRNA (Adopted from Reynolds et al. 2004)

silencing (Yamada et al. 2007) silenced only one out of the three α-amylase genes in *Aspergillus oryzae* and found merely 10% silencing, which is not sufficient to enter the therapeutic phase. Hence, designing common siRNA for all the genes targeting an enzyme becomes a very arduous task. Chitin synthase is a well-accepted antifungal drug target, yet it cannot be used to design siRNA(s) for *R. oryzae* because genome level analysis has revealed that 23 genes can encode chitin synthase (Ibrahim et al. 2009). The prospects of selecting the conserved regions of an enzyme are lucrative, as a single siRNA might be effective against several fungi causing a particular disease, but the enzyme should be present exclusively in fungi. Hence, target selection is critical toward its gene nature and diversity.

12.3.1.4 Resources and Tools

There are several tools available to design the siRNA, few of them are described in Table 12.2.

siVirus It is an effective siRNA designing tool for several divergent viral genomes such as human immunodeficiency virus (HIV), *hepatitis C virus* (HCV), influenza virus, and *SARS coronavirus*, based on various siRNA designing guidelines such as Ui-Tei et al. (2004), Reynolds et al. (2004), and Amarzguioui and Prydz (2004), while the off-targeting siRNA sequences were identified using siDirect tool. Out of 35 predicted siRNAs, 31 sequences of siRNA against conserved region of HIV-1 genome have shown inhibition of viral replication, which conclusively supports the designing capacity of these tools (Naito et al. 2006).

RNA Rule Set 1.0 It is a Java-based program to predict the efficient siRNA design. Basically siRNArules 1.0 operates with two scoring sums: one for the positive rules deduced from the set of the best siRNAs and one for the negative rules deduced from the set of the worst siRNAs. The rank is calculated by a simple sum of these two values. How these rules should be weighted to obtain the most efficient algorithm is a challenge for the open-source community (Holen 2006).

DSIR This tool was developed by Vert et al. (2006) which uses only two parameters for designing of siRNA, i.e., position-specific thermodynamic stability and non-position-specific motif formation. Their composite score decides which

Table 12.2 List of selected siRNA design tools

S. no	siRNA designing tool	Website
1	OligoWalk	http://rna.urmc.rochester.edu/cgi-bin/server_exe/oligowalk/oligowalk_form.cgi
2	BLOCK-iT™ RNAi designer	https://rnaidesigner.thermofisher.com/rnaiexpress/rnaiExpress.jsp
3	RNAi design tool	http://eu.idtdna.com/Scitools/Applications/RNAi/RNAi.aspx?source=menu
4	siDESIGN center	http://dharmacon.gelifesciences.com/design-center/
5	siRNA at whitehead	http://sirna.wi.mit.edu/
6	TROD	http://www.unige.ch/sciences/biologie/bicel/websoft/RNAi.html
7	AsiDesigner	http://sysbio.kribb.re.kr:8080/AsiDesigner/menuDesigner.jsf
8	GenScript's siRNA design center	http://www.genscript.com/design_center.html
9	SiDirect	http://sidirect2.rnai.jp/
10	Side	http://predictor.nchu.edu.tw/side/about_siDE.php
11	siVirus	http://sivirus.rnai.jp/
12	DSIR	http://biodev.extra.cea.fr/DSIR
13	SiSearch	http://www.biolabprotocols.com/details/2985/siSearch.html

siRNA is better through lasso regression technique. It also provides the options to design 19–21 bp siRNA. The effectiveness of this tool is comparable and evaluated as one of the best tools by Matveeva et al. (2007) while comparing with other tools.

12.4 RNAi Delivery

In silico siRNA designing is followed by its efficient delivery into an organism of choice for knockdown of the target gene. siRNA delivery strategies employed in fungi mainly comprise of soaking approach using chemically synthesized siRNA or inserting inverted repeat transgenes (IRT) into the desired plasmid using long-hairpin RNA (lhRNA). Table 12.3 lists some of the fungal species in which these strategies have been performed.

12.4.1 Soaking

This technique employs chemically synthesized siRNA(s) which are absorbed by the organism. Briefly, it involves annealing of the siRNA strands, quantification of the siRNA duplex, and soaking of the organism with these double-stranded siRNA (s). This protocol has been followed in protozoan parasite *Entamoeba histolytica* (Solis and Guillén 2008) and schistosomes (Ndegwa et al. 2007). In the model

Table 12.3 Silencing by exogenous siRNA

Presence of RNAi machinery in fungi	Gene targeted	Strategy	Inhibition	Refs.
Ascomycetes				
Aspergillus nidulans	*ODC* – fungal ornithine decarboxylase gene	Soaking method	Inhibition of 10%, 26%, 33%, 33% observed with 10 nM, 15 nM, 20 nM, 25 nM siRNA, respectively	Khatri and Rajam (2007)
Aspergillus nidulans	Bristle brlAβ – developmental regulatory gene	Inverted repeat of alcA (alcohol dehydrogenase) promoters flanking 498 bp brlAβ upstream	Expression was 3–4 fold less abundant on threonine than that on glucose	Barton and Prade (2008)
Aspergillus oryzae	brlA and α-amylase genes	Hairpin RNA cassette of 556 bp brlA, 1656 bp, and 750 bp α-amylase	Decreased signal for brlA gene, reduction in α-amylase activity	Ibrahim et al. (2009)
Aspergillus fumigatus	ALB1/PKSP – polyketide synthase (melanin biosynthesis pathway) and FKS1 β-(1,3) glucan synthase (cell wall polysaccharide)	500 bp inverted repeats of ALB1, FKS, and FKS/ALB1	29% of pALB1 transformants showed 5% (white colonies) or 24% (light-green colonies) reduction in ALB1 expression, 1% of pFKS1 transformants showed complete RNAi phenotype	Mouyna et al. (2004)
Aspergillus niger	xlnR encodes xylanases and cellulases	Inverted repeat of 834 bp of *xlnR*	12% transformants showed decreased activities	Oliveira et al. (2008)
Aspergillus parasiticus and *A. flavus*	AflR – transcription factor for expression of aflatoxin biosynthetic genes *VerA* – gene which is a part of aflatoxin biosynthetic pathway	Inverted repeat of 670 bp of *aflR*	*aflR* IRT transformants expressed *verA* at levels below the detection limits by N.Blot	McDonald et al. (2005)
Fusarium graminearum	Tri6 – transcription factor regulating expression of trichothecene (mycotoxin) biosynthetic genes	Inverted repeat constructs of tri6 ORF with 602 nt in sense and 588 nt in antisense direction	Wheat head blight did not spread to neighboring spikelets	McDonald et al. (2005)

Fusarium verticilloides (or *F. moniliforme*)	*gus* gene-encoding beta-glucuronidase	Inverted repeat of 627 bp of *gus* gene	Two transformed colonies showed a reduction of 62% and 96% in the *gus* gene expression	Tinoco et al. (2010)
Neurospora crassa	Albino gene *al-1* involved in carotenoid biosynthesis	Full length *al-1* gene (1971 bp)	24% of transformants showed albino phenotype	Cogoni and Macino (1997)
Magnaporthe oryzae (M. grisea)	Enhanced GFP (eGFP)	Constructs with sense-sense, antisense-antisense, and sense-antisense orientation of ~700 nt of eGFP	40 out of 80 IRT transformants emitted 20% fluorescence relative to original GFP	Kadotani et al. (2003)
Schizosaccharomyces pombe	Enhanced GFP (eGFP)	Inverted repeat of 760 bp of eGFP	Gfp mRNA reduced more than twofold	Sigova et al. (2004)
Histoplasma capsulatum	AGS1 or α-(1,3)-glucan synthase, cell wall polysaccharide	Inverted repeat of 678 bp of coding sequence of AGS1	Loss of α-(1,3)-glucan synthase from cell walls	Rappleye et al. (2004)
Cryptococcus neoformans (JEC21 strain)	CAP59 – capsule synthesis ADE2 – adenine biosynthesis	520 nt inverted repeats of CAP59 and ADE2	~25% reduction giving dull (CAP59suppression) and pink (ADE2 suppression) colonies	Liu et al. (2002)
Bipolaris oryzae	PKS1 – polyketide synthase gene, involved in melanin biosynthesis pathway	Inverted repeats of PKS1 gene fragment (756 bp)	70% of the transformants showed a melanin-deficient (white color) phenotype	Moriwaki et al. (2007)
Sclerotinia sclerotiorum	*rgb1* encoding PP2A (type 2A phosphoprotein phosphatase) B regulatory subunit	Inverted repeat of 1.1 kbp of *rgb1* under the control of *A. nidulans* promoter and terminator	*rgb1* suppression inhibited sclerotial maturation and caused reduced pathogenesis	Erental et al. (2007)
Colletotrichum lagenarium	Enhanced GFP (eGFP)	Construct consisted of 0.72 kb eGFP under the control of *A. nidulans* promoter and terminator	GFP fluorescence reduced to less than 20% of the parent strain	Nakayashiki et al. (2005)
Basidiomycetes				
Coprinus cinereus	recA-like recombinase in meiosis – Lim15/Dmc1	Vector construct with 750 nt antisense and 650 nt sense strand	23% having meiotic defects, i.e., fruiting body having white cap	Namekawa et al. (2005)

(continued)

Table 12.3 (continued)

Presence of RNAi machinery in fungi	Gene targeted	Strategy	Inhibition	Refs.
Schizophyllum commune	Structural gene – SC15	334 nt hairpin construct of SC15	80% reduction in aerial hyphae formation and attachment	Jong et al. (2006)
Zygomycetes				
Mortierella alpina	Δ12-desaturase desaturates oleic acid to linoleic acid	467 nt inverted repeat of Δ12-desaturase	25–88% decrease in fatty acid production	Takeno et al. (2005)
Mucor circinelloides	*carB* – encodes phytoene dehydrogenase enzyme, involved in carotenoid biosynthesis pathway	pMAT647 with 2483 nt of *carB*	pMAT647 and pMAT754 gave a silencing frequency of 3.16% and 3.12% with a complete albino phenotype	Nicolás et al. (2003)
		pMAT754 with complete cDNA sequence (2334 nt) of *carB*		

filamentous fungus, *Aspergillus nidulans*, 10 nM, 15 nM, 20 nM, and 25 nM siRNA were used, and a growth inhibition of 10%, 26%, 33%, 33% was observed, respectively, along with a 50% reduction in germ tube length following treatment with 10 nM siRNA-treated samples (Khatri and Rajam 2007; Amarzguioui et al. 2004). The soaking method results in transient expression making it less potent than direct microinjection (Schepers 2005), and therefore its efficiency in targeting the fungal thick cell walls consistently is not much significant.

12.4.2 Transgenic RNAi

The soaking method, transient RNAi, has limitations in terms of duration and experimental variability of the RNAi effect. The expression of inverted repeat (IR) transgenes, transgenic RNAi, has been shown to induce stable RNAi in fungi (Liu et al. 2002). However, variability, often, has been observed in the gene silencing effect by dsRNA-producing transgenes depending on the target gene, the type of construct, site of integration, and transgene copy number. An IRT is constructed using a target gene sequence, which is incorporated in an organism-specific plasmid. The plasmid containing IRT is transformed into the organism that upon transcription forms a hairpin loop which is cleaved by endogenous Dicer to generate siRNA(s). The generated small interfering RNAs (siRNAs) cleave the endogenous mRNA(s) with the help of RISC. Considering that there are several hairpin-expressing RNA systems available for robust RNA silencing using a tissue-specific RNA pol II promoter for tissue-specific expression of dsRNA and that hairpin RNA assures efficient formation of dsRNA, it is the frequently used method for induction of RNA silencing (Paddison and Vogt 2008). One type of IRTs is a long-hairpin RNA (lhRNA) which generally consists of more than 300 bp open reading frame, a spacer of approximately 250–500 oligonucleotides and an inverted repeat of the gene sequence. The IRT construct is inserted into the plasmid specific for a particular fungus and is transformed into the respective fungi either by protoplast formation or electroporation. But the transformation/cloning efficiency for long inverted repeats of a cDNA is very low, and the selection process is very time-consuming. The construction of lhRNA(s) is difficult as it involves several rounds of PCR and cloning steps, and moreover, it's necessary that vectors/plasmids should be available for each species to carry out siRNA delivery. Generally the longer size of spacer helps in the easy cloning of the inverted repeat and improves the efficiency of RNAi. Here, the use of IR constructs containing a short spacer (20–50 bp) in the middle of an inverted repeat improves the cloning efficiency (Yu et al. 2004). Moreover, the presence of introns as a spacer in the constructs also improves their effectiveness and enhances cloning efficiency [48. 49] (Wesley et al. 2001, *Kalidas and Smith* 2002). Nevertheless, these strategies have been successfully used to trigger silencing in various fungal systems like *Bipolaris oryzae* (Moriwaki et al. 2007) in which an IRT was constructed consisting of sense and antisense polyketide synthase gene of 756 bp encoding for melanin production separated by a spacer of cutinase gene intron obtained from a study on

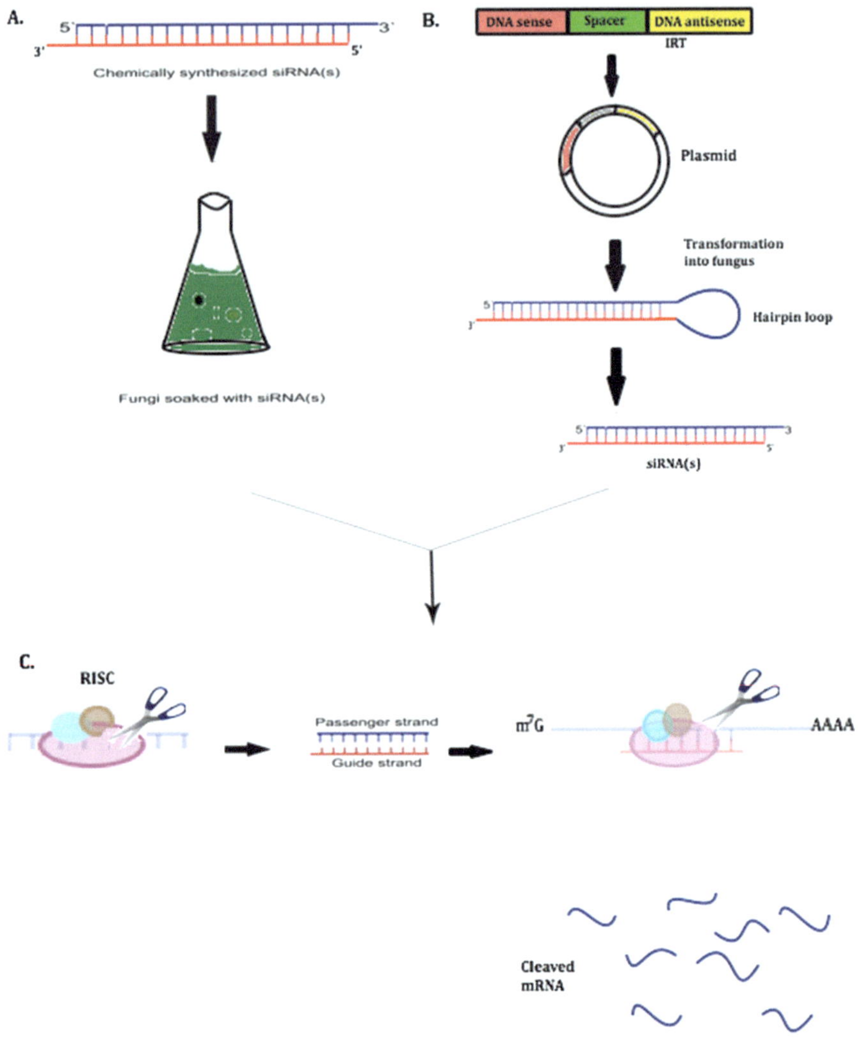

Fig. 12.2 A schematic representation of the two delivery methods of siRNA is shown. (**a**) Soaking method which involves soaking of fungi with siRNA. (**b**) IRT consists of the sense and antisense orientation of the gene separated by a spacer. It is incorporated into plasmid and transformed into fungi. Endogenous Dicer cleaves the hairpin loop which is formed upon transcription of IRT and siRNA(s) is generated. (**c**) The passenger strand is cleaved, and the guide strand of siRNA enters the RISC where it cleaves the target mRNA

Magnaporthe oryzae (Kadotani et al. 2003), and this yielded melanin-deficient (white) phenotype in 70% transformants. Another study reported only 12% transformants of *xlnR* encoding xylanases and cellulases of *A. niger* (Oliveira et al. 2008) showing decreased activities compared to the control strain. The reason was attributed to either rearrangement in silencing constructs or apparent decrease in xylanolytic activities or may be absence of "true" co-transformants.

Another prominent IRT successfully used in mammalian systems is the short-hairpin RNA (shRNA) which consists of a 19 bp siRNA sense and antisense sequence separated by a spacer of 9 nucleotides where the siRNA sequence should be 100% homologous to the target mRNA. The shRNA expression vectors available for mammalian systems require RNA polymerase III system as shown in Fig. 12.2.

12.5 Validation of Gene Silencing

A preliminary way to investigate the knockdown of a gene and percentage of silencing in fungus is to measure the mycelial growth of fungus by either colony diameter method or mycelial dry weight method (radial growth assays). Blotting techniques, fluorescence imaging, and biochemical assays are the further confirmatory tests which are used to analyze silencing at the RNA and protein levels. A study was carried out by Kadotani et al. (2003) in the blast fungus *M. oryzae* (Holen 2006) in which they investigated RNA silencing using enhanced green fluorescent protein (eGFP). Sense-sense, antisense-antisense, and sense-antisense IRT constructs of eGFP separated by a partial sequence of β-glucuronidase gene as internal spacer were employed. Significant silencing was induced only by sense-antisense IRT construct as detected by the loss of GFP fluorescence using an image analyzer. Several studies have employed Northern blot analysis for studying gene silencing in various fungi (Yamada et al. 2007; Kadotani et al. 2003; Hammond et al. 2007, 2008; Hammond and Keller 2005; Segers et al. 2006; Janus et al. 2007). The captured fluorescence images of plasmids (IRTs) containing fluorescence tags like GFP (Segers et al. 2006; Janus et al. 2007) assist in better understanding of protein localization, confirm the presence of plasmid in the cell, and also facilitate in differentiating the silenced from the non-silenced transformants. Real-time PCR and the reverse transcription PCR (Khatri and Rajam 2007; Oliveira et al. 2008; Janus et al. 2007; Liu et al. 2002) setup are also employed using appropriate gene-specific primers to detect and quantify the gene transcript levels. Liu et al. conducted a study (Liu et al. 2002) on RNA silencing in the pathogenic fungus *Cryptococcus neoformans* where they constructed IRTs on genes related to capsule synthesis and adenine biosynthesis pathway. They observed 25% reduction in the gene expression imparting the transformants with dull and pink colonies, respectively, and silencing analyzed by employing reverse transcription PCR with gene-specific primers and PCR product stained with SYBR Green I nucleic acid gel stain for analysis of DNA-associated fluorescence. Janus et al. investigated a reporter system *DsRed* to identify silencing transformants by cosilencing *DsRed* with the protein isopenicillin N synthase (*pcbC*) involved in cephalosporin C biosynthesis in the filamentous fungus *Acremonium chrysogenum* and employed immunoblotting using a polyclonal antibody against the isopenicillin N synthase to assess the downregulation of the *pcbC* gene (Janus et al. 2007). Biochemical assays can be performed as well which are specific for the enzyme/protein of interest to evaluate silencing of the target protein by specific siRNA(s). Hammond and Keller in 2005 executed the norsolorinic acid (NOR) analysis by thin-layer chromatography and observed whether the *A. nidulans aflR* IRTs can suppress NOR production (Hammond and Keller 2005). Research discoveries in vitro need to be accompanied by establishment of the potentiality of siRNA as a therapeutic. The antifungal drug dosages within the therapeutic window have been tested and determined on animal models, mostly murine models (Ibrahim et al. 2008a, b) for fungal diseases like zygomycosis, aspergillosis, candidiasis, cryptococcosis; but no pharmacological studies have yet been conducted on animal models employing siRNA therapy for fungal infections. Thus, dedicated efforts are needed to assess siRNA therapy for mycoses in various animal models and further investigate in humans through clinical trials.

12.6 Conclusion

RNAi has been explored in fungi in the post-genomics era, not only to understand the RNA silencing machinery but also as a tool to treat fungal diseases which cannot be effectively managed by conventional drugs. For instance, the past few decades have seen a rise in the opportunistic invasive fungal infections (mycoses) due to significant increase in the population of immunocompromised patients. The limitations of existing antifungal therapies have provided impetus toward exploring RNAi as a therapeutic option. *N. crassa* was the organism in which the first RNAi gene was discovered which was succeeded by numerous in vitro studies to achieve improved silencing in fungi and promising RNAi as a therapeutic tool. Identification of a potential target gene is necessitated against which siRNA is designed in silico, and recent whole-genome studies in fungi have greatly expedited the novel drug discovery process. Even though RNAi-related research discoveries have made rapid progress from in vitro to in vivo and clinical trials for neurodegenerative diseases and cancer, proper pharmacokinetic parameters of safety and efficacy for RNAi still remain to be answered for mycoses. RNAi as a therapeutic approach is relatively at a nascent stage, and uninterrupted probing should prove useful to drive it as a preferred option for the treatment of mycoses in the long run.

References

Amarzguioui M, Prydz H (2004) An algorithm for selection of functional siRNA sequences. Biochem Biophys Res Commun 316(4):1050–1058

Barton L, Prade R (2008) Inducible RNA interference of *brlAβ* in *Aspergillus nidulans*. Eukaryot Cell 7:2004–2007

Bhanderi B et al (2009) Antifungal drug resistance – concerns for veterinarians. Veterinary World 2:204–207

Chamilos G, Kontoyiannis D (2005) Update on antifungal drug resistance mechanisms of *Aspergillus fumigatus*. Drug Resist Updat 8:344–358

Cogoni C, Macino G (1997) Isolation of quelling-defective (qde) mutants impaired in posttranscriptional transgene-induced gene silencing in *Neurospora crassa*. Proc Natl Acad Sci U S A 94:10233–10238

Drinnenberg I et al (2009) RNAi in budding yeast. Science 326:544–550

Dujon B et al (2004) Genome evolution in yeasts. Nature 430:35–44

Erental A et al (2007) Type 2A phosphoprotein phosphatase is required for asexual development and pathogenesis of *Sclerotinia sclerotiorum*. MPMI 20:944–954

Fire A et al (1998) Potent and specific genetic interference by double-stranded RNA in *Caenorhabditis elegans*. Nature 391:806–811

Fulci V, Macino G (2007) Quelling: post-transcriptional gene silencing guided by small RNAs in *Neurospora crassa*. Curr Opin Microbiol 10:199–203

Hammond T, Keller N (2005) RNA silencing in *Aspergillus nidulans* is independent of RNA-dependent RNA polymerases. Genetics 169:607–617

Hammond T et al (2007) *Aspergillus* Mycoviruses are targets and suppressors of RNA silencing. Eukaryot Cell 7:350–357

Hammond T et al (2008) RNA silencing gene truncation in the filamentous fungus *Aspergillus nidulans*. Eukaryot Cell 7:339–349

Holen T (2006) Efficient prediction of siRNAs with siRNArules 1.0: an open-source JAVA approach to siRNA algorithms. RNA 12(9):1620–1625

Ibrahim A et al (2008a) Comparison of lipid amphotericin B preparations in treating murine zygomycosis. Antimicrob Agents Chemother 52:1573–1576

Ibrahim A et al (2008b) Combination echinocandin-polyene treatment of murine mucormycosis. Antimicrob Agents Chemother 52:1556–1558

Ibrahim ML et al (2009) Genomic analysis of the basal lineage fungus *Rhizopus oryzae* reveals whole-genome duplication. PLoS Genet 5:e1000549

Janus D et al (2007) An efficient fungal RNA-silencing system using the *DsRed* reporter gene. Appl Environ Microbiol 73:962–970

Jinek M, Doudna J (2009) A three-dimensional view of the molecular machinery of RNA interference. Nature 457:405–412

Jong J et al (2006) RNA-mediated gene silencing in monokaryons and dikaryons of *Schizophyllum commune*. Appl Environ Microbiol 72:1267–1269

Kadotani N et al (2003) RNA silencing in the phytopathogenic fungus *Magnaporthe oryzae*. MPMI 16:769–776

Kalidas S, Smith DP (2002) Novel genomic cDNA hybrids produce effective RNA interference in adult Drosophila. Neuron 33(2):177–184

Kawasaki K et al (2003) Design and synthesis of novel benzofurans as a new class of antifungal agents targeting fungal N-myristoyltransferase. Part 3. Bioorg Med Chem Lett 13:87–91

Khatri M, Rajam M (2007) Targeting polyamines of *Aspergillus nidulans* by siRNA specific to fungal ornithine decarboxylase gene. Med Mycol 45:211–220

Krcmery V (1996) Emerging fungal infections in cancer patients. J Hosp Infect 33:109–117

Leuschner P et al (2006) Cleavage of the siRNA passenger strand during RISC assembly in human cells. EMBO Rep 7:314–320

Li L et al (2010) RNA interference pathways in filamentous fungi. Cell Mol Life Sci 67:3849–3863

Liu H et al (2002) RNA interference in the pathogenic fungus *Cryptococcus neoformans*. Genetics 160:463–470

Matveeva O et al (2007) Comparison of approaches for rational siRNA design leading to a new efficient and transparent method. Nucleic Acids Res 35:e63

McDonald T et al (2005) RNA silencing of mycotoxin production in *Aspergillus* and *Fusarium* species. MPMI 18:539–545

Moriwaki A et al (2007) RNA-mediated gene silencing in the phytopathogenic fungus *Bipolaris oryzae*. FEMS Microbiol Lett 269:85–89

Moussian B (2008) The role of GlcNAc in formation and function of extracellular matrices. Comp Biochem Physiol 149:215–226
Mouyna I et al (2004) Gene silencing with RNA interference in the human pathogenic fungus *Aspergillus fumigatus*. FEMS Microbiol Lett 237:317–324
Mukherjee P et al (2005) Combination treatment of invasive fungal infections. Clin Microbiol Rev 18:163–194
Münsterkötter M, Mannhaupt G (2008) The posttranscriptional machinery of *Ustilago maydis*. Fungal Genet Biol 45:S40–S46
Naito Y et al (2006) siVirus: web-based antiviral siRNA design software for highly divergent viral sequences. Nucleic Acids Res 34((Web Server issue)):W448–W450
Nakayashiki H (2005) RNA silencing in fungi: mechanisms and applications. FEBS Lett 579:5950–5957
Nakayashiki H et al (2005) RNA silencing as a tool for exploring gene function in ascomycete fungi. Fungal Genet Biol 42:275–283
Namekawa S et al (2005) Knockdown of *LIM15/DMC1* in the mushroom *Coprinus cinereus* by double-stranded RNA-mediated gene silencing. Microbiology 151:3669–3678
Ndegwa D et al (2007) Protocols for gene silencing in schistosomes. Exp Parasitol 117:284–291
Nicolás F et al (2003) Two classes of small antisense RNAs in fungal RNA silencing triggered by non-integrative transgenes. EMBO J 22:3983–3991
Oliveira J et al (2008) Efficient cloning system for construction of gene silencing vectors in *Aspergillus niger*. Appl Microbiol Biotechnol 80:917–924
Paddison PJ, Vogt PK (2008) RNA interference, Current topics in microbiology and immunology, vol 320. Springer-Verlag, Berlin
Pauw B, Picazo J (2008) Present situation in the treatment of invasive fungal infection. Int J Antimicrob Agents 32:S75–S79
Rappleye C et al (2004) RNA interference in *Histoplasma capsulatum* demonstrates a role for α-(1,3)-glucan in virulence. Mol Microbiol 53:153–165
Reynolds A et al (2004) Rational siRNA design for RNA interference. Nat Biotechnol 22:326–330
Ribes J et al (2000) Zygomycetes in human disease. Clin Microbiol Rev 13:236–301
Richardson M (1991) Opportunistic and pathogenic fungi. J Antimicrob Chemother 28:1–11
Rogers T (2008) Treatment of zygomycosis: current and new options. J Antimicrob Chemother 61: i35–i39
Romano N, Macino G (1992) Quelling: transient inactivation of gene expression in *Neurospora crassa* by transformation with homologous sequences. Mol Microbiol 6:3343–3353
Ruddon R (2007) Cancer biology, 4th edn. Oxford University Press, New York
Schepers U (2005) RNA interference in practice: principles, basics, and methods for gene silencing in *C. elegans*, drosophila, and mammals, 1st edn. Wiley-VCH Verlag GmbH & Co. KGaA, Weinheim
Segers G et al (2006) Hypovirus papain-like protease p29 suppresses RNA silencing in the natural fungal host and in a heterologous plant system. Eukaryot Cell 5:896–904
Sigova A et al (2004) A single Argonaute protein mediates both transcriptional and posttranscriptional silencing in *Schizosaccharomyces pombe*. Genes Dev 18:2359–2367
Singh S et al (2005) New fungal metabolite geranylgeranyltransferase inhibitors with antifungal activity. Nat Prod Res 19:739–747
Siomi H, Siomi M (2009) On the road to reading the RNA-interference code. Nature 457:396–404
Solis C, Guillén N (2008) Silencing genes by RNA interference in the protozoan parasite *Entamoeba histolytica*. In: Barik S (ed) RNAi: design and application. Humana Press, Totowa, pp 113–128
Song J et al (2004a) The *Candida albicans* lanosterol 14-α Demethylase (ERG11) gene promoter is maximally induced after prolonged growth with antifungal drugs. Antimicrob Agents Chemother 48:1136–1144
Song J et al (2004b) Crystal structure of Argonaute and its implications for RISC slicer activity. Science 305:1434–1437

Takeno S et al (2005) Improvement of the fatty acid composition of an oil-producing filamentous fungus, *Mortierella alpina* 1S-4, through RNA interference with Δ12-desaturase gene expression. Appl Environ Microbiol 71:5124–5128

Tinoco M et al (2010) In vivo trans-specific gene silencing in fungal cells by *in planta* expression of a double-stranded RNA. BMC Biol 8:27

Tuschl T et al (1999) Targeted mRNA degradation by double-stranded RNA in vitro. Genes Dev 13:3191–3197

Ui-Tei K et al (2004) Guidelines for the selection of highly effective siRNA sequences for mammalian and chick RNA interference. Nucleic Acids Res 32(3):936–948

Vert JP et al (2006) An accurate and interpretable model for siRNA efficacy prediction. BMC Bioinformatics 30(7):520

Wapinski I et al (2007) Natural history and evolutionary principles of gene duplication in fungi. Nature 449:54–61

Warnock D, Richardson M (1991) Fungal infection in the compromised patient, 2nd edn. Wiley, Chichester

Wesley SV et al (2001) Construct design for efficient, effective and high-throughput gene silencing in plants. Plant J 27(6):581–590

Yamada O et al (2007) Gene silencing by RNA interference in the Koji Mold *Aspergillus oryzae*. Biosci Biotechnol Biochem 71:138–144

Yiu S et al (2005) Filtering of ineffective siRNAs and improved siRNA design tool. Bioinformatics 21:144–151

Yu J et al (2004) Transgenic RNAi-mediated reduction of MSY2 in mouse oocytes results in reduced fertility. Dev Biol 268(1):195–206

Genome-Wide Essential Gene Identification in Pathogens 13

Budhayash Gautam, Kavita Goswami, Satendra Singh, and Gulshan Wadhwa

Abstract

Genome-wide, a gene can be designated as indispensable for the survival of a cell or an organism, and its interruption can lead to the malfunctioning or death of the organism. Due to its essentiality for survival, these could be proposed as novel and promising candidates for broad-spectrum drug targets, if these are conserved across a genus. Identification of essential gene has been done in many organisms, and interestingly, most of them were pathogenic in nature. The genome-scale elucidation of essential genes plays an important role in development and complete genome availability. At large scale, gene-inactivation technologies such as targeted gene inactivation, genetic footprinting, and transposon-based mutagenesis are controlled by essential genes. In silico, numerous strategies and tools also have been developed, such as subtractive genomics, essentiality base mapping, and target identification using phylogenetic profiling. Bioinformatic approaches can also be used to analyze experimentally generated data. This chapter is referred to provide an overview of some of these methodologies which are often used to identify essential genes and their functions and discuss advantage and drawbacks of the methods.

B. Gautam (✉) · S. Singh
Department of Computational Biology and Bioinformatics, Jacob Institute of Biotechnology and Bioengineering, Sam Higginbottom University of Agriculture, Technology and Sciences, Allahabad, Uttar Pradesh, India
e-mail: budhayash.gautam@shiats.edu.in

K. Goswami
Plant RNAi Biology Group, International Center for Genetic Engineering and Biotechnology, New Delhi, India

G. Wadhwa
Department of Biotechnology, Apex Bioinformatics Centre, Ministry of Science & Technology, New Delhi, India

G. Wadhwa et al. (eds.), *Current trends in Bioinformatics: An Insight*,
https://doi.org/10.1007/978-981-10-7483-7_13

Keywords
Essential gene · DEG · Genome-wide identification · Subtractive genomics · Essentiality-based mapping · In silico drug target identification

13.1 Introduction

The revolutionary advancement in the sequencing technologies has made possible the whole-genome sequencing of large genomes in the biomedical research field. *Haemophilus influenzae* was the first bacterial genome to be sequenced in 1995. Currently, >1400 bacterial and eukaryotic genomes have been completely sequenced which are publicly available in databases such as DDBJ (DNA Data Bank of Japan), GenBank, and EMBL (European Molecular Biology Laboratory) Nucleotide Sequence Database (Fleischmann et al. 1995). These databases provide substantial amounts of sequences and vitally influenced the research in every field of biomedical science. The emergence of antibiotics resistance in various pathogens and their ability to acquire and spread resistance are creators of urgent requirement for the development of broad-spectrum drug targets and antibiotics (Barrett and Stanberry 2009). Vast amounts of large genomic data require desirable approaches for the prediction, identification, and selection of a target for the probe. However, essentiality of genes and their protein products can provide a tool to solve the problem; in fact, these have been used very often recently (Deng et al. 2010). Gene essentiality refers to the involvement in survival of a cell, and its deletion and disruption are lethal for an organism under certain environmental conditions (Date and Marcotte 2003).

Thus, essential gene-related encoded functions are considered a basis of life (Koonin 2000). An organism comprises a set of minimal genes which are required and adequate to maintain the functioning of a cell, and identification of these minimal genes and their function is very difficult (Lehoux et al. 2001). Essential genes have several important applications; these genes are not only necessary for the survival of the organism, but also these are evolutionarily conserved than nonessential genes; therefore these genes are good candidates for drug targets for any pathogenic disease because most of the drugs often target the genes performing critical cellular processes (Huynen and Bork 1998. Essential genes that are conserved in a number of different genera would be good candidates for broad-spectrum drug targets rather than species specific identification of essential genes in pathogenic organisms and their involvement in pathways and how it control pathogens by identifying potential targets and further use to develop antimicrobial drug (Lin and Zhang 2011). It can also reveal the pathogenic organism's relationships during evolution (Cooper and Duffield 2011).

Essential gene identification has been done recently by utilizing several methods, but broadly these methodologies can be categorized in two groups:

(1) identification by experimental methodologies and (2) identification by computational methodologies. Both groups have been very efficient in identifying essential genes. However, there are certain pros and cons to each technique which are discussed in the following sections.

13.2 Database of Essential Genes (DEG)

On the basis of identification of essential genes in various organisms, a database of essential genes was constructed known as DEG (Luo et al. 2013). Originally it contains essential gene information from seven organisms (DEG 1.0, October 21 2002); the most current version (DEG 13.3, January 7, 2016) contains 49,702 essential genes and 646 essential noncoding sequences (Fig. 13.1). DEG 11.5 was the last updated version, available with 23,378 essential genes and 680 essential noncoding sequences in 64 organisms, 40 prokaryotes including 39 bacteria and 1 archaea and 24 eukaryotes (November 24, 2015) (Table 13.1) (http://www.essentialgene.org/). This version has more numbers of noncoding genes than the latest version of DEG. Another database is available for predicted essential genes named pDEG 1.0 (December 8, 2010) which contains 5880 essential genes out of 11,703 total available genes from 16 mycoplasma organisms (http://tubic.tju.edu.cn/pdeg/).

The DEG database is divided in two subclasses, having essential gene data of prokaryotes and eukaryotes. For each prokaryote and eukaryote organism, DEG contains the DNA and protein sequences. Every gene has a unique identification number gene reference number and name and gene function with DEG entry. Prokaryotic essential genes are also linked to NCBI (National Center for Biotechnology Information) to the Clusters of Orthologous Groups (COGs) (http://www.ncbi.nlm.nih.gov/). Publications on essentiality are also available, linked to the related organism, and related PubMed link is also provided. Apart from browsing the gene records and searching essential genes by their accession numbers, gene names, functions, and organisms, additionally, BLAST (Basic Local Alignment

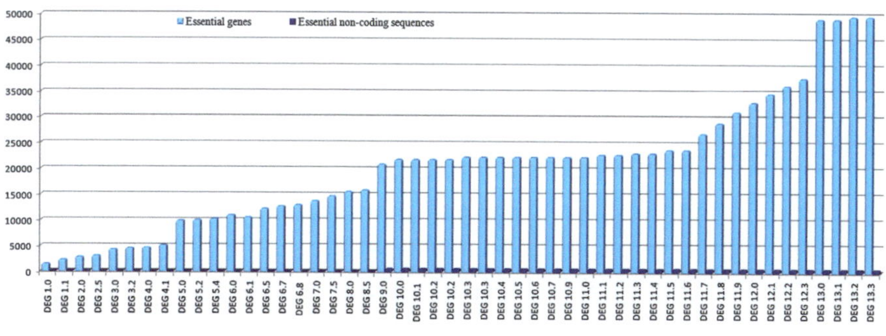

Fig. 13.1 Status of essential gene (cyan color) and the noncoding gene (blue color) in DEG database

Table 13.1 Essential genes of various organisms currently available in Deg 11.5

S. No.	Kingdom	Organism	Essential genes	Total genes	%Essential gene	Median total length (Mb)
1.	Prokaryote	*Acinetobacter baylyi* ADP1	499	3426	14.57	3.9
2.	Prokaryote	*Bacillus subtilis* 168	271	4673	5.80	4.09642
3.	Prokaryote	*Bacteroides fragilis* 638R	547	4469	12.24	5.30944
4.	Prokaryote	*Bacteroides thetaiotaomicron* VPI-5482	325	7838	4.15	6.23288
5.	Prokaryote	*Burkholderia pseudomallei* K96243	505	5953	8.48	7.12696
6.	Prokaryote	*Burkholderia thailandensis* E264	406	5721	7.10	6.69622
7.	Prokaryote	*Campylobacter jejuni* subsp. *jejuni* NCTC 11168 = ATCC 700819	228	3413	6.68	1.67782
8.	Prokaryote	*Caulobacter crescentus*	480	14,708	3.26	4.23667
9.	Prokaryote	*Escherichia coli* MG1655 I	609	5049	12.06	5.16825
10.	Prokaryote	*Escherichia coli* MG1655 II	296	322	91.93	5.16825
11.	Prokaryote	*Francisella novicida* U112	392	1782	22.00	1.86692
12.	Prokaryote	*Haemophilus influenzae* Rd. KW20	642	1795	35.77	1.85826
13.	Prokaryote	*Helicobacter pylori* 26,695	323	3297	9.80	1.63321
14.	Prokaryote	*Mycobacterium tuberculosis* H37Rv	614	12,689	4.84	4.38293
15.	Prokaryote	*Mycobacterium tuberculosis* H37Rv II	84	147	57.14	NA
16.	Prokaryote	*Mycobacterium tuberculosis* H37Rv III	73	112	65.18	NA
17.	Prokaryote	*Mycoplasma genitalium* G37	381	530	71.89	0.579677
18.	Prokaryote	*Mycoplasma pulmonis* UAB CTIP	310	817	37.94	0.963879
19.	Prokaryote	*Porphyromonas gingivalis* ATCC 33277	281	3407	8.25	2.33474
20.	Prokaryote	*Pseudomonas aeruginosa* PAO1	336	18,541	1.81	6.56769
21.	Prokaryote	*Pseudomonas aeruginosa* UCBPP-PA14	335	6051	5.54	6.56769
22.	Prokaryote	*Rhodopseudomonas palustris* CGA009	522	5093	10.25	5.43588
23.	Prokaryote	*Salmonella enterica* serovar *Typhi*	353	20,135	1.75	4.76352
24.	Prokaryote	*Salmonella enterica* serovar *Typhi* Ty2	358	9322	3.84	4.76352
25.	Prokaryote	*Salmonella enterica* serovar *Typhimurium* SL1344	353	4921	7.17	4.76352

26.	Prokaryote	*Salmonella enterica* subsp. *enterica* serovar *Typhimurium* str. 14028S	105	5634	1.86	4.76352
27.	Prokaryote	*Salmonella typhimurium* LT2	230	11,666	1.97	4.76352
28.	Prokaryote	*Shewanella oneidensis* MR-1	403	4917	8.20	5.13142
29.	Prokaryote	*Sphingomonas wittichii* RW1	535	5462	9.79	5.62889
30.	Prokaryote	*Staphylococcus aureus* N315	302	2851	10.59	2.86686
31.	Prokaryote	*Staphylococcus aureus* NCTC 8325	351	2979	11.78	2.86686
32.	Prokaryote	*Streptococcus pneumoniae*	244	57,548	0.42	2.0988
33.	Prokaryote	*Streptococcus pyogenes* MGAS5448	227	44,923	0.51	1.83015
34.	Prokaryote	*Streptococcus pyogenes* NZ131	241	1843	13.08	1.83015
35.	Prokaryote	*Streptococcus sanguinis*	218	2505	8.70	2.34878
36.	Prokaryote	*Synechococcus elongatus* PCC 7942	682	2806	24.31	2.71926
37.	Prokaryote	*Vibrio cholerae* N16961	779	4203	18.53	4.01969
38.	Prokaryote	*Methanococcus maripaludis* S2	519	1804	28.77	1.7467
39.	Eukaryote	*Arabidopsis thaliana*	356	71,611	0.50	97.3801
40.	Eukaryote	*Aspergillus fumigatus*	35	15,981	0.22	29.0541
41.	Eukaryote	*Caenorhabditis elegans*	294	54,250	0.54	100.286
42.	Eukaryote	*Danio rerio*	315	84,612	0.37	1391.74
43.	Eukaryote	*Drosophila melanogaster*	339	40,217	0.84	148.504
44.	Eukaryote	*Homo sapiens*	2196	744,812	0.29	2994.61
45.	Eukaryote	*Mus musculus*	2114	203,040	1.04	2541.65
46.	Eukaryote	*Saccharomyces cerevisiae*	1110	97,127	1.14	12.3018
47.	Eukaryote	*Schizosaccharomyces pombe* 972 h-	1260	13,022	9.68	NA

Search Tool) searches against DEG entries are also available with whole-genome annotation for prokaryotes and multi-gene search for eukaryotes (http://tubic.tju.edu.cn/deg/blast_genome_anno.php?db=a).

All these functions have to be done separately in both the databases, because DEG is divided into prokaryote and eukaryote sub-databases. The whole database can also be downloaded. DEG is a freely accessible database which is available at http://tubic.tju.edu.cn/deg or http://www.essentialgene.org.

13.3 Identification of Essential Genes Using Experimental Methods

At present, complete genomes of various species have been available for the prediction of function of uncharacterized genomes that can be made by comparative genomics and other methods with the help of available genetic and biochemical information. The availability of complete genomic sequences raised an important question on essential gene quantity, means what are the number of genes which are essential for cellular life or what is the minimal genome of organism?

Although bacterial genomes differ in their sizes and gene repertoires, every genome can contain an essential gene either it is large or small; they must contain all the information related to the cell which allows the cell to perform many essential (housekeeping) functions (Juhas et al. 2012). This gene enables the cell to maintain "reproduction, metabolic homeostasis and evolve"; these are the main properties of living cells. Cells are capable of importing metabolites, but for the functional proteins, they have to rely on their own essential gene which takes part in the production of protein (Akerley et al. 1998). Necessarily, the protein-coding gene has to determine the minimal set of genes to maintain a living cell; the minimal gene set is becoming an appealing issue; it is defined as a "group of genes which would be enough to maintain the functioning of cellular life under the most favorable conditions imaginable, that is, in the availability of essential nutrients and in the deficiency of environmental stress." Recently, many research groups have tried to define the essential gene set in bacteria using various experimental and computational techniques, discussed in the following section in brief (Gil et al. 2004).

13.3.1 Transposon-Based Strategies for Gene Essentiality

Transposons provide an alternative method for defining essential genes. Different available approaches can be utilized depending on the type of information that we obtained and is available in the genetic systems of the organism of interest (Devine and Boeke 1994). Transposons, segments of DNA flanked by sequences which can move from one location to another in the genome, are repeatedly inverted. That can be recognized by a protein, transposase, an enzyme which allows transposon to transpose. The movement of transposons from one location to another depends on the transposase as it recognizes and cleaves sequences which could be any undetermined nuclear sequence (Bardarov et al. 2002). Mutagenesis in transposon results in interruption in the genomic region where it has to insert. If an insertion takes place within a predicted ORF, it allows the resulting strain to be form a colony, leads to improbable that is essential for ORF for viability under those conditions (Judson and Mekalanos 2000).

As per reports, the essential genes of a bacterial chromosome can be identified in two ways:

1. The *negative approach* – it is an approach that is unable to identify essential genes as it isolates the regions which are not essential while presuming the rest to be essential (Fig. 13.2).
2. The *positive approach* – it identifies essential genes by generating a conditional mutation and screening a lethal phenotype (Fig. 13.3).

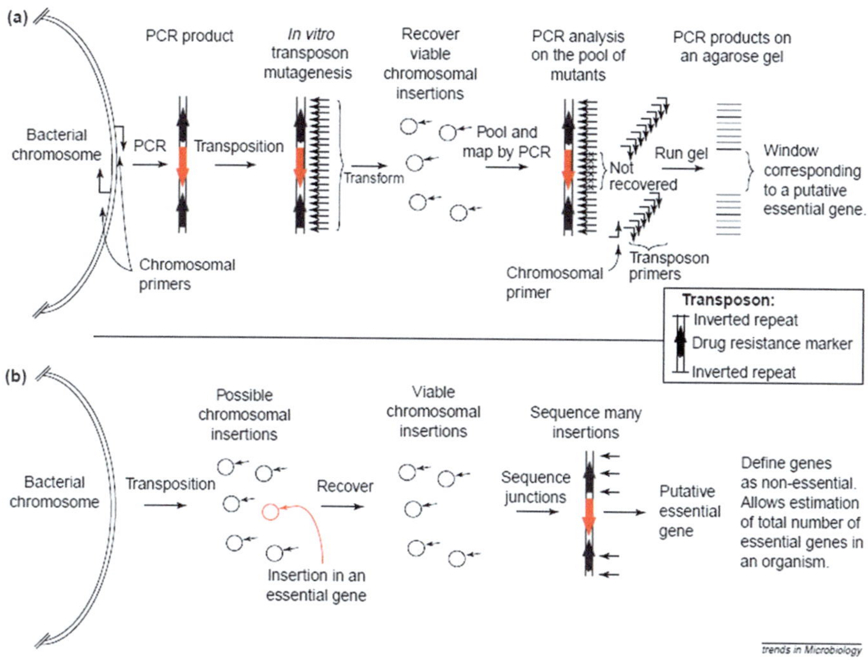

Fig. 13.2 Negative approaches to identify essential bacterial genes: identify essential genes by identification of regions that are not essential. Regions of the chromosome that cannot be disrupted are presumed essential. (**a**) PCR-mapping approaches. A specific region of the bacterial chromosome is amplified by PCR and is subjected to in vitro mutagenesis with a transposon containing a drug-resistance marker. The small horizontal arrows signify some of the possible locations of insertions in the PCR product. Genes are designated by vertical black (nonessential) or red (essential) arrows. Insertions that allow viable colonies to form are pooled and analyzed by PCR. Upon analysis, those regions that do not allow insertions (crossed-out arrows) show up as empty regions on the agarose gel and are presumed to define essential genes. (**b**) Global transposon mutagenesis. The whole bacterial chromosome is the target for transposon mutagenesis. A large number of viable insertions are analyzed by sequencing. This method requires many insertions to be sequenced before statistically significant conclusions can be drawn

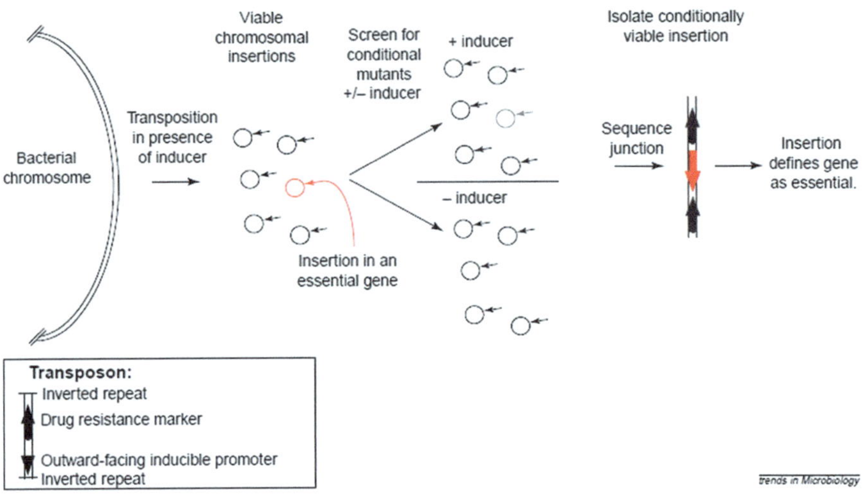

Fig. 13.3 The positive approach to identify essential genes: identify directly genes that are essential. Transposition with a transposon containing an outward-facing inducible promoter at one edge in the presence of the inducer results in many possible transposon insertions. The horizontal arrows signify possible insertion locations on the bacterial chromosome. Screening identifies insertions that disrupt the promoter region of an essential gene (red arrow). The strain generated by such an insertion is dependent on the inducer for viability. The insertional junction is sequenced, allowing the identification of the downstream essential gene

This question comes to mind: why is this anti-correlated approach being used to identify the essential genes? A problem in identifying essential genes was that a knockout of an essential gene was lethal.

To avoid this problem, the negative approach can be used to assign a location of transposon insertions which facilitate the observation of transposon insertion in a particular region; unobserved insertion seems to be essential (Hayes 2003). In any case either using transposons or not, it is difficult to identify the essential region of a chromosome because of the transposons' nature. At genomic level, analysis or identification of essential gene needs extensive sequencing or PCR. Using PCR mapping, a large data set of specific regions of genome targeted by transposon mutagenesis can be pooled together; this is a significant approach which helps to generate possible colonies of desired data set (Ivics et al. 2009).

Resolved PCR product analysis an agarose gel visualizing the "windows" which corresponds to essential ORFs. The large saturated studies allow drawing a significant conclusion about the importance of multiple ORFs within a coding region. Global transposon mutagenesis is not able to generate a sufficient number of insertions that are required for equilibration of depleting inserted transposons in particular ORFs (Kang and Fisher 2005). In genome-wide studies, more experiments need to be performed to provide a definitive proof of a specific gene which has not been disrupted essentially. There is a disadvantage of negative approach that is does not express the severity of gene while increased number in the closely related genes showed that it cannot disrupted the not essential gene. Biological experiment to check the essentiality of the insertion is still under demonstration; a prescribed possibility is that these genes are nonessential. Insertion is allowed only at some particular site that is conserved and can apply to the essential gene (Kleckner 1981). A quality that permits insertions in a couple of areas inside of the coding locale could even now be vital. Thus, the negative approach recognizes qualities that ought to be examined, assisted by practical studies.

An advantage of the positive approach for distinguishing the essential gene is that it does not need any development of a complementing plasmid to point out that an ORF is essential, since it uses a built-in test for the identification of biological function of genes. Any contingent insertion that is created can be utilized for biochemical examination of the quality item being referred to without further strain development, and crucial ORFs with insignificant 5′ areas that are tolerant for transposon insertions will at present produce a restrictive phenotype (Salama et al. 2004). Some disadvantages of this approach are that an inducible promoter might not provide enough expression to overcome the defect created by knocking out the natural promoter or, conversely, basal expression of an inducible promoter might be too high to allow identification of essential genes for which only minute amounts of gene product are required.

It likewise has a few advantages such as an inducible promoter would not give enough expression to thrash the desert made by knocking out the characteristic promoter or, on the other hand, basal expression of an inducible promoter may be too high to permit distinguishing proof of crucial qualities for which just moment measures of quality item are required (O'sullivan et al. 2006). The size of the objective within which an associated insertion can happen fluctuates relying upon the sequence quality and can usually be small for the most part; it may be tough or not possible to hit a few genes. DNA structures pose issues for all methodologies (Touchon et al. 2009).

Polar impacts from the presence of an associated insertion may be severe, limiting insertions upstream of a necessary gene. In spite of the fact that the positive approach would possibly defeat this confinement for a few genes, if interpretation of a downstream gene is coupled to a gene that is discontinuous by the nearness of the insertion, these insertions won't be found (Luisi et al. 2002). On the off chance there's not coupled translation to the essential gene and could have a few qualities, downstream of the inducible promoter, which requires affirmation of the proposed gene by complementation studies (Ishikawa and Hori 2013).

Although simple in concept, essentiality of gene is a difficult question to address experimentally. It is possible to perform experiments under defined in vitro conditions (Akerley et al. 1998). However, this does not simplify the genetic interplay of essential processes that occurs in bacteria. Many gene distractions will result into intermediate phenotypical changes which mark them as a tedious task in classifying a gene as "essential" or "nonessential" (Jothi et al. 2007). This is compounded by the fact that many genes have some degree of redundancy, with

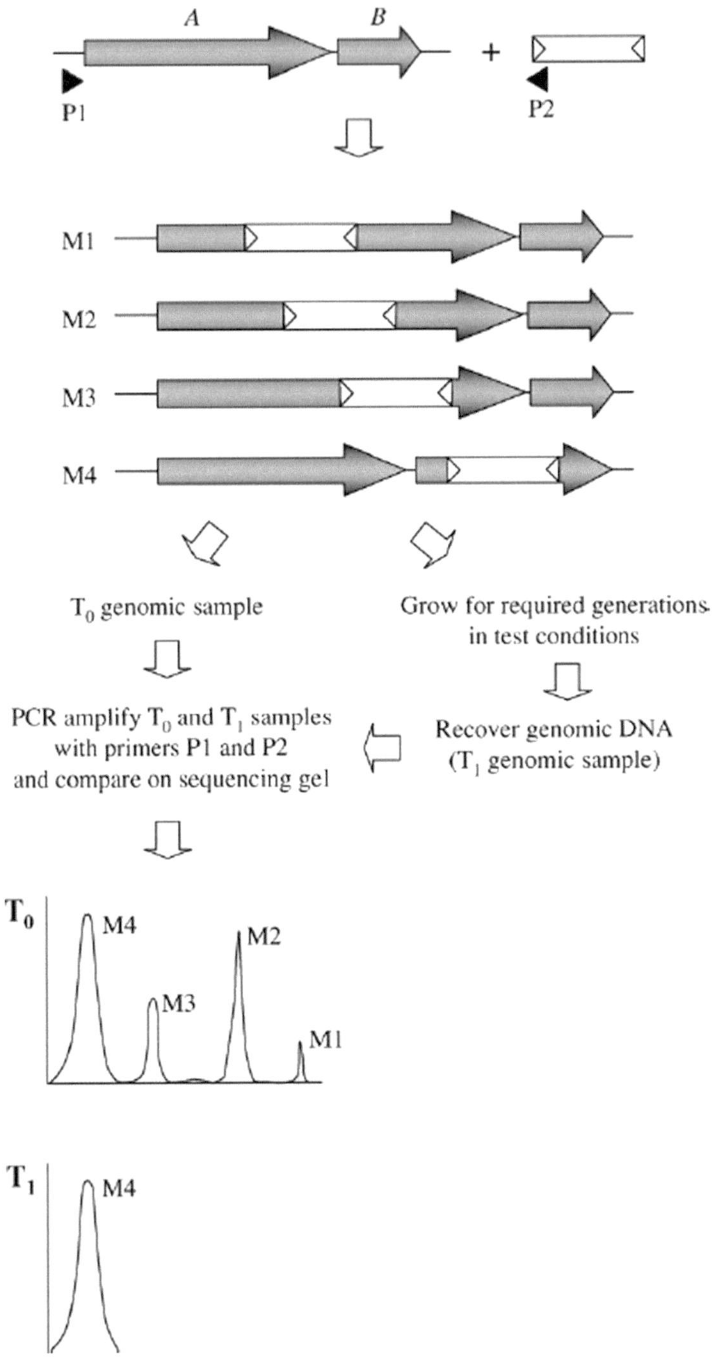

Fig. 13.4 Genetic footprinting with transposons. A transposon (open box) is used to comprehensively mutagenize the microorganism of interest. Four transposon insertions are shown: three of

other pathways that are able to compensate a certain extent for the mutation in question (Kensche et al. 2008). However, transposon-based approaches, which identify the function of genes, should be the focus of researchers, to further our understanding of the basic biological processes in a cell.

Although easy in thought, vitality is a troublesome issue to address tentatively. It is conceivable to perform tests under characterized conditions by utilizing characterized media or development conditions; in any case, this doesn't alter the genetic interaction of essential processes that are occurring inside of a bacterium (Grünblatt et al. 2014).

Many gene interruptions will bring about a middle phenotype that makes it difficult to characterize a quality as "essential" or "nonessential." This is often combined with the setting wherein a lot of genes have some level of excess, with different pathways ready to remunerate to a specific degree for the mutation being referred to (Langille et al. 2013). However, transposon-based approaches recognized genes that scientists ought to concentrate on to advance our comprehension of fundamental natural procedures and also the method of how these procedures communicate (Frøkjær-Jensen et al. 2010).

13.3.2 Genetic Footprinting with Transposons

Genetic footprinting is a molecular approach that helps in deciding the function of a sequenced gene, using an assembly of a substantial pool of transposon insertion mutants and the consequent recognizable gene that gets to be drained when this pool is proliferated under certain natural conditions (Huynen and Bork 1998).

The initial phase in genetic footprinting is the era of an exhaustive transposon arbitrary insertion pool that ideally will incorporate mutants with insertions at completely different locations within the gene of interest. The second step includes outgrowth of the mutagenized pool under the applicable test conditions for a particular selected variety of generations (Jimenez-Ruiz et al. 2014). At last, genomic DNA is recovered from the pool previously and then after the fact development. These DNA area units are subjected to PCR utilizing a couple of groundwork, one that strengthens the transposon close to the gene and one that primes inside or adjacent the gene of interest (Fig. 13.4).

Fig. 13.4 (continued) these insertions (M1–M3) are at different locations in an essential gene A, whereas the fourth insertion (M4) is located in an inessential gene B. Genomic DNA is prepared from the entire mutant pool (T0 genomic sample). The mutant pool also is grown under appropriate experimental conditions, e.g., in minimal growth medium, for a specific number of generations, after which genomic DNA is prepared (T1 genomic sample). Primers P1 and P2, which anneal upstream of gene A and within the transposon, respectively, are used in separate amplification reactions with the T0 and T1 DNAs. The products are analyzed on a sequencing gel. As A is an essential gene under the test conditions, the amplification products M1–M3 that are present in the T0 sample are absent from the T1 sample owing to depletion of the corresponding inviable mutants during growth. In contrast, the M4 amplification product is still evident at T1 because the corresponding transposon mutation is in a dispensable gene

The consumption of the PCR product taking place after development under trial conditions is indicative of a decrease in the quantity of cells within the population that harbor transposon insertions in the test gene, which is essential for viability under these conditions (Wong and Mekalanos 2000). In the event that the development of mutants containing a disturbance of the genes of interest is healthy under the test conditions, no critical distinction in the example of PCR items prior and then afterward outgrowth will be ascertained. Individual genes of interest may be tested by genetic footprinting, or a range of quality genes may be analyzed using aliquots of victimization DNA derived from a solitary outgrowth experiment (Smith et al. 1995).

13.3.3 Targeted Gene Deletion

An allele is an alternative form of gene, and its replacement helps to achieve targeted gene deletion which can be studied by a single gene or by whole genome where each gene is available. There are some reports on targeted gene deletion such as Joshi et al. (2002) which reported that targeted gene deletion has been used to obtain Leishmania mutants that are unable to express leishmanolysin L genes (gp63 genes 1–6). Distribution of targeted deletion gene is based on the mutant type; either it is marked mutant or unmarked mutant. Marked mutant: target gene deletion mutation primarily involves conferring antibiotic resistance to the mutant as a marker that is surrounded by resolvase recognition target sites and has a polar effect (Sassetti et al. 2003). Due to this antibiotic resistance in marked mutant, insertion of additional mutation becomes complicated; on the other hand unmarked mutant, also known as null or in-frame mutation, does not have involvement of any antibiotic resistance marker resulting in allowing the additional mutation. It requires target gene deletion in open reading frame of desired chromosome which leads to mutant phenotype (Singh et al. 1997).

13.4 Identification of Essential Genes Using Computational Methods

The computation method known as in silico analysis is a time-saving and easy approach to identify the essential genes as compared to the experimental approach which is time consuming and highly labor intensive. This computational approach is also useful for genome sequencing which is essential for genome-wide gene studies, although experimental techniques are being upgraded in identifying the essential genes. Lin and Zhang (2011) reported that only 15 bacterial genomes are available till now related to the genome-wide gene essentiality, while around 1400 genomes have been sequenced and ~4000 bacterial genomes are under construction or underway. So for the prediction of essential genes from the sequenced data, in silico approach is very important (Xiong et al. 2006).

A number of algorithm pipelines have been designed for essential gene identification; most of them are based on the genomic features such as growth-related sequenced data or any other feature which includes connectivity in evolutionary rate (Yaveroglu and Can 2009), Protein-protein interaction network, variation in mRNA expression, GC content, phylogenetic conservation, codon adaptation index (CAI), predicted subcellular localization and codon usages (Zhang and Ren 2015).

13.4.1 Subtractive Genomics Approach

Development of strain-specific drugs and effective therapies against pathogens is a serious challenge in the medical field. Subtractive genomics approach has been established as an application to evaluate the property of possible targets; this evaluation is performed using two ways: "essentiality" and "selectivity". It includes selection of the essential gene for replication, growth and development, capable or essential for the survival of the microorganism (Hosen et al. 2014). Essentiality of a gene refers to intolerance to mutational inactivation of cell, and its status is confirmed using restricted lethal mutants (Jordan et al. 2002). To identify the appropriate drug target, firstly cellular processes and involved essential genes unique to the parasite have to be identified; the rest of the genes that are nonhomologous can be removed to make the analysis easy (Forsyth et al. 2002). The technique has been used in a number of studies to predict putative drug targets in a number of bacterial species (Gautam et al. 2012), although few of these have been tested experimentally (Sarangi et al. 2009).

13.4.2 Essentiality-Based Mapping Approach

Mapping is an in silico approach to treat the target genes of microorganisms which depends on the accessibility and the similarity of the genes. Identification of the target should be assembled in the direction in which they can communicate individually to the microorganism and should have the potential to classify so that designing of pathogen-specific drug can be facilitated without making any damage to normal flora and without development of any undesired drug resistance (Xu et al. 2011). For the identification of pathogen-selective target genes, a "set of rules based" on the process has to be created so that, after an appropriate filtering, a set of the most important target genes can be selected (Holman et al. 2009). The selected gene of a desired organism can then be mapped to the essential gene of known organisms. This mapping has been done to identify or to know the important genes for growth and conserve the degenerate genome which indicates that nonessential functions have been selectively lost due to divergent bacterium. Therefore, conservation of the essential gene depends on the availability of the related genome; the resulting gene can then be later ranked by scoring system using BLAST, a homology search tool that presents genes related to the desired gene (Song and Ko 2008).

13.4.3 Target Identification Using Phylogenetic Profiling and In Silico Validation

Phylogenetic analysis is the study of ancestrally related sequences, which helps in profiling a set of gene or protein sequences of reference genomes to check the relation among them. Homology search between gene products is used to analyze the function of genes at the basis of similarity search. The greater similarity coverage is the likelihood of proteins to share membership in the same pathway or cellular system (Snitkin et al. 2006). This sharing of similarity with known proteins can help to assign a putative function to uncharacterized proteins. At genome-wide scale, comparative analysis of the entire genome has the power to reveal functional linkages known as "interactomes," elucidating both known and novel pathways and cellular systems (Pellegrini 2012).

Phylogenetic analysis has been established as a widely known technique for identifying the biological roles for unknown proteins. This algorithm capitalizes on the tendency of proteins to work together in the same cellular system and to travel together in evolutionary processes of speciation, lateral transfer, and gene loss and helps to build a phylogenetic tree or structure of related data sets to find out the specific gene or gene family (Ranea et al. 2007). This system may constitute of multi-subunit protein complexes, biochemical pathways, protein-modifying enzymes with their targets, etc. Taxonomic distribution of a given protein showed the reflected evolutionary history of its family (Psomopoulos et al. 2013). Therefore, analysis of the protein family abundant across large numbers of genomes may provide information about the protein given that, if one protein is linked to another protein, they may be involved in the same system (Mikkelsen et al. 2005). A phylogenetic analysis is performed using the simplest form of binary characters 1s or 0s which indicate or characterize the presence or absence of a protein or marker (a segment of DNA sequence with known location on the chromosome) across defined taxonomic groups (Basu et al. 2011). This phylogenetic profile generates a data set to identify the homology in order to come up with statistical evidence, and a score is generated for each protein sequence against other protein families in other species which represents the similarity level of a sequence (Plaimas et al. 2010).

This technique is based on some algorithms which help to find out the relationship among the species, but there are some boundaries in that the members of comparative data set should be functionally equivalent. Nonetheless, phylogenetic profiling can be very successful in identifying the ancestor relationship between protein families and their association with the function of the known protein; they likely cooperate in the same cellular processes (Freilich et al. 2009). For validation of the predicted function of an unknown such as biological roles or molecular functions, it needs to be combined with other types of evidence incorporated in various genome-scale data (Gaasterland and Ragan 1998).

Therefore, phylogenetic profiling serves as a computational tool for the annotation of uncharacterized genes by covering many functional modules. Genome availability of many organisms will support the discovery of novel pathways and help to understand the physiology of organisms (Tatusov et al. 1997). There are some well-characterized organisms, but their involvement in any of these pathways is still undiscovered, which could be important for the bacterium (Schmidt and Oliver 1989). Phylogenetic profiling approach can help to reveal the functional information of genes of these organisms which can help to assume their involvement in some pathways. This approach can also be used in target drug identification for pathogens, by selecting the responsible gene for the disease without affecting the host gene. As identification of the essential gene by excremental approach takes time, phylogenetic profiling is helpful to identify the probable genes; afterward, the identified gene can be tested and validated experimentally. An in silico approach is a computation methodology which is used to validate the experimentally derived data of essential genes by comparing with predicted data. There are many experiments that have been performed to identify the lethal genes that could not be validated (Rusmini et al. 2014). Many computational methods have been established to predict essential genes which include identification of orthologs, analysis of genomic inheritance, network analysis, ancestral gene conservation, and machine-learning-based integrative approaches. The most commonly used approach for the prediction of essential gene is phylogenetic conservation (Sakharkar et al. 2004).

13.5 Conclusions

Essential gene search has been a challenge for the researcher; it is defined as one whose loss is lethal under a certain environmental condition. The identification of essential genes in microorganisms guarantees to (i) distinguish crucial gene and pathways for controlling pathogenic microorganisms by recognizing potential target for antimicrobial drug development, (ii) uncover the negligible gene set for living organism and to reveal insight into the inception of life, and (iii) reveal the relationship among bacteria during evolution. A few vast mutant libraries of model organisms have been built. In spite of the fact that these libraries are precious for research in systems biology, they demonstrate conflicting essential gene results, notwithstanding for closely related strains. This absence of agreement has forestalled reliable prediction of essential gene or pathways in species that have not yet been inspected. The genome-scale explanation of essential gene has potential for advancement and accessibility of completed genome sequences and substantial scale gene-inactivation technologies, for example, transposon-based mutagenesis, target gene inactivation, and genetic footprinting. Essentiality base mapping, subtractive genomics, and phylogenetic profiling for target identification are in silico strategies and tools that have been developed for the analysis of essential genes. These bioinformatics tools can also be used along with experimental methods. Although, computationally generated data must be valid through experimental methodologies, however, there are some disadvantages of these approaches. In silico identification of essential gene is very important because excremental identification of essential gene is exceedingly labor intensive and time consuming; however, genome sequencing speed also so much outpaces those of other methods which is important for the genome-wide study of essential gene. With the expanding capacity for genome sequencing, the in silico prediction of essential genes is going to be a whole lot necessary.

Acknowledgment The authors are grateful to the Sam Higginbottom University of Agriculture, Technology and Sciences, Allahabad, for providing the facilities and support to complete the present research work.

References

Akerley BJ et al (1998) Systematic identification of essential genes by in vitro mariner mutagenesis. Proc Natl Acad Sci 95:8927–8932

Bardarov S et al (2002) Specialized transduction: an efficient method for generating marked and unmarked targeted gene disruptions in Mycobacterium tuberculosis, M. bovis BCG and M. smegmatis. Microbiology 148:3007–3017

Barrett AD, Stanberry LR (2009) Vaccines for biodefense and emerging and neglected diseases. Academic, Amsterdam

Basu MK et al (2011) ProPhylo: partial phylogenetic profiling to guide protein family construction and assignment of biological process. BMC Bioinformatics 12:1

Cooper I, Duffield M (2011) The in silico prediction of bacterial essential genes. In: Méndez-Vilas A (ed) Science against microbial pathogens: communicating current research and technological advances. FORMATEX Microbiology Series N° 3, vol 1. Formatex Research Center, Badajoz

Date SV, Marcotte EM (2003) Discovery of uncharacterized cellular systems by genome-wide analysis of functional linkages. Nat Biotechnol 21:1055–1062

Deng J et al (2010) Investigating the predictability of essential genes across distantly related organisms using an integrative approach. Nucleic Acids Res 39(3):795–807

Devine SE, Boeke JD (1994) Efficient integration of artificial transposons into plasmid targets in vitro: a useful tool for DNA mapping, sequencing and genetic analysis. Nucleic Acids Res 22:3765–3772

Fleischmann RD et al (1995) Whole-genome random sequencing and assembly of Haemophilus influenzae Rd. Science 269:496–512

Forsyth R et al (2002) A genome-wide strategy for the identification of essential genes in Staphylococcus aureus. Mol Microbiol 43:1387–1400

Freilich S et al (2009) Stratification of co-evolving genomic groups using ranked phylogenetic profiles. BMC Bioinformatics 10:1

Frøkjær-Jensen C et al (2010) Targeted gene deletions in C. elegans using transposon excision. Nat Methods 7:451–453

Gaasterland T, Ragan MA (1998) Microbial genescapes: phyletic and functional patterns of ORF distribution among prokaryotes. Microb Comp Genomics 3:199–217

Gautam B et al (2012) Metabolic pathway analysis and molecular docking analysis for identification of putative drug targets in Toxoplasma gondii: novel approach. Bioinformationtics 8:134–141

Gil R et al (2004) Determination of the core of a minimal bacterial gene set. Microbiol Mol Biol Rev 68:518–537

Grünblatt E et al (2014) Imaging genetics in obsessive-compulsive disorder: linking genetic variations to alterations in neuroimaging. Prog Neurobiol 121:114–124

Hayes F (2003) Transposon-based strategies for microbial functional genomics and proteomics. Annu Rev Genet 37:3–29

Holman AG et al (2009) Computational prediction of essential genes in an unculturable endosymbiotic bacterium, Wolbachia of Brugia malayi. BMC Microbiol 9:1
Hosen MI et al (2014) Application of a subtractive genomics approach for in silico identification and characterization of novel drug targets in Mycobacterium tuberculosis F11. Interdiscip Sci Comput Life Sci 6:48–56
Huynen MA, Bork P (1998) Measuring genome evolution. Proc Natl Acad Sci 95:5849–5856
Ishikawa M, Hori K (2013) A new simple method for introducing an unmarked mutation into a large gene of non-competent Gram-negative bacteria by FLP/FRT recombination. BMC Microbiol 13:86
Ivics Z et al (2009) Transposon-mediated genome manipulation in vertebrates. Nat Methods 6:415–422
Jimenez-Ruiz E et al (2014) Advantages and disadvantages of conditional systems for characterization of essential genes in Toxoplasma gondii. Parasitology 141:1390–1398
Jordan IK et al (2002) Essential genes are more evolutionarily conserved than are nonessential genes in bacteria. Genome Res 12:962–968
Joshi PB et al (2002) Targeted gene deletion in Leishmania major identifies leishmanolysin (GP63) as a virulence factor. Mol Biochem Parasitol 120:33–40
Jothi R et al (2007) Discovering functional linkages and uncharacterized cellular pathways using phylogenetic profile comparisons: a comprehensive assessment. BMC Bioinformatics 8:1
Judson N, Mekalanos JJ (2000) Transposon-based approaches to identify essential bacterial genes. Trends Microbiol 8:521–526
Juhas M et al (2012) High confidence prediction of essential genes in Burkholderia cenocepacia. PLoS One 7:e40064
Kang DC, Fisher PB (2005) Complete open reading frame (C-ORF) technology: simple and efficient technique for cloning full-length protein-coding sequences. Gene 353:1–7
Kensche PR et al (2008) Practical and theoretical advances in predicting the function of a protein by its phylogenetic distribution. J R Soc Interface 5:151–170
Kleckner N (1981) Transposable elements in prokaryotes. Annu Rev Genet 15:341–404
Koonin EV (2000) How many genes can make a cell: the minimal-gene-Set concept 1. Annu Rev Genomics Hum Genet 1:99–116
Langille MG et al (2013) Predictive functional profiling of microbial communities using 16S rRNA marker gene sequences. Nat Biotechnol 31:814–821
Lehoux DE et al (2001) Discovering essential and infection-related genes. Curr Opin Microbiol 4:515–519
Lin Y, Zhang RR (2011) Putative essential and core-essential genes in Mycoplasma genomes. Sci Rep 1
Luisi PL et al (2002) The notion of a DNA minimal cell: a general discourse and some guidelines for an experimental approach. Helv Chim Acta 85:1759–1777
Luo H et al (2013) DEG 10, an update of the database of essential genes that includes both protein-coding genes and noncoding genomic elements. Nucleic Acids Res 42(Database issue):D574–D580
Mikkelsen TS et al (2005) Improving genome annotations using phylogenetic profile anomaly detection. Bioinformatics 21:464–470
O'sullivan GJ et al (2006) Potential and limitations of genetic manipulation in animals. Drug Discov Today Technol 3:173–180
Pellegrini M (2012) Using phylogenetic profiles to predict functional relationships. Bacterial Mol Netw Methods Protoc 804:167–177
Plaimas K et al (2010) Identifying essential genes in bacterial metabolic networks with machine learning methods. BMC Syst Biol 4:1
Psomopoulos FE et al (2013) Detection of genomic idiosyncrasies using fuzzy phylogenetic profiles. PLoS One 8:e52854
Ranea JA et al (2007) Predicting protein function with hierarchical phylogenetic profiles: the Gene3D Phylo-Tuner method applied to eukaryotic genomes. PLoS Comput Biol 3:e237

Rusmini R et al (2014) A shotgun antisense approach to the identification of novel essential genes in Pseudomonas aeruginosa. BMC Microbiol 14:1
Sakharkar KR et al (2004) A novel genomics approach for the identification of drug targets in pathogens, with special reference to Pseudomonas aeruginosa. In Silico Biol 4:355–360
Salama NR et al (2004) Global transposon mutagenesis and essential gene analysis of Helicobacter pylori. J Bacteriol 186:7926–7935
Sarangi AN et al (2009) Subtractive genomics approach for in silico identification and characterization of novel drug targets in Neisseria meningitidis serogroup B. J Comput Sci Syst Biol 2:255–258
Sassetti CM et al (2003) Genes required for mycobacterial growth defined by high density mutagenesis. Mol Microbiol 48:77–84
Schmidt M, Oliver D (1989) SecA protein autogenously represses its own translation during normal protein secretion in Escherichia coli. J Bacteriol 171:643–649
Singh IR et al (1997) High-resolution functional mapping of a cloned gene by genetic footprinting. Proc Natl Acad Sci 94:1304–1309
Smith V et al (1995) Genetic footprinting: a genomic strategy for determining a gene's function given its sequence. Proc Natl Acad Sci 92:6479–6483
Snitkin ES et al (2006) Comparative assessment of performance and genome dependence among phylogenetic profiling methods. BMC Bioinformatics 7:420
Song JH, Ko KS (2008) Detection of essential genes in Streptococcus pneumoniae using bioinformatics and allelic replacement mutagenesis. Microb Gene Essentiality Protoc Bioinformatics 416:401–408
Tatusov RL et al (1997) A genomic perspective on protein families. Science 278:631–637
Touchon M et al (2009) Organised genome dynamics in the Escherichia coli species results in highly diverse adaptive paths. PLoS Genet 5:e1000344
Wong SM, Mekalanos JJ (2000) Genetic footprinting with mariner-based transposition in Pseudomonas aeruginosa. Proc Natl Acad Sci 97:10191–10196
Xiong J et al (2006) Genome wide prediction of protein function via a generic knowledge discovery approach based on evidence integration. BMC Bioinformatics 7:268
Xu P et al (2011) Genome-wide essential gene identification in Streptococcus sanguinis. Sci Rep 1:125
Yaveroglu ON, Can T (2009) Predicting protein-protein interactions from protein sequences using phylogenetic profiles. Int J Comput Electr Autom Control Inf Eng 3:1971–1977
Zhang Z, Ren Q (2015) Why are essential genes essential? – the essentiality of Saccharomyces genes. Microbial Cell 2:280–287

Disease Informatics

14

Sayak Ganguli and Abhijit Datta

Abstract

The field of disease informatics around the world has focused on the application of information technology to understand and prevent disease outbreaks. The recent *Ebola* outbreak has again pointed out our deficiencies in the proper management and vaccination practices around the world in situations of a pandemic disease. However analyses at the molecular level targeting the proteins and other immune system components provide us with the opportunity to identify effective targets for small molecule-based targeting as well as to understand the biology of the disease. Several therapeutic approaches around the world are being explored. Starting from traditional chemical medicine to ethnomedicinal practices, small molecule compound libraries are being screened using virtual screening procedures for the quest to identify and predict lead molecules of the future having limited side effects but increased efficacy. Apart from small molecule-based therapeutic strategies, oligonucleotide- and aptamer-based strategies are also being explored which enables us to directly interfere with the genome function of a particular pathogen. Combinatorial libraries and high-throughput practices such as next-generation sequencing have also accelerated the discovery of information and genetic medicine or personalized medicine which looked like a distant dream a few years back but is gradually transforming into reality. In this era of information generation, at the big data level, it is imperative that informatics-based strategies be explored and utilized fully so as to manage and analyze information in real time. Bioinformatics and clinical informatics approaches are continuously being utilized for providing patient health-care support, and scientists around the globe are working round the

S. Ganguli
Theoretical and Computational Biology Division, Amplicon Institute of Interdisciplinary Science and Technology, Palta, West Bengal, India

A. Datta (✉)
Department of Botany, Jhargram Raj College, Medinipur, West Bengal, India

G. Wadhwa et al. (eds.), *Current trends in Bioinformatics: An Insight*,
https://doi.org/10.1007/978-981-10-7483-7_14

clock to tackle diseases from the epidemiological, molecular, and post-medical phases. Basic science research investigating the biology of the diseases is also being funded since it forms the stepping stones on which the entire discipline of disease informatics stands firm. This chapter deals with three different diseases (Parkinson's disease, influenza, and AIDS (caused by HIV 1)) and how various bioinformatics approaches help us to understand the biology of the disease and its effects. Each case study provides a putative translational output of the disease management which can be utilized by researchers in the clinical trial phase.

Keywords
Diseases · microRNAs · Ebola · Parkinson's Disease · Natural Products

14.1 Introduction

A major conglomerative approach toward the management, i.e., to prevent, detect, respond, and manage infectious disease outbreaks of plants, animals, and humans, requires the role of information systems (Damianos et al. 2002; Buehler et al. 2004). Currently, various laboratories, health-care providers, and government agencies at local, state, national, and international levels have realized the need for accumulating data and development of integrated management systems which provide access to data, perform analytics, as well as provide predictions to such diseases. Prime examples of such an effort are the US Department of Agriculture's (USDA) effort toward certain animal diseases (e.g., mad cow disease and foot-and-mouth disease) and the US Geological Survey's (USGS) enterprise toward the management of databases for wildlife diseases through its National Wildlife Health Center (NWHC) and numerous partners. A wide array of analytical tools has been developed by researchers and public agencies working on health issues for proper prediction of outbreaks as well as general surveillance. For instance, such models have been employed to predict outbreaks of the West Nile virus (WNV) (Eidson 2001; Eidson et al. 2001; Julian et al. 2002; Guptill et al. 2003) and of influenza (Hyman and LaForce 2003).

14.2 Bioinformatics Approaches to Deal with Disease Information

The main aim of bioinformatics is to integrate information which remains scattered around the World Wide Web and in various niches of bibliographic information systems. This is solved by the development of databases and data warehouses which may contain specific information regarding a particular disease or may be a collection of multiple disease phenotypes as well as the causal changes at the genome level.

Table 14.1 New disease-related databases

Database name	URL	Brief description
ArrayMap	http://www.arraymap.org	Gene copy number profiling in human cancers
GRASP	http://apps.nhlbi.nih.gov/grasp/	GWAS results
sc-PDB	http://bioinfo-pharma.u-strasbg.fr/scPDB/	Potential drug-binding sites in protein structures from the PDB
ADReCS	http://bioinf.xmu.edu.cn/ADReCS	Adverse drug reaction classification system
AHTPdb	http://crdd.osdd.net/raghava/ahtpdb/	Antihypertensive peptides database
BCCTBbpa	http://bioinformatics.breastcancertissuebank.org/	Breast cancer campaign tissue bank bioinformatics portal
Cancer3D	http://www.cancer3d.org	Mapping of cancer mutations to protein structures
CancerPPD	http://crdd.osdd.net/raghava/cancerppd/	Experimentally validated anticancer peptides
Candidate cancer gene database	http://ccgd-starrlab.oit.umn.edu/	Cancer genes identified in transposon-based genetic screens in mice
CMPD	http://cgbc.cgu.edu.tw/hmpd	Cancer mutant proteome database
DDMGD	http://www.cbrc.kaust.edu.sa/ddmgd/	Associations between gene methylation and disease
EHFPI	http://biotech.bmi.ac.cn/ehfpi/	Essential host factors for pathogenic infection
EpilepsyGene	http://122.228.158.106/EpilepsyGene	Genes and mutations related to epilepsy
MethHC	http://MethHC.mbc.nctu.edu.tw	DNA methylation in human cancer
MoonProt	http://www.moonlightingproteins.org/	Moonlighting proteins
Organ System Heterogeneity DB	http://mips.helmholtz-muenchen.de/Organ_System_Heterogeneity/	Phenotypic effects of diseases and drugs on different organs
Platinum	http://structure.bioc.cam.ac.uk/platinum	Experimentally measured effects of mutations on protein-ligand complexes
PubAngioGen	http://www.megabionet.org/aspd	Public angiogenesis research portal
PyIgClassify	http://dunbrack2.fccc.edu/PyIgClassify/default.aspx	Clusters of conformations of antibody complementarity-determining regions
VADE	http://bmi-tokai.jp/VaDE/	VarySysDB disease edition: disease-associated genomic polymorphisms
ViRBase	http://www.rna-society.org/vhncrnadb/	Virus-host interaction-associated ncRNAs

Adapted from Galperin et al. (2015)

Every year the Nucleic Acids Research database issue has a collection of the new databases that became available for public usage on the web, and many of them are disease-related ones (Table 14.1).

In this era of big data, numerous high-throughput methods are continuously contributing to the data deluge along with the numerous ongoing and completed genome projects. Thus now not only are the researchers faced with the problem of understanding the complexity of the disease, but also they are now in a quandary to properly streamline the available data. This not only initiates the need for proper ontology development but also necessitates the automation of the analytical platforms.

Many genes and pathways are involved in complex diseases such as Parkinson's disease, and many high-throughput experimental approaches are generally applied for the analyses of these complex disease profiles. Bioinformatics approaches in combination with systems biology approaches need to be utilized to explore the molecular mechanisms of complex diseases (cardiovascular diseases, cancers, and mental disorders). This includes the integration of relevant multiple data sources, derived database construction, if possible candidate gene selection, and finally network and pathway analysis of the disease. The main focus should be on integrating information on dense overlapping regulons, protein interactions with ligands, proteins and DNA/RNA, and mapping of pathways for complex diseases such that they can be utilized for the prediction of candidate biomarkers and aid in genome-wide association studies.

Two essential and conserved regulators of gene expression are transcription factors and microRNAs, and data suggest that that there is coregulation of these transcription factors and microRNAs as well as other noncoding RNAs which may have implications in disease phenotypes. Data mining presents another important challenge in bioinformatics, and real-time data mining platforms are the order of the day to deal with the fast-changing landscape of integrated data. High-throughput sequencing experiments and increase in analyses efficiency of automated pipelines have contributed toward the increase in interests toward feature analysis, evolution, and comparative genomics of specific gene and gene families (Fig. 14.1).

14.3 Combating Parkinson's Disease

Selective demise of the neurons of the substantia nigra pars compacta is one of the key characteristics of Parkinson's disease. This phenomenon leads to progressive motor dysfunction. Levy bodies which are nothing but cytoplasmic ubiquitinated protein deposits accumulate inside the neuron. Often these inclusions are thread-like and are referred to as Lewy neurites.

The accumulation of these deposits causes cell death.

The major component of these proteinaceous deposits has been identified to be a 140-residue protein α-synuclein (aS) through extensive biochemical analyses (Cordato and Chan 2004).

The protein synuclein is a vertebrate-specific protein which is coded by three genes. They belong to a closely knit cluster of presynaptic protein family. They are

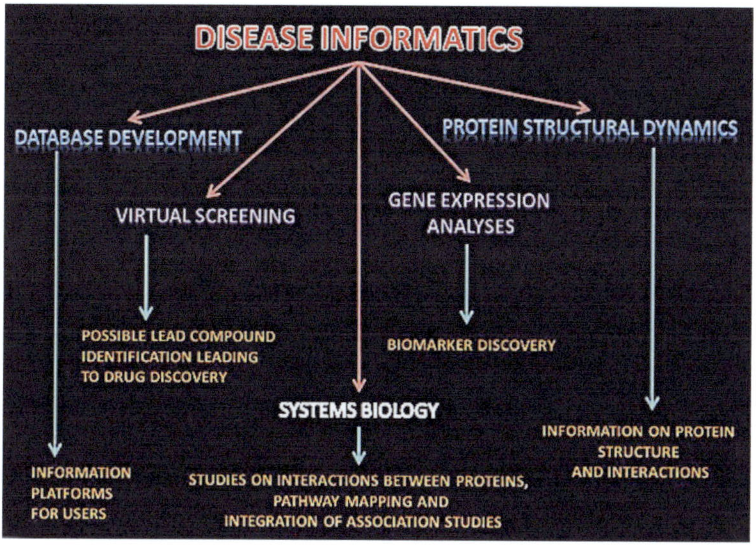

Fig. 14.1 Bioinformatics approaches to deal with disease-related data

small and abundantly found in neural tissues. One of the essential features of synuclein is the presence of acidic stretches within the COOH terminal region and a repetitive degenerative motif KTKEGV spread over the first 87 residues. The family of synuclein proteins includes α-synuclein, which was also called non-amyloid component precursor protein, or synelfin; β-synuclein, also referred to as phospho-neuro protein 14 (Dev et al. 2003; Diamandis et al. 2000); and γ-synuclein, also known as breast cancer-specific gene 1 or persyn. Furthermore, gene triplication event of alpha synuclein along with several candidate mutations in the protein such as A30P, E46K, and A53T has been directly correlated to familial PD. Misfiling leading to aggregation of synuclein is a very common indicator of PD, and this is progressive with age (Doss-Pepe et al. 2005).

This work focuses on the identification of functional partners of alpha synuclein (Fig. 14.2) and screening an inhibitor ligand capable of specific binding with the non-amyloid plaque region of the molecule which is responsible for aggregation into Lewy bodies.

The current study demonstrates the ubiquitous presence of α-synuclein in mammalian members indicative of orthologue distribution; the associated interacting proteins of α-synuclein however showed differential occurrence in members having α-synuclein orthologues (Fig. 14.3) but were found to be strictly coherent with organismic complexity. The identified ligand pyrroloquinoline quinone should serve as an effective inhibitor for plaque formation by α-synuclein and thus aid in future Parkinson disease therapy.

Fig. 14.2 Interacting partners of alpha synuclein

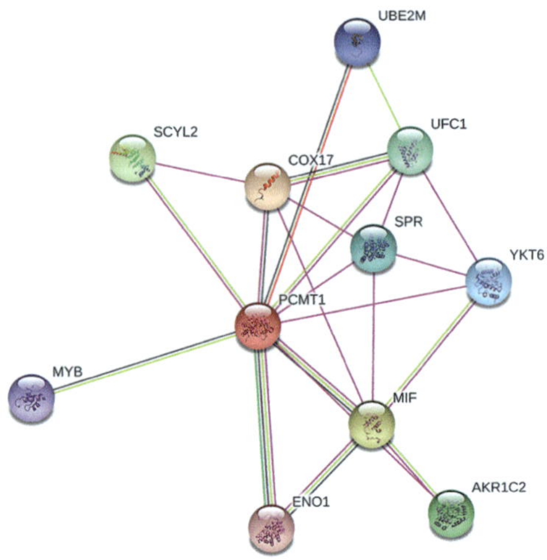

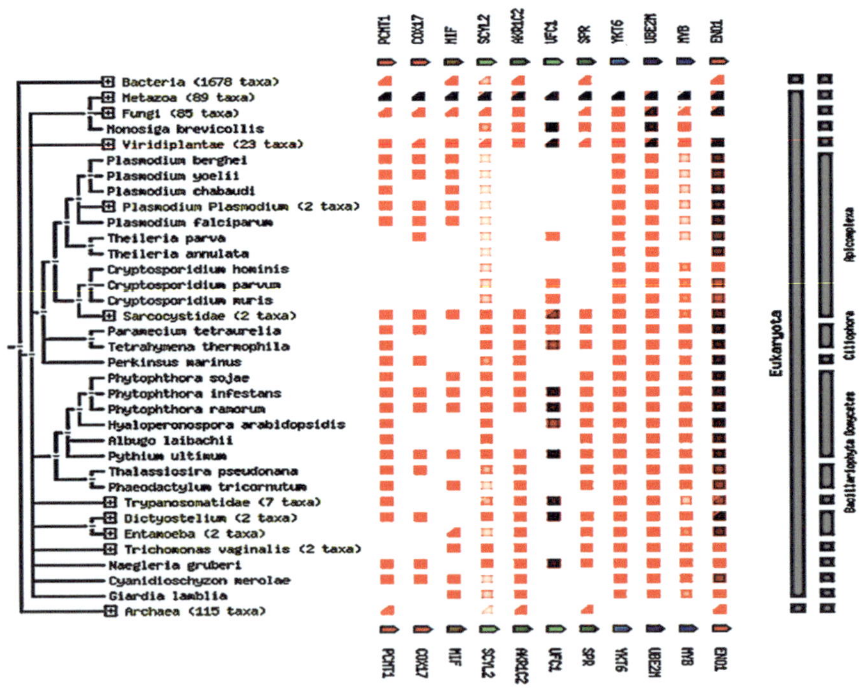

Fig. 14.3 Phylogenetic occurrence of alpha synuclein

14.4 Analysis of Parkin

Mutation in the parkin gene causes autosomal recessive juvenile Parkinsonism. This highlights that ubiquitin-mediated proteolysis may play an important role in the pathobiology of PD. The wild and mutant forms of parkin (R42P, R42H, R42C etc.) protein were used in this study, and the interacting proteins were studied for their binding efficiencies with parkin. Then results of mutation, common ligand-binding sites, and docking were then correlated with each other. Phosphorylation sites and the accessible surface area of the proteins were also compared.

The study comprehensively analyzes the structural variations that are arising from the various mutations reported in literature in the parkin protein. The results obtained show that the mutations do not impart any notable structural change in the protein nor do they have any possible impact on the surface properties.

Ligand and substrate interactions with the normal and mutant proteins and the most common substrates show that there are no significant changes in the docking scores. However as parkin mutations have been reported to be involved in familial Parkinsonism, our study does not reveal any possible alteration in binding of substrate moieties (Fig. 14.4). Only in case of CDCrel-1, the actual sites of binding with the normal were found to be different with that of the mutants, and there was distinct loss of hydrogen bonding among the residues (Arg 42 and Asp 114).

This further proves that the reported mutations (Arg42Pro, Arg42His, and Arg42Cys) in parkin disrupts its progressive binding with CDCrel and leads to a loss of function (Fig. 14.5).

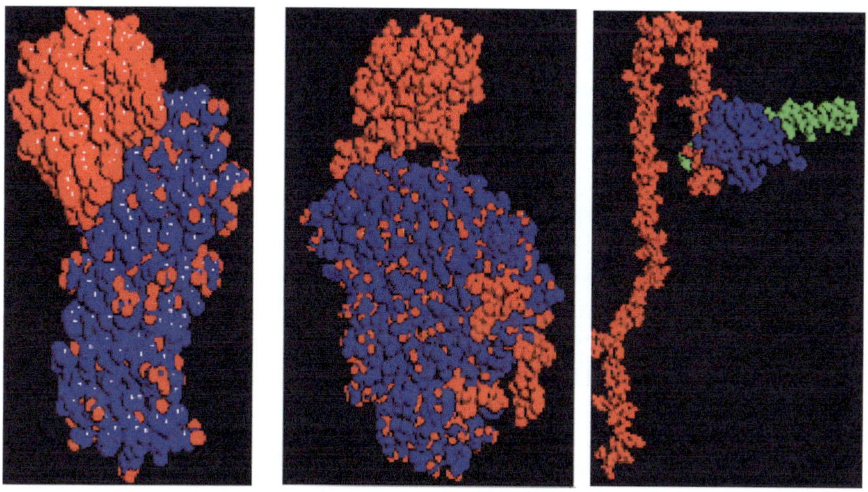

Fig. 14.4 Docking interactions of the different proteins with parkin

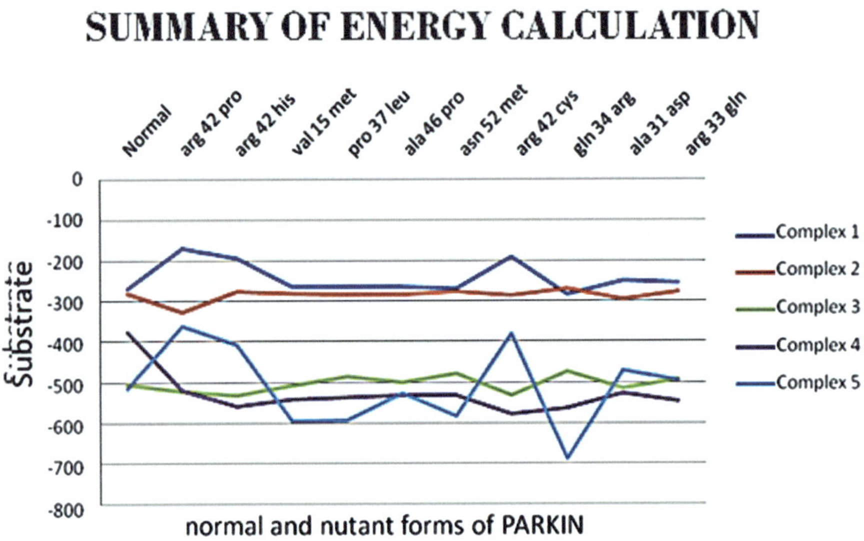

Fig. 14.5 Energy calculations of the different interacting partners

14.5 Analyses of UCHL1

After getting the crystallographic structure of UCHL-1 from PDB, the mutation I93M was done in silico. This specific mutation was done according to the established fact that this mutation shifts the hydrolase activity of UCHL-1 to ligase activity which ultimately hampers the ubiquitin proteasome system (UPS). Then the comparative in silico studies (position-wise and distance-wise) of probable phosphorylated residues were done. Structures of probable substrates were obtained either from pdb or by homology modeling starting with raw sequence. Those structures were then used as ligands, and structure of UCHL-1 was used as receptor for docking. After docking (both with wild-type and mutant UCHL-1), energy parameters of shape, force, and total energy for each probable substrate were compared (Figs. 14.6 and 14.7).

No relative change of position or distance of probable phosphorylated residues were observed after comparing wild-type and mutant UCHL-1.Comparative docking result showed some relative changes of energy values between wild-type and mutant UCHL-1 which could be significant for further studies.

14.6 Analysis of PINK 1

The study focused on the analysis of structural complexity of the protein PINK 1 which has been reported to be associated with Parkinson's disease. Three identified mutations were tested to locate the phosphorylation sites which were then compared

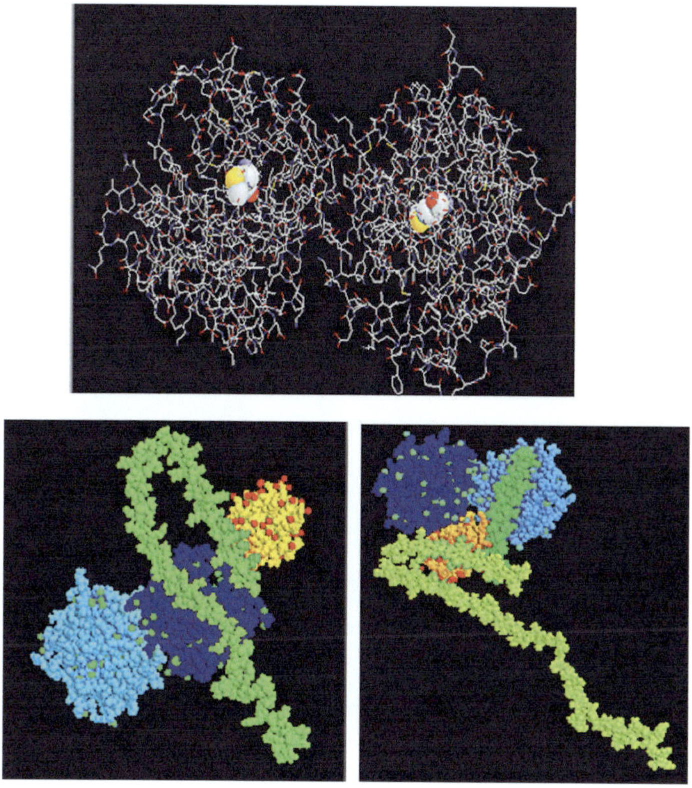

Fig. 14.6 Docking interactions of the different proteins with UCHL1

with those of the normal PINK 1. However, no detectable differences were identified among the members. When protein-protein interactions were studied then, it was found that the mutants showed greater binding energies than the wild type thus leading to the conclusion that mutations affected the stability of the complex.

14.7 Fighting Influenza

The development of oseltamivir resistance was long predicted and now clinically proven as the virus had for long showed signs of bypassing certain structural attributes of oseltamivir binding by causing site-altering mutations. These mutations have allowed neuraminidase to function resulting in the spread of resistant viruses which both survive and propagate. The spread of the virus is accentuated by the fact that neuraminidase is responsible for the release of these new virus particles. An inhibitor against neuraminidase should be able to mimic the sialic acid receptors which serve as natural targets of the neuraminidase. A potent inhibitor should have

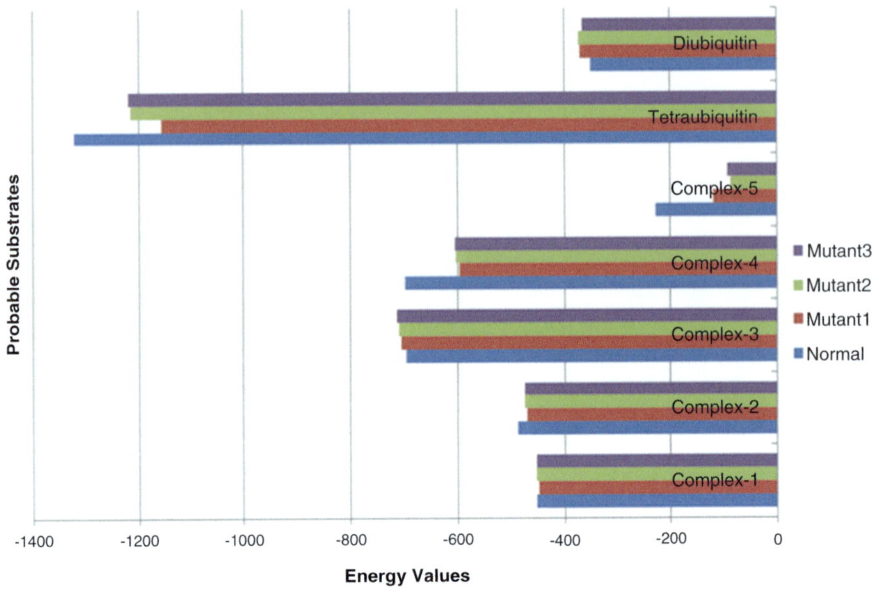

Fig. 14.7 Energy calculations of the different interacting partners of UCHL1

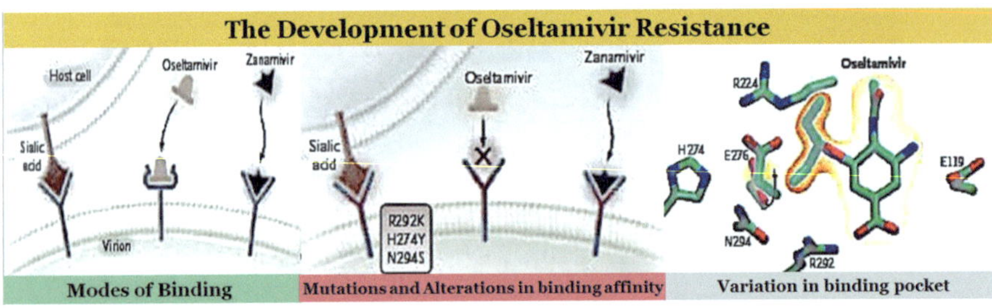

Fig. 14.8 The neuraminidase active site is dynamic as it interacts with oseltamivir by creating a pocket, though it can directly interact with zanamivir. This pocket formation is eliminated by the various reported mutations. The rotation of E276 and bonding of the amino acid to R224 results in pocket formation and is prevented by R292K, N294S, and H274Y mutations. These confer resistance to oseltamivir. Further an E119V mutation allows a water molecule to bind to the space that is created by the structurally small valine which in turn interferes with oseltamivir binding

the property of interacting with the neuraminidase active site and prevent any further interaction of the protein (Fig. 14.8).

Thus the requirement for better and effective lead compound becomes a necessity to combat this pandemic disease. If we go through the practices of drug discovery and design over the years, we find that drugs were discovered and formulated using long-drawn processes involving several biochemical and chemical experiments leading to synthesis and screening against numerous biological screens. Promising

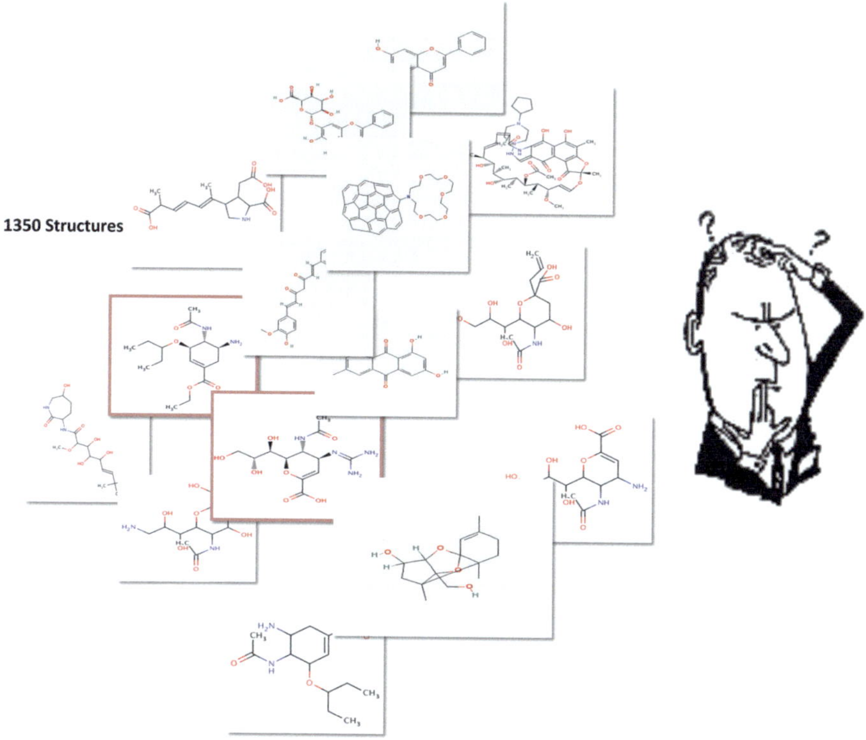

Fig. 14.9 The initial pool of natural compounds

compounds were then investigated further for their ADMET properties. Computational screening has made this entire process cost-effective and faster.

Here we study the most talked-about herbal lead compounds and their potential binding affinity to the effector molecules of major disease-causing agents, H5N1 and H1N1(neuraminidase) (Fig. 14.9). The work also encompasses the screening of the nanoparticle compound fullerene which has been reported to have anti-HIV activity. Further studies were also performed with telomerase which has been the target of numerous anticancer experiments.

Initially a large pool of compounds were assembled which according to medicinal literature showed effectivity against flu. They were screened using the following steps to eventually yield the final working set:

- Search algorithm for new inhibitors.
- Starting with known lead compounds, a database is searched to create a pool of putative drugs (Thimm et al. 2004).
- These compounds are compared to known inhibitors and noninhibitors, and drugs with similarities to inactive structures are removed from the list of proposed inhibitors (Chen and Reynolds 2002).

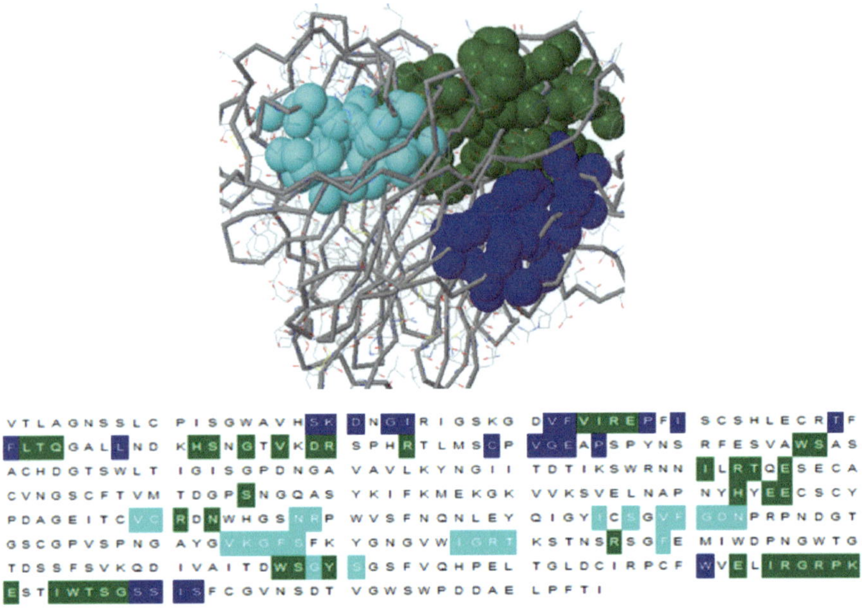

Fig. 14.10 Ligand-binding pockets in the neuraminidase of H5N1 and H1N1 subtypes

- Combining structural features of ineffective substances with property filtering rules allows the exclusion of further candidates. Drugs surpassing this sieve are proposed as new inhibitors (Lipinski 2000).

The results revealed that most herbal lead compounds screened (Yim et al. 1999; Liontas and Yeger 2004; Cui et al. 2006) were effective targets against H5N1 neuraminidase; namely baicalein (Gao et al. 1996; Matsuzaki et al. 1996) was found to be an effective target for inhibiting the neuraminidase of H1N1 virus by binding to its most active pocket. Telomerase was found to be effectively bound by curcumin (Youssef and Sherbeny 2005) at the RNA-binding interface. The study with nanoparticle fullerene (Bakry et al. 2007) also showed that it has binding affinity with the catalytic core of the protein of H1N1 but not H5N1 and thus should serve as effective delivery system for small molecule drugs in cases of H1N1 infection (Figs. 14.10 and 14.11).

Our study confirms that in silico drug screening is an effective alternative for identification of lead compounds. Several natural lead compounds were identified and tested using molecular docking for their effectiveness against major molecules of interest for diseases such as avian and swine influenza and cancer. Baicalein, emodin, and resveratrol were identified to be effective lead compounds that had the ability to bind to neuraminidase. Their binding energies were also found to be lower than most approved drugs.

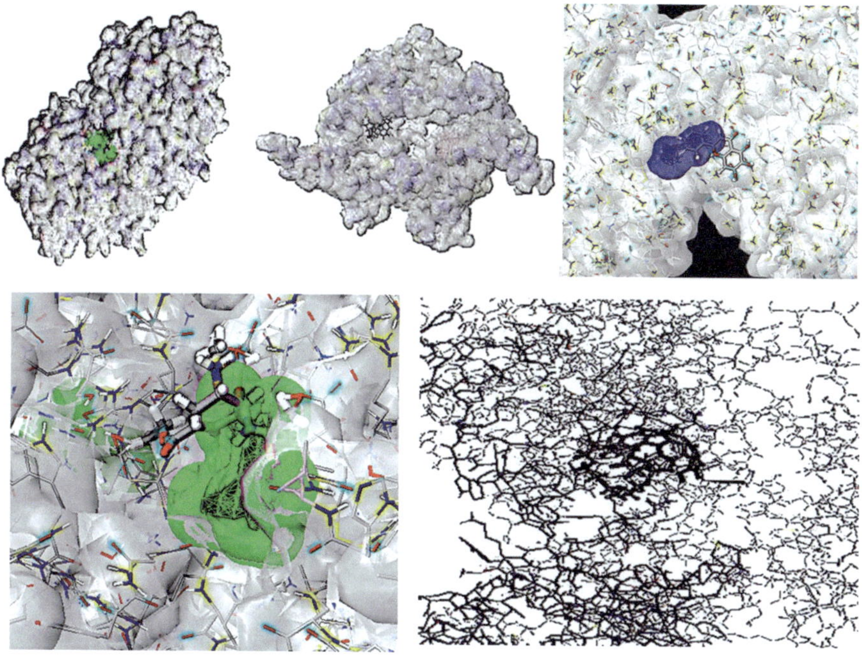

Fig. 14.11 Interactions of fullerene with neuraminidase (H5N1 and H1N1) and telomerase

14.8 Combating HIV 1 by Targeting RNA Regulatory Elements in the HIV 1 Genome

AIDS has spread over the years throughout the world and probably through the routes of human migration and settlement. The genome of HIV-1 has also been studied in detail to discover the chink in the armor of this deadly pathogen. The genome-encoded proteins have also been isolated, recombined, and expressed in various systems to study the possible kinetics and discover routes of inhibition. Nucleoside and non-nucleoside reverse transcriptase inhibitors were the first line of anti-HIV drugs that were produced in the market, but as our understanding of the complex genome has increased, the therapeutic strategies have evolved along with the virus virulence. Thus currently HAART utilizes combinatorial therapy to affected individuals. Still the element of control is lacking as is evident from the persistent number of cases of HIV-1 being reported from around the world. The virus is a very smart organism which keeps on mutating key residues in its genome rendering the drug molecules ineffective thus the premise of the work is justified since the targets are genomic landmarks in the RNA genome mainly RNA secondary structural elements which influence the processes of translation or any other form of gene expression by binding to host cell proteins. Inhibition of these elements

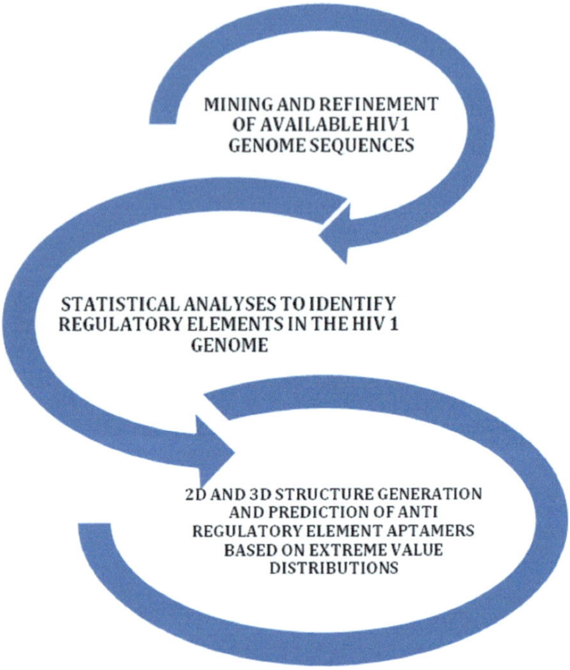

Fig. 14.12 The design cycle of for SECIS element detection and designing of anti-SECIS aptamers

using aptamers can pave the way for next-generation therapeutics based on RNA interference (Banik et al. 2013).

The identification and prediction of these regulatory elements were done using a modified Bayesian analyses (Roy et al. 2011) (Fig. 14.12).

Once the regulatory motifs have been identified, then their secondary structures would be obtained using a modified Zuker's algorithm which when fed into a symbolic programming-based module would result in the generation of three-dimensional structures suitable for performing molecular dynamic simulations. The anti-regulatory aptamers would be designed using an algorithm which focuses on base pairing (Ganguli et al. 2011) (Fig. 14.13).

14.9 Conclusion

The use of bioinformatics for the analyses and integration of disease-related information is steadily increasing. Efforts are on to create a virtual cell which would enable us to understand and visualize the nature of interactions that takes place in real time. Mathematical and statistical validations of established models as well as generation of robust algorithms which would associate all influential parameters affecting a disease phenotype are the major challenges that bioinformaticians face in the years to come.

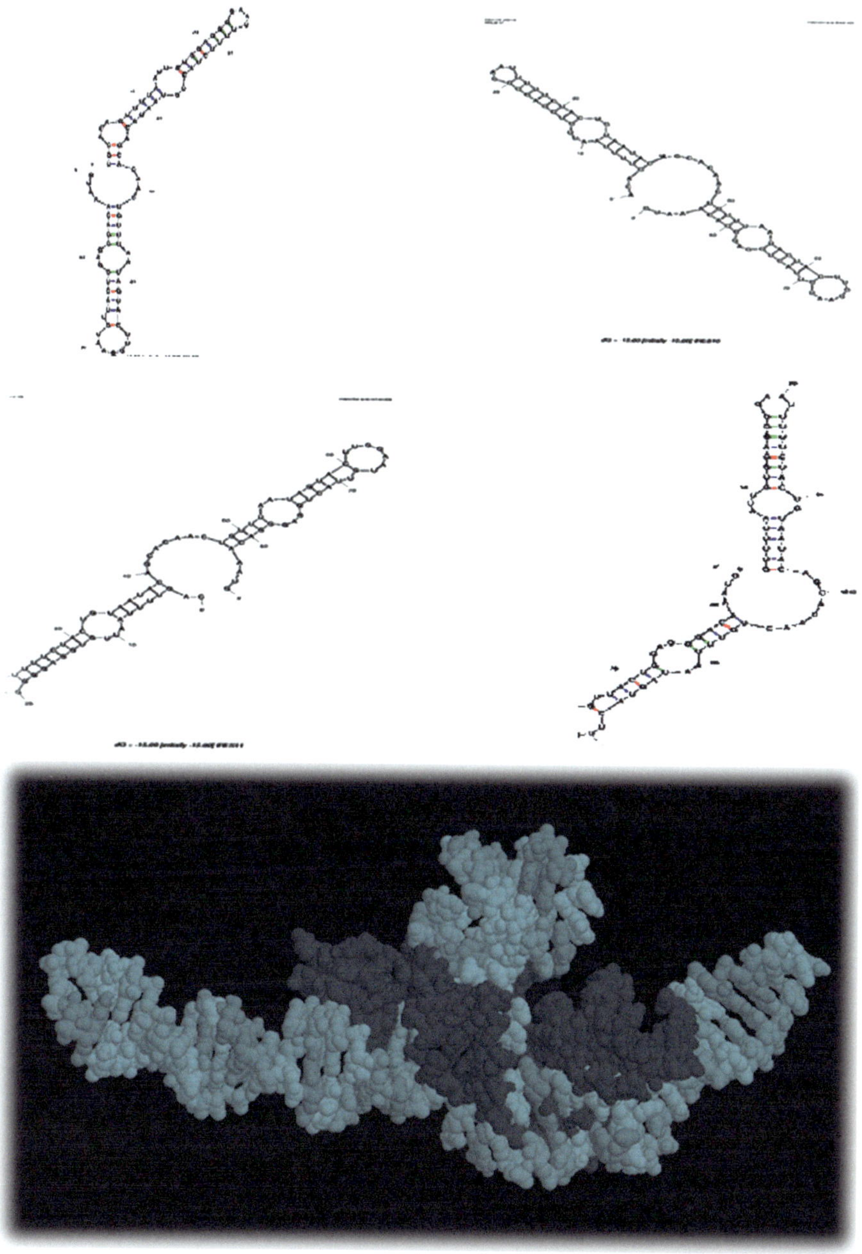

Fig. 14.13 Results obtained following analyses. (**a**) Secondary structures of identified SECIS elements in the HIV-1 genome. (**b**) Interactions of modeled SECIS element and human selenocysteine tRNA

References

Bakry R et al (2007) Medicinal applications of fullerenes. Int J Neuromed 2(4):639–649

Banik R, Ganguli S, Datta A (2013) HIV-1 genome analyses reveals conserved Musashi binding elements (MBE) – possible roles in glioblastoma multiforme. Int J Comput Bioinforma In Silico Model 2(6):293–296

Buehler J et al (2004) Framework for evaluating public health surveillance systems for early detection of outbreaks: recommendations from the CDC Working Group. Morb Mortal Wkly Rep 53(RR-5):1–13

Chen X, Reynolds CH (2002) Performance of similarity measures in 2D fragment-based similarity searching: comparison of structural descriptors and similarity coefficients. J Chem Inf Comput Sci 42:1407–1414

Cordato DJ, Chan DK (2004) Genetics and Parkinson's disease. J Clin Neuro Sci 2:119–123

Cui SX et al (2006) Curcumin inhibits telomerase activity in human cancer cell lines. Int J Mol Med 18:227–231

Damianos L et al (2002) MiTAP for bio-security: a case study. AI Mag 23(4):13–29

Dev KK, Hofele K, Barbieri S, Buchman VL, van der Putten H (2003) Part II: alpha-synuclein and its molecular pathophysiological role in neurodegenerative disease. Neuropharmacology 45:14–44

Diamandis EP, Yousef GM, Luo LY, Magklara A, Obiezu CV (2000) The new human kallikrein gene family: implications in carcinogenesis. Trends Endocrinol Metab 11:54–60

Doss-Pepe EW, Chen L, Madura K (2005) Alpha-synuclein and parkin contribute to the assembly of ubiquitin lysine 63-linked multi ubiquitin chains. J Biol Chem 17:16619–16624

Eidson M (2001) Neon needles in a haystack: the advantages of passive surveillance for West Nile virus. In: White DJ, Morse DL (eds) West Nile virus: detection, surveillance, and control. New York Academy of Sciences, New York, pp 38–53

Eidson M et al (2001) Dead crow densities and human cases of West Nile Virus, New York State, 2000. Emerg Infect Dis 7:662.664

Galperin YM et al (2015) The 2015 nucleic acids research database issue and molecular biology database collection. Nucl Acids Res 43(Database issue):D1–D5

Ganguli S, Mitra S, Datta A (2011) Antagomirbase: a putative antagomir database. Bioinformation 7(1):41–43

Gao D, Sakurai K, Katoh M (1996) Inhibition of microsomal lipid peroxidation by baicalein: a possible formation of an iron-baicalein complex. Biochem Mol Biol Int 39:215

Guptill SC et al (2003) Early-season avian deaths from West Nile virus as warnings of human infection. Emerg Infect Dis 9:483.484

Hyman J, LaForce T (2003) Modeling the spread of influenza among cities. In: Banks H, Castillo-Chavez C (eds) Bioterrorism: Mathematical modeling applications in homeland security Chapter 10. Society for Industrial and Applied Mathematics, p 211–236

Julian KG et al (2002) Early season crow mortality as a sentinel for West Nile virus disease in humans, northeastern United States. Vector Borne Zoonotic Dis 2:145–155

Liontas A, Yeger H (2004) Curcumin and resveratrol induce apoptosis and nuclear translocation and activation of p53 in human neuroblastoma. Anticancer Res 24:987–998

Lipinski CA (2000) Drug-like properties and the causes of poor solubility and poor permeability. J Pharmacol Toxicol Methods 44:235–249

Matsuzaki Y et al (1996) Cell death induced by baicalein in human hepatocellular carcinoma cell lines. Jpn J Cancer Res 87:170–177

Roy P et al (2011) Structural analysis of predicted hiv-1 SECIS elements. World J AIDS 1:208–218

Thimm M, Goede A, Hougardy S, Preissner R (2004) Comparison of 2D similarity and 3D superposition. Application to searching a conformational drug database. J Chem Inf Comput Sci 44:1816–1822

Yim H et al (1999) Emodin, an anthraquinone derivative isolated from the rhizomes of *Rheum palmatum*, selectively inhibits the activity of casein kinase II as a competitive inhibitor. Planta Med 65:9–13

Youssef KM, El-Sherbeny MA (2005) Synthesis and antitumor activity of some curcumin analogs. Arch Pharm (Weinheim) 338:181–189

Development in Malaria and Anemia Screening: Medical Imaging Informatics Approach

15

Dev Kumar Das, Chandan Chakraborty, Rashmi Mukherjee, and Ashok K. Maiti

Abstract

Medical imaging informatics (MII) includes problems of image data representation and abstraction. This provides immense help not only in standardization and interoperability but also enhances image data usability for data mining, decision support, and visual modeling and simulation. Hematological research has been significantly substantiated with the advancement of medical informatics approach. Among various hematological disorders, malaria and anemia are very common diseases that affect the human population as major health burden. This book chapter focuses on the quantitative evaluation of erythrocytes (red blood cells, RBCs) for characterization of malaria parasites and its differential infections. Anemic erythrocytes have also been recognized from light microscopic images with respect to their shape, size, and other quantitative attributes.

Keywords

Blood pathology · Malaria · Anaemia · Microscopic image analysis · Classification models

D. K. Das · C. Chakraborty (✉)
School of Medical Science & Technology, Indian Institute of Technology Kharagpur, Kharagpur, West Bengal, India

R. Mukherjee
RNLKWC, Vidyasagar University, Midnapur, West Bengal, India

A. K. Maiti
Medipath Clinic (P) Ltd., West Medinipur, West Bengal, India

G. Wadhwa et al. (eds.), *Current trends in Bioinformatics: An Insight*,
https://doi.org/10.1007/978-981-10-7483-7_15

15.1 Background

Medical imaging informatics [MII] provides a clinically relevant computing and visualization platform that improves the performance of medical imaging services within the health-care system (Branstetter 2007a, b). It investigates about how medical image information is extracted, analyzed, enhanced, and exchanged.

The modern clinical practice still relies on visual inspection of blood smears, cytopathological and radiological images, etc., along with biochemical and biophysical parameters. These combined data contain the critical information required for clinical practice, research, and education of medicine and health care. This necessitates the use of computer-based imaging, its related field of graphics and visualization, as well as pattern recognition tools to come out with the proper interpretation of the integrated biochemical, biophysical, and image data (Mukherjee et al. 2010). Earlier, there was recognition of the need for computer-based knowledge representations, not only of anatomy at the tissue level but also at the multiple underlying biological levels (cellular, molecular, atomic) leading to the definition of structural informatics (Brinkley 1991). Meanwhile, knowledge-based representations by integration of multimodal image interpretations (Hohne et al. 1995), segmentation, and visualization systems (Robb and Hanson 1995; Ghosh et al. 2010) have assisted in bridging the gap between imaging and mainstream informatics work. Productive collaborations among researchers in imaging and informatics may help in providing new insights to different pathological conditions.

15.2 Medical Imaging Informatics

Medical informatics researchers have contributed considerably to the development of medical imaging methods and systems (Sinha et al. 2002). Biomedical imaging and its interdisciplinary areas are proliferated by quantitative models of multimodal imaging techniques. In modern times, clinics require huge storage and retrieval systems due to large numbers of patient image studies. This initiated the development of picture archiving and communication system (PACS). In addition, annotated medical image databases in combination with improved 2D and 3D reconstruction and visualization algorithms are helping in making medical imaging results more accessible to clinicians. In the recent times, the Visible Human Project and Human Brain Project brought forward significant informatics challenges of knowledge representation, modeling, and information retrieval for dealing with large-scale repositories of such multimodal digital image sets (Branstetter 2007a, b). These issues encouraged medical informatics researchers in imaging problems to develop what can be described as a medical imaging informatics subspecialty (Mejino et al. 2001; Sinha et al. 2002).

MII includes problems of image data representation and abstraction. This provides immense help not only in standardization and interoperability but also enhances image data usability for data mining, decision support, and visual modeling and simulation. Applicability of MII involves elementary issues in the standardization of image information transfer, underlying knowledge representation, coding, electronic medical record, computerized guidelines for health care, information compression, efficient indexing of image databases, security, and confidentiality of imaging records, surgery, etc. (Taylor et al. 1998). In this chapter application of MII for malaria and anemia screening is discussed elaborately.

15.3 Computerized Detection of Malaria and Anemia from Blood Smear Images

In modern medical science, microscopic imaging technology has been progressing significantly for disease detection because of quantitative characterization of structural, textural, and intensity information at tissue and cell levels. Toward this direction, microscopic evaluation of peripheral blood smear has an enormous contribution toward generating the effective pathological images, which stand as the basis of making the better diagnostic decision for characterizing hematological disorders, e.g., malaria, anemia, leukemia, diabetes, cancer, AIDS, psoriasis, etc. Among these, malaria and anemia are the most common of hematological disorders in which erythrocytes are affected. Among Southeast Asian countries, India has got maximum malaria patients, approximately, two-thirds of confirmed malaria patients. The mortality due to malaria is higher than anemia. It causes 1.5–2.7 millions of death per annum (Raviraja et al. 2007, 2008). In India, ~ 50–60% of malaria patients are affected by *Plasmodium vivax (P. vivax)*, and 40–50% of them are infected by *Plasmodium falciparum* (*P. falciparum*) which is reported in the National Vector Borne Disease Control Program (NVBDCP-2011). *P. falciparum* infection is more harmful than *P. vivax* infection where its infection rate is rapidly increasing over Indian population. Mortality rate due to anemia and malaria reduction is possible by making an appropriate diagnostic tool which will diagnose correctly and timely. A peripheral blood smear is commonly considered for identifying abnormal erythrocytes associated with anemia or malaria.

The conventional approach for detecting and confirming the type/stage of malaria and anemia is a subjective method of the stained thin smears under microscope by hematopathologists. In this evaluation process, the pathologist examines abnormal erythrocytes under the light microscope subjectively. Such evaluation is time-taking and error-prone, which leads to interobserver variation.

Under such circumstances, quantitative microscopic imaging informatics plays a very important role not only in better visualization and characterization of erythrocytes but also automated screening of malaria and anemia diseases along with its substages (Das et al. 2015a, b). The informatics approach consists of mainly microscopic imaging under proper magnification, microscopic image analysis, and pattern recognition methods.

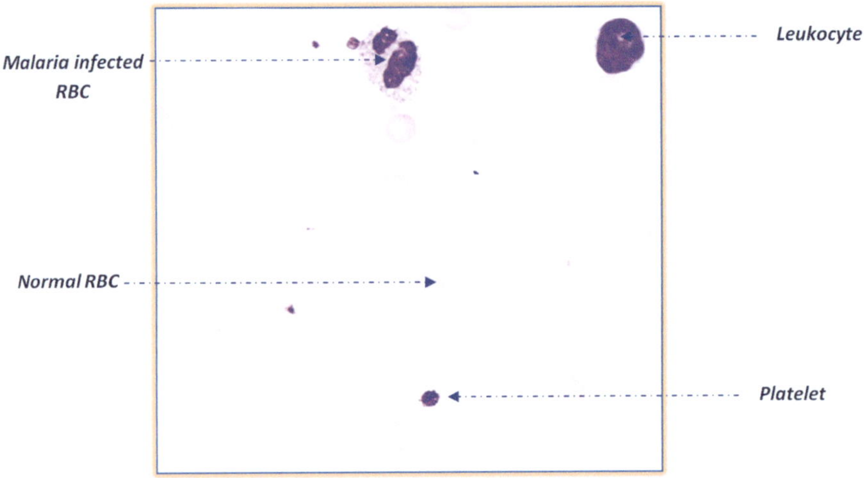

Fig. 15.1 Microscopic image view [H & E stained] of three types of blood cells

15.3.1 Microscopic Imaging of Blood Smears

A blood smear provides shape information along with color differentiation of blood cells. It consists of blood cells, viz., erythrocyte, leukocyte, and platelet and plasma. Figure 15.1 shows the blood smear image which contains normal and malaria-infected erythrocytes, leukocytes, as well as platelets. Table 15.1 summarizes the morphological changes of malaria-infected RBCs as seen during microscopic evaluation of peripheral blood smear.

Malaria Malaria is a parasite-infected disease where *Plasmodium* species, viz., *P. falciparum*, *P. vivax*, *P. malariae*, and *P. ovale*, are mainly responsible for malaria infection. These *Plasmodium* species are basically transmitted from one infected human to another human via the bite of infected *Anopheles sp.* mosquitoes. In the Indian subcontinent, although *P. vivax* infection is higher than *P. falciparum*, however, mortality due to *P. falciparum* infection is more than *P. vivax*. There are three life stages of malaria parasite, viz., early ring trophozoite, schizont, and gametocyte, which are visible under the microscope. In the case of *P. falciparum* infection, typical ring and gametocyte stages are visible under the microscope, but in the case of *P. vivax* infection, all the three stages are visible.

Anemia Anemia is one of the most common disorders of the blood where hemoglobin (Hb) and total erythrocyte counts are the most important parameters for evaluating anemia diseases. Erythrocytes contain Hb protein inside the cell which contains iron and transport oxygen. Instead of Hb and total erythrocytes, there are other types of blood parameters, which are also considered for anemia diagnosis.

Table 15.1 Morphological changes of RBCs during different stages of *P. vivax* and *P. falciparum* infection

Infection types		Class			
		P. falciparum		*P. vivax*	
RBC	Size	Normal size		Enlarged shape	
	Shape	Round, crenated		Round or oval shaped, bizarre	
Stippling spots		Maurer's spots		Schuffner's spot	
Pigmentation		Black or dark brown pigmentation		Golden brown pigmentation	
Ring trophozoite stage		Smallest, sometimes two chromatin dots, multiple rings	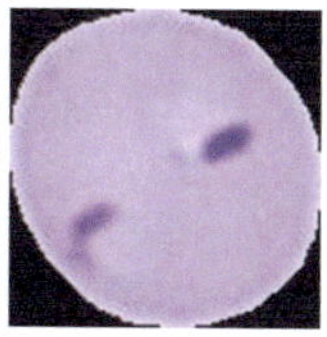	Large, one chromatin dot, often two rings	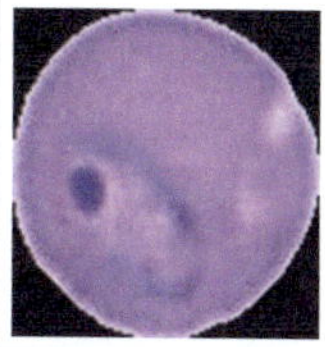
Schizont stage		Medium size, numerous chromatin masses, course pigments	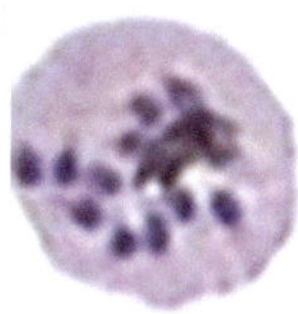	Large, numerous chromatin masses, fine pigments	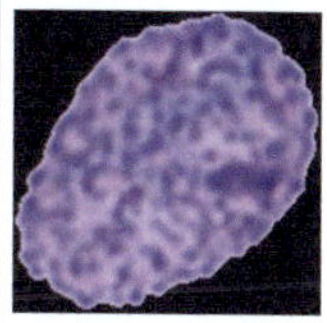
Gametocyte stage		Hemispherical shaped, larger and slim, central chromatin	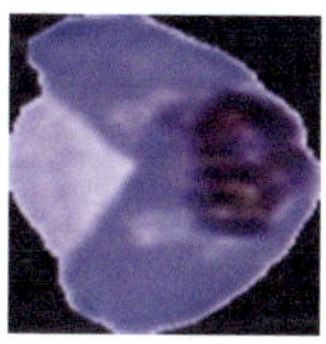	Spherical compact	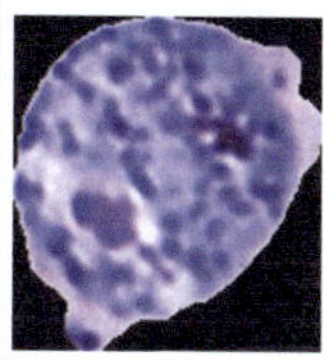

Table 15.2 Altered characteristics of anemic RBCs

Anemic RBC	Shape parameters
Sickle cell	It is crescent-/sickle-shaped and appears when the oxygen concentration is reduced
Teardrop	Drop-shaped, elongated-like structure commonly found in microcytic hypochromic diseases
Acanthocyte	Irregular-shaped with long spiny projection commonly occurring due to the low density of lipoprotein
Echinocyte	Crenated-shaped sharp-pointed projection evenly distributed on the cell membrane
Elliptocyte	Elliptical-, cigar-, or egg-shaped commonly found in megaloblastic anemia cases

Table 15.2 summarizes the altered topological characteristics of different types of anemic RBCs. As such, blood smear examination is a must for identification of various anemia types. In the microscopic evaluation process, an expert examines

morphologic changes in the erythrocytes for characterizing anemia. Several morphological changes occur during anemic condition. Here, five types (teardrop, sickle cell, echinocyte, acanthocyte, elliptocyte) of morphological disordered erythrocytes are considered for characterizing using machine learning approach. Here we have described an application of microscopic imaging informatics approach for automated screening of malaria and anemia using images of peripheral blood smears. Figure 15.2 shows the overall MII approach for malaria and anemia screening.

15.3.2 Erythrocyte Segmentation by Boundary Detection

Malaria and anemia are characterized based on morphological and intensity variation of erythrocytes. In view of this, segmentation of erythrocytes is very important from the whole image. Erythrocyte recognition leads to segmentation of all erythrocytes preserving their shapes from the blood smear images, where these shapes are helpful in disease characterization (Ghosh et al. 2011). Rule-based (Kumar et al. 2006), watershed segmentation, and Chan-Vese segmentation (Purwar et al. 2011) approaches were applied for segmenting the erythrocytes.

15.3.2.1 Rule-Based Method

Kumar et al. (2006) framed a set of rules for clumped cell splitting. Sio et al. (2007) adopted using edge detection and edge linking for segmenting *P. falciparum* infection (Kumar et al. 2006; Sio et al. 2007). Edge detection and edge linking were considered to enhance the segmentation result. The rule-based approach for robust clump splitting (see Fig. 15.3) consists of the following steps:

Step1. Edge detection of RBCs using edge correlation coefficient criterion.
Step2. Then apply edge linking for reducing edge discontinuities.
Step3. *Rule-based approach:*

(a) Convex hull and boundary arc detection
(b) Candidate split line selection
(c) Best split line selection

15.3.2.2 Watershed Segmentation Method

The marker-controlled watershed algorithm is a semiautomated segmentation (Gonzalez and Woods 2008) method consisting of region-growing and edge detection method for partitioning an image into catchment basins and watershed lines. Here, an image is considered as a topographic surface where the watershed

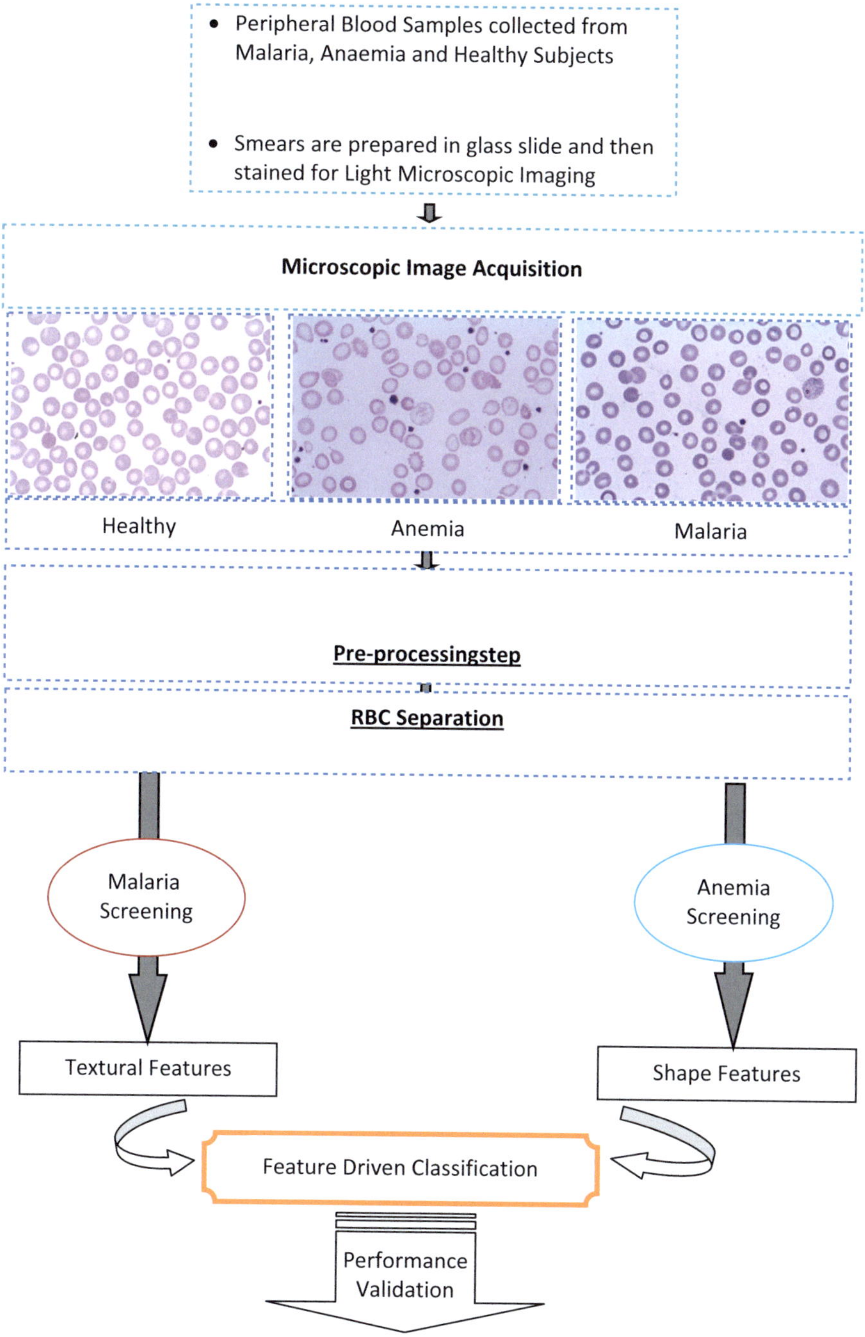

Fig. 15.2 Overall approach for malaria and anemia imaging informatics

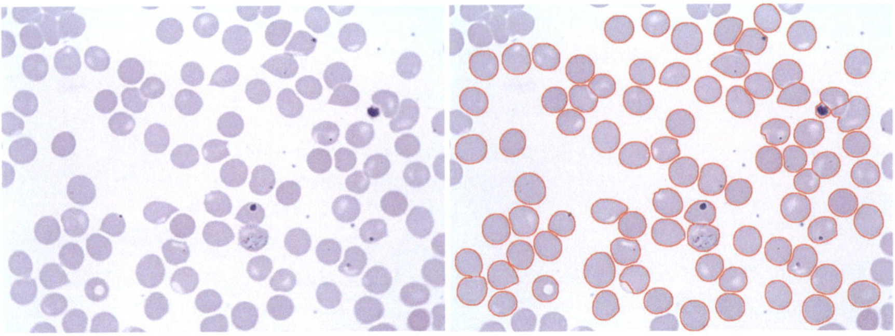

Fig. 15.3 Rule-based approach **a** original image, **b** delineated RBCs with recognized boundary (Das et al. 2015a, b)

transform refers to a flooding process. In this case, a separator line is drawn whenever two bodies of water from neighborhood regions meet each other. Simultaneously two catchment basins or watershed are constructed in the neighborhood seed regions. The segmentation procedure consists of the following steps:

Step 1: Morphological gradient calculation.
Step 2: Foreground object selection.
Step 3: Background object selection.
Step 4: Distance transform calculation.
Step 5: Apply watershed transform.
Step 6: Label each erythrocyte and to the feature extraction module.

It can be observed in Fig. 15.4 that this technique leads to over-segmentation of the cells where parasite-infected region may not be always properly identified.

15.3.2.3 Cell Boundary Detection Using Chan-Vese Method

An active contour-based method, called as Chan-Vese segmentation, is considered to fit into the boundaries of the foreground region. Contour evolution is performed by using the level set method.

From Fig. 15.5, it can be observed that Chan-Vese segmentation technique does not perform well for overlapping cell segmentation. Rule-based approach provides a better result for segmenting overlapping erythrocytes, but sometimes erythrocyte's cell loses its shape due to over-segmentation. However, in these cases, marker-controlled watershed segmentation method provides a better result for segmenting overlapping erythrocyte's cell.

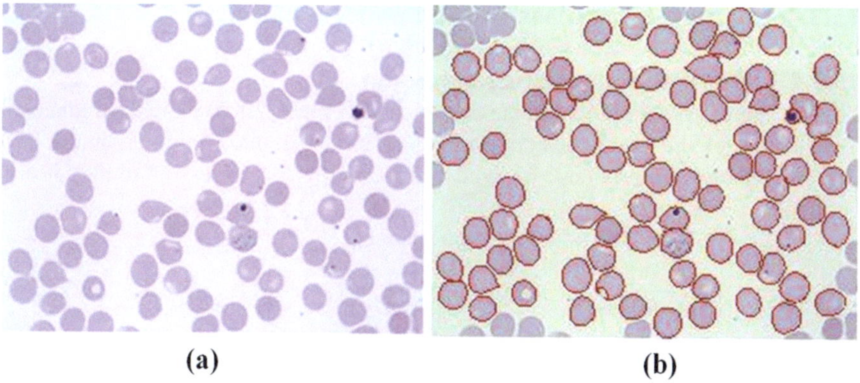

Fig. 15.4 Watershed segmentation (**a**) original image, (**b**) boundary tracked cells

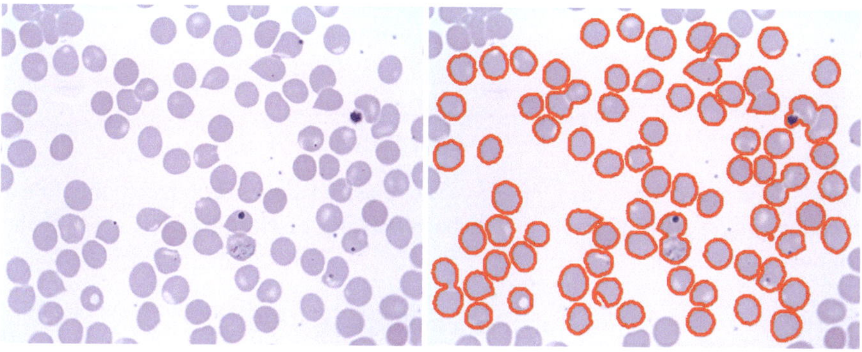

Fig. 15.5 Results showed the detected RBCs inscribed by the cell boundaries evolved from active contour models

15.3.3 Recognition and Extraction of Erythrocyte Features

In the conventional evaluation process, a pathologist observes color variation for malaria diagnosis. In the case of anemia diagnosis, anemia-affected erythrocytes become irregular in terms of morphological nature. So for characterizing anemic condition, morphological features are important. Here, a total of 80 textural (statistical and nonstatistical) and 16 morphological features (shape features and Hu's moment) were considered to discriminate five types of malaria infection stages and non-infected erythrocytes. In order to discriminate abnormal erythrocytes for the anemic (especially thalassemia) patient, morphological and Hu's moment-based features are extracted. In this approach, 18 geometrical and 7 invariant moment-based features were extracted from five types of abnormal erythrocytes.

15.3.4 Computerized Classification and Validation

15.3.4.1 Multivariate Logistic Regression

Logistic regression is a probabilistic classification (Webb 2003) which is frequently used as one of the good predictive models. Here, the study involves multiclass pattern classification problem; we used multiclass logistic regression model. It can be defined for C classes as:

$$\log\left(\frac{p(x|w_k)}{p(x|w_C)}\right) = \alpha_{k0} + \alpha_k^T x, \quad k = 1, 2, 3 \ldots\ldots. C - 1$$

Posterior probabilities are obtained using logit models (Das et al. 2013) and it is defined as:

$$P[w_k|x] = \frac{\exp\left(\alpha'_{k0} + \alpha_k^T x\right)}{1 + \sum_{k=1}^{C-1} \exp\left(\alpha'_{k0} + \alpha_k^T x\right)}$$

$$\cdot$$

$$\cdot$$

$$P[w_C|x] = \frac{1}{1 + \sum_{k=1}^{C-1} \exp\left(\alpha'_{k0} + \alpha_k^T x\right)}$$

Here $\alpha'_{k0} = \alpha_{k0} + \log(p(w_k)/p(w_C))$. The classification depends on the linear functions $\alpha'_{k0} + \alpha_m^T x$. Variable x will be consisting of class w_m if it satisfies the following condition:

$$\max\left\{\alpha'_{k0} + \alpha_k^T x\right\} = \alpha'_{k0} + \alpha_k^T x > 0$$

Otherwise, transfer x to class w_c. Maximum likelihood (ML) estimation is used to estimate regression parameters α by maximizing the log-likelihood. It can be defined as:

$$\begin{aligned}\log L(\alpha) &= \arg\max_{\alpha} \sum_{i=1}^{n} \log P(x|\alpha) \\ &= \arg\max_{\alpha} \sum_{i=1}^{n} \log \frac{1}{1 + \exp(-\alpha^T x)}\end{aligned}$$

15.3.4.2 Multilayer Perceptron (MLP) Network Model

A multilayer perceptron (MLP) model is a feed-forward neural network consisting of multiple layers of nodes (Duda et al. 2001). The basic model of MLP model is shown in Fig. 15.6 below. Here a single hidden layer is considered where d-dimensional feature space $X = [x_1, x_2, x_4, \ldots\ldots\ldots\ldots\ldots x_d]$and k classes or output $z = [z_1, z_2, z_3, \ldots\ldots z_k]$are provided. Each hidden layer j receives the weighted sum of its input and is denoted as *net*:

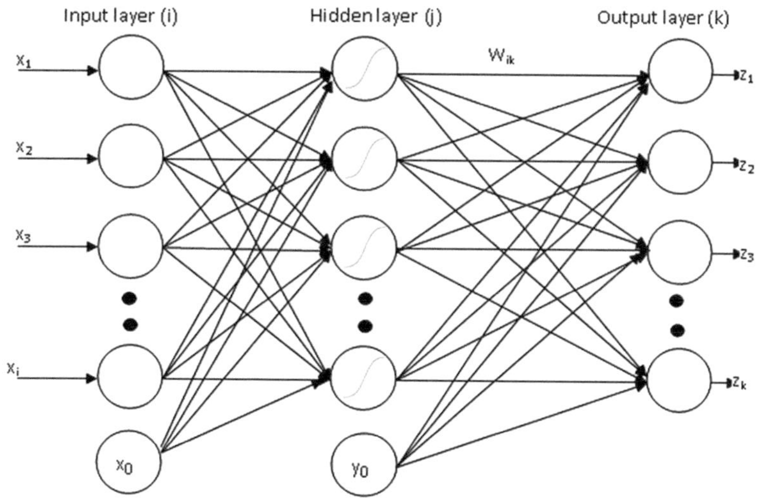

Fig. 15.6 Multilayer perceptron network architecture for data classification

$$\text{net}_j = \sum_{i=1}^{d} x_i w_{ji} + w_{j0}$$

where i and j mark the input and hidden layer indices. w_{ji} provides the weights from mapping input to the hidden layer. $w_{j0} = x_0 = 1$ is considered as the bias input, and output in the hidden layer is defined as $y_j = f(\text{net}_j)$. Hence, the net sum input in the output layer node is defined as:

$$\text{net}_k = \sum_{j=1}^{m} y_j w_{\text{kj}} + w_{k0}$$

Here, k and m denote output layer and number of hidden nodes. The hidden nodes were considered using the following formula as:

$$m = \frac{\text{No of features} + \text{No of classes}}{2}$$

The output of the output layer is defined as $z_k = f(\text{net}_k)$, where $f(net)$ is an activation function, and it is defined by $f(\text{net}) = \frac{1}{1+e^{-\text{net}}}$. Here, output error should be minimized by updating weights of the input where the training error is defined by:

$$J(w) = \frac{1}{2} \sum_{k=1}^{c} (t_k - z_k)^2$$

where target and network outputs are indicated by t_k and z_k. c denotes the number of nodes in the output layer. Firstly weights are randomly initialized in the network. The weight updation is done by gradient descent method (DevKumar et al. 2012). This is defined as:

$$\Delta w = -\eta \frac{\partial J}{\partial w} \quad \text{or} \quad \Delta w_{\mathrm{pq}} = -\eta \frac{\partial J}{\partial w}$$

η denotes learning rate. Here $\eta = 0.3$was considered. Final weights are updated as:

$$w_{ji} = w_{ji} + \Delta w_{ji}$$

$$w_{kj} = w_{kj} + \Delta w_{kj}$$

Algorithms for Multilayer Perceptron Neural Network
Step1: Weight (w) and threshold (θ) initialization.
Step2: Given input and expected output.
Step3: True output calculation; $Z_k = f(\mathrm{net}_k)$.
Step4: Weight updation by rule: $w = w + \Delta w$.
Step5: Stop iteration till absolute difference between subsequent weights is lesser than the predefined small positive value or $\|J(w)\| < \theta$.

Extracted textural and morphological features were analyzed for malaria and anemia characterization. All significant features were divided into a different subset of features. Finally, each subset of features is trained to MLP and logistic regression model where performances were evaluated with respect to sensitivity, specificity, PPV, and overall accuracy. MLP provides comparatively better performance (specificity $= 96.1 \pm 0.21\%$; sensitivity $= 93.78 \pm 0.18\%$) for malaria-infected erythrocyte recognition. In anemia detection, MLP and logistic regression provide equivalent performances for effective screening.

15.4 Conclusion

As observed, this chapter addresses the computerized detections of malaria and anemia condition. Its practicality is addressed by citing the issues of imaging such as staining variation at different imaging setups or differences during sample preparations. As it can be seen from the above, it is still early days for MII, and there is increasing urgency to standardize the exchange of medical image data because of its increasing operability and usability for both clinical practice and research. MII described here is merely the beginning of a challenging scientific and technological aspect of bioinformatics. Extensive research is required in improving the underlying integration of basic science, cognitive science, and computational modeling of the imaging modalities. Significant efforts will be expected from medical informatics community as there are many challenges ahead. Some of the other informatics challenges that require deeper scientific, technical, and clinical research include developing better techniques for automatic segmentation and registration of medical images in 3D and 2D as well as improving results from integrating automatic image segmentation with expert-derived manual segmentations and annotations. Research is also needed to develop advanced imaging systems in more general health-care delivery software and evaluate their results within health-care environments with effective feedback for learning how to modify and update the imaging methods in different clinical contexts.

References

Branstetter BF (2007a) Basics of imaging informatics. Part 1. Radiology 243(3):656–667

Branstetter BF (2007b) Basics of imaging informatics: Part 2. Radiology 244(1):78–84

Brinkley JF (1991) Structural informatics and its applications in medicine and biology. Acad Med 66(10):589–591

Das DK et al (2013) Quantitative microscopy approach for shape-based erythrocytes characterization in anaemia. J Microsc 249(2):136–149

Das DK et al (2015a) Automated system for characterization and classification of malaria-infected stages using light microscopic images of thin blood smears. J Microsc 257(3):238–252

Das DK et al (2015b) Computational microscopic imaging for malaria parasite detection: a systematic review. J Microsc 260(1):1–19

DevKumar D et al (2012) Textural pattern classification of microscopic images for malaria screening. Advances in therapeutic engineering. CRC Press, Boaco Raton, pp 419–446

Duda RO et al (2001) Pattern classification. Wiley, New York

Ghosh M et al (2010) Automated leukocyte recognition using fuzzy divergence. Micron 41 (7):840–846

Ghosh M et al (2011) Development of Renyi's entropy based fuzzy divergence measure for leukocyte segmentation. J Med Imaging Health Inform 1:334–340

Gonzalez RC, Woods RE (2008) Digital image processing. Prentice Hall, New York

Hohne KH et al (1995) A new representation of knowledge concerning human anatomy and function. Nat Med 1(6):506–511

Kumar S et al (2006) A rule-based approach for robust clump splitting. Pattern Recogn 39 (6):1088–1098

Mejino JLV et al (2001) Symbolic modeling of structural relationships in the foundational model of anatomy. KR 2004 Workshop on formal biomedical knowledge representation

Mukherjee R et al (2010) Clinical biomarker for predicting preeclampsia in women with abnormal lipid profile: statistical pattern classification approach. Systems in Medicine and Biology (ICSMB)

Purwar Y et al (2011) Automated and unsupervised detection of malarial parasites in microscopic images. Malar J 10:364

Raviraja S et al (2007) In: Ibrahim F, Osman NAA, Usman J, Kadri NA (eds) Analysis of detecting the malarial parasite infected blood images using statistical based approach, 3rd Kuala Lumpur international conference on biomedical engineering 2006: Biomed 2006, 11 â€ 14 December 2006 Kuala Lumpur, Malaysia. Springer Berlin Heidelberg, Berlin/Heidelberg, pp 502–505

Raviraja S et al (2008) In: Abu Osman NA, Ibrahim F, Wan Abas WAB, Abdul Rahman HS, Ting H-N (eds) A novel technique for malaria diagnosis using invariant moments and by image compression, 4th Kuala Lumpur international ronference on biomedical engineering 2008: BIOMED 2008 25â€ 28 June 2008 Kuala Lumpur, Malaysia. Springer Berlin Heidelberg, Berlin/Heidelberg, pp 730–733

Robb R, Hanson D (1995) The analyze software system for visualization and analysis in surgery simulation. In: Lele SR, Richtsmeier JT (eds) Computer integrated surgery. MIT Press, Cambridge, MA, pp 175–190

Sinha U et al (2002) A review of medical imaging informatics. Ann N Y Acad Sci 980:168–197

Sio SW et al (2007) Malaria count: an image analysis-based program for the accurate determination of parasitemia. J Microbiol Methods 68(1):11–18

Taylor RH et al (1998) Computer-integrated surgery. Technol Clin Appl Clin Orthop Relat Res 354:5–7

Webb AR (2003) Introduction to statistical pattern recognition. Statistical pattern recognition. Wiley, pp 1–31

16 Role of Bioinformatics in Drug Resistance Prediction for HIV/AIDS

Jayakanthan Mannu and Premendu P. Mathur

Abstract

The successful treatment of human immunodeficiency virus (HIV) infection is majorly affected by development of viral drug resistance. This complicates physician to choose the right choice of drugs. In such a scenario, a series of bioinformatics software tools and databases have been developed for predicting drug resistance, and responses to combination therapy from viral genotype have been developed to support physician. In this paper, we provided an up-to-date review on current treatment options, exploring the potential of novel targets and developed computational tools and databases for current HIV therapy in viral drug resistance.

Keywords

Combination therapy · Drug resistance · HAART · HIV databases · Genotypic resistance testing · Phenotypic resistance testing

16.1 Introduction

Human immunodeficiency virus-1 (HIV-1) is one of the important targets for clinical research that causes AIDS in human. This RNA virus replicates inside the host cell by integrating its genetic material with the host cell genome (Terwilliger et al. 1990). Several researches are happening worldwide to neutralize the activity of this virus for curing HIV-1-infected patients. There are no effective

J. Mannu
Department of Plant Molecular Biology and Bioinformatics, Centre for Plant Molecular Biology and Biotechnology, Tamil Nadu Agricultural University, Coimbatore, Tamil Nadu, India

P. P. Mathur (✉)
Department of Biochemistry and Molecular Biology, School of Life Sciences, Pondicherry University, Pondicherry 605014, India

G. Wadhwa et al. (eds.), *Current trends in Bioinformatics: An Insight*,
https://doi.org/10.1007/978-981-10-7483-7_16

drugs yet to treat the patients because this virus is highly resistant to the drugs (Kozal 2009; Sluis-Cremer et al. 2015). Hence a monotherapy is not advisable for AIDS treatment. Highly active antiretroviral therapy (HAART) is widely in practice with the coadministration of two or more drugs from different classes of HIV drugs simultaneously to treat AIDS patients (Sterne et al. 2005; Gange et al. 2002; Zeldin and Petruschke 2004). Today, totally 25 anti-HIV drugs (Table 16.1) are on the market, in which most of the drugs that interfere with the replication cycle of the virus are currently approved for use in clinical practice. These drugs can be classified into different types based on their target of action.

Table 16.1 US FDA-approved drugs in HIV treatments

Drug class	Generic name (acronym)	Brand name	Target enzymes for metabolism (substrate (S), inhibitor (I), or inducer (A))
Non-nucleoside reverse transcriptase inhibitors	Delavirdine (DLV)	Rescriptor	CYP3A4(S,I), CYP2D6(S,I), CYP3A5(I), CYP3A7(I), CYP2C9(I), CYP2C19(I), CYP1A2(I), CYP2C8(I)
	Rilpivirine (RPV)	Edurant	CYP3A(S)
	Efavirenz (EFV)	Sustiva	CYP2C19 (I), CYP2C9 (I), CYP2B6 (S, I, A), CYP3A4 (S, I, A), CYP1A2 (I), CYP2D6 (I)
	Etravirine (ETR)	Intelence	CYP3A (S,A), CYP2C9(S,I), CYP2C19(S,I)
	Nevirapine (NVP)	Viramune	CYP2B6(S,A), CYP3A4(S,I,A), CYP3A5(S), CYP2C9(S,I), CYP2A6 (S), CYP2D6(S,I)
Nucleoside reverse transcriptase inhibitors	Abacavir (ABC)	Ziagen	Alcohol dehydrogenase 6 and UDP-glucuronosyltransferase 1-1
	Didanosine (ddl)	Videx, Videx EC (enteric-coated)	No cytochrome P450 interactions
	Emtricitabine (FTC)	Emtriva, Coviracil	No cytochrome P450 interactions
	Lamivudine (3TC)	Epivir	No cytochrome P450 interactions
	Stavudine (d4T)	Zerit	No cytochrome P450 interactions
	Tenofovir DF (TDF)	Viread	CYP1A2 (I)
	Zidovudine (ZDV, AZT)	Retrovir	CYP2A6 (S), CYP2C19 (S), CYP2C8 (S), CYP2C9 (S) and CYP3A4 (S)
Protease inhibitors	Amprenavir (APV)	Reyataz	CYP3A4(S,I), CYP2B6(I), CYP2C19 (I), CYP2C8(S), CYP2C9(S), CYP2D6(S), CYP3A5(S)
	Atazanavir (ATV)	Reyataz	CYP3A4(S,I), CYP2C9(S)

(continued)

Table 16.1 (continued)

Drug class	Generic name (acronym)	Brand name	Target enzymes for metabolism (substrate (S), inhibitor (I), or inducer (A))
	Darunavir (DRV)	Prezista	CYP3A4(S,I)
	Fosamprenavir (FPV)	Lexiva	CYP3A4(S,I)
	Indinavir (IDV)	Crixivan	CYP3A4(S,I), CYP3A7(S,I), CYP3A5(S,I), CYP2C19(I), CYP2C9 (I), CYP2D6(I)
	Lopinavir (LPV)	Kaletra	CYP3A4(S,I), CYP2D6(I), CYP2C19 (I,A), CYP1A2(I), CYP2B6(I), CYP2C9(I)
	Nelfinavir (NFV)	Viracept	CYP2B6(I), CYP2C19(S,I), CYP3A4 (S,I), CYP3A7(S,I), CYP3A5(S,I), CYP1A2(I), CYP2C9(I), CYP2D6(I)
	Ritonavir (RTV)	Norvir	CYP2C19(I,A), CYP2B6(S,I,A), CYP2C8(I,A), CYP3A7(S,I), CYP3A5(S,I), CYP3A4(S,I,A), CYP2D6(S,I), CYP2C9(I,A), CYP1A2(S,A), CYP2E1(I)
	Saquinavir (SQV)	Invirase	CYP3A7(S,I), CYP3A5(S,I), CYP3A4(S,I), CYP2C19(I), CYP2C8 (I), CYP2C9(I), CYP2D6(S,I),
	Tipranavir (TPV)	Aptivus	CYP3A4(S,I), CYP2D6(S,I), CYP2C19(S,I)
Fusion inhibitors	Enfuvirtide (T-20)	Fuzeon	No cytochrome P450 interactions
CCR5 antagonists	Maraviroc (MVC)	Selzentry	CYP3A4(S)
Integrase inhibitors	Raltegravir (RAL)	Isentress	UDP-glucuronosyltransferase 1-1 (S)
	Dolutegravir (DTG)	Tivicay	UDP-glucuronosyltransferase 1-1 (S), CYP3A(S)
	Elvitegravir	Vitekta	CYP3A4(S), UGT1A1(S)

1. Nucleoside reverse transcriptase inhibitors (NRTIs), which are structurally similar to deoxynucleosides, inhibit the functions of reverse transcriptase by incorporating into the nascent DNA chain during DNA elongation.
2. Non-nucleoside reverse transcriptase inhibitors (NNRTIs), which are another class of drugs that inhibits reverse transcriptase, bind to the active site of reverse transcriptase and disrupt DNA polymerization reaction.
3. HIV-1 protease enzyme is essential for the life cycle of this virus (Davies 1990; Brik and Wong 2003). The virus synthesizes its proteins as polyprotein precursor form and should be cleaved to transform into mature, fully functional proteins to infect the cells. HIV-1 protease cleaves the precursor form of viral polyprotein to

transform into the mature functional proteins (Lambert et al. 1992; Rose et al. 1995). Inactivation of HIV-1 protease by specific chemical compounds should render the virus noninfectious (Kohl et al. 1988). The compounds such as ritonavir, amprenavir, tipranavir, indinavir, saquinavir, nelfinavir, lopinavir, fosamprenavir, darunavir, and atazanavir are the US FDA-approved drugs available for HIV-1 protease inhibition (Table 16.1). Finally, entry inhibitors block the entry of virus virions into their target cells (Flexner 2007). The entry of the virus is mediated by sequential interactions of the viral proteins gp120 and gp41 with host CD4 receptor and a coreceptor, usually CCR5 or CXCR4 (Berger et al. 1999; Dimitrov et al. 1998).

16.2 HIV Drug Resistance Testing

Drug resistance is a major factor contributing to therapy failure (Vandamme et al. 1999). The occurrence of high mutation and high replication is the genetic basis for drug resistance of HIV. Hence to identify the potential combination of drugs for HIV treatment, it is important to carry out the resistance testing. Two types of resistance testing are available; one is genotypic resistance testing, in which it screens the viral genome for resistance-associated mutations (Dunn et al. 2011) (AIDS info Clinical Guidelines Portal updated on May 1, 2014, http://www.fda.gov/oashi/aids/virals.html). In phenotypic resistance testing, viral activity will be measured in cell culture assays in presence or absence of viral drugs (MacArthur 2009). Although, these approaches give possible mutations for drug resistance, it is very difficult for clear understanding of results due to existence of vast number of mutations and mutation patterns that confer the resistance. Therefore, the development of computational methods that correlate specific genotypes with resistant phenotypes would allow development of effective HIV therapy against drug-resistant HIV variants (Zazzi et al. 2004; Beerenwinkel et al. 2005).

16.3 Geno2pheno

The geno2pheno is a freely accessible online tool (http://www.geno2pheno.org/). This tool helps physician or laboratory technician to input relevant portions of the viral genome sequence that was sequenced using patient's blood sample into the server (Fig. 16.1).

The basic requirement of the tool is to include complete coding region of the protease enzyme and also 660 coding bases of reverse transcriptase as input sequence. The server matches the input genome against reference strain HXB2 for sequence alignment. The mutations present in the genome will be identified from the alignment results, and then statistical learning methods will be applied to estimate drug resistance against virus (Beerenwinkel et al. 2003). The geno2pheno

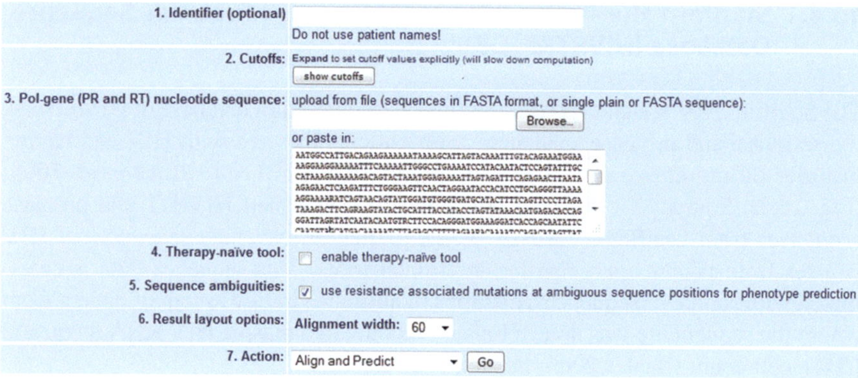

Fig. 16.1 HIV-1 gene DNA sequence submission page of geno2pheno server for the prediction of phenotypic resistance of the respective virus to 15 antiretroviral drugs

web service is available at http://www.genafor.org. For a comprehensive review on the prediction one may refer to Lengauer and Sing (2006).

16.4 HIV Drug Resistance Databases

HIV drug resistance databases were developed to store and analyze genomic data of different HIV strains and associated sequence variations, which are responsible for HIV drug resistances. Currently, there are two major databases on HIV drug resistance available to help researchers and clinicians with vast array of data of HIV variation. These databases are not only helpful for drug design and discovery but also make new dimensional view into the molecular and cellular biology, immunology, and other fields of medical research. These two databases are:

1. The Stanford HIV RT and Protease Sequence Database allows users to identify resistance-conferring mutations by screening their input DNA sequences against stored sequences in the database.
2. The Los Alamos HIV Sequence Database allows users to access all published genomic DNA, coding protein sequences of HIV with subtype, and sequence annotations. This database also offers multiple tools and programs for their analysis.

The review of Kuiken et al. (2003) has explained about the uses of these two databases (Kuiken et al. 2003). In brief, the uses of these databases are given below.

16.4.1 Stanford HIV Reverse Transcriptase and Protease Sequence Database (HIVRT&PrDB)

The Stanford HIV RT and Protease Sequence Database is a database of HIV reverse transcriptase and protease sequences from patients infected with HIV and having histories of antiretroviral treatment (http://hivdb.stanford.edu) (Rhee et al. 2003) (Fig. 16.2). Source of the database includes all published HIV RT and protease sequences from GenBank, EMBL, DDBJ, journal articles, and sequences of HIV isolates from persons participating in clinical trials. This sequence data is cross-linked with source of sequences, histories of antiretroviral treatment of patient from whom the sequencing was done, clinical outcome like plasma HIV RNA level, and CD4+ cell count (Shafer 2006; Kuiken et al. 2003).

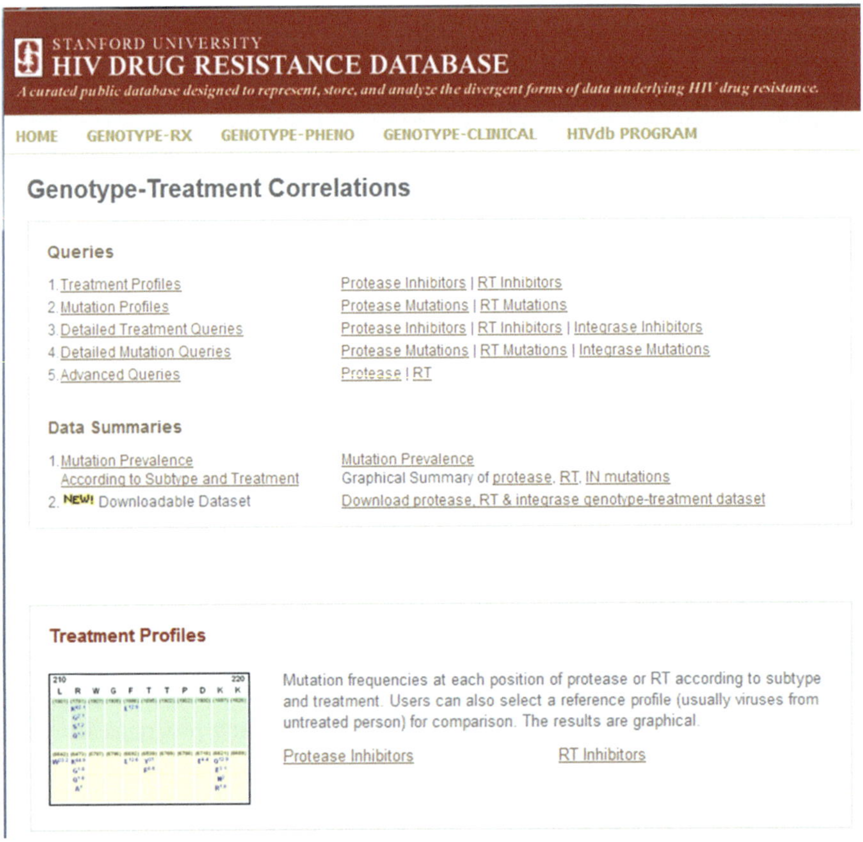

Fig. 16.2 Home page of HIV Drug Resistance Database used to retrieve data on treatment profile for various anti-HIV drugs

16.4.1.1 Database Search Interface

The search of genotype-treatment correlation data permits users to view treatment options associated with specific genotype of HIV. The data shown are for both protease and reverse transcriptase. The occurrence of mutation frequencies at each position of protease or reverse transcriptase with respect to subtype and treatment options can be retrieved from treatment profiles page. In this "Protease inhibitor" and "Reverse Transcriptase inhibitor" pages, users can opt to select number of protease inhibitors/reverse transcriptase from 1 to 8 as combination of drugs, to review mutation frequencies associated with specific subtype. Figure 16.3 shows the treatment profile for protease inhibitor nelfinavir (NFV). The result of

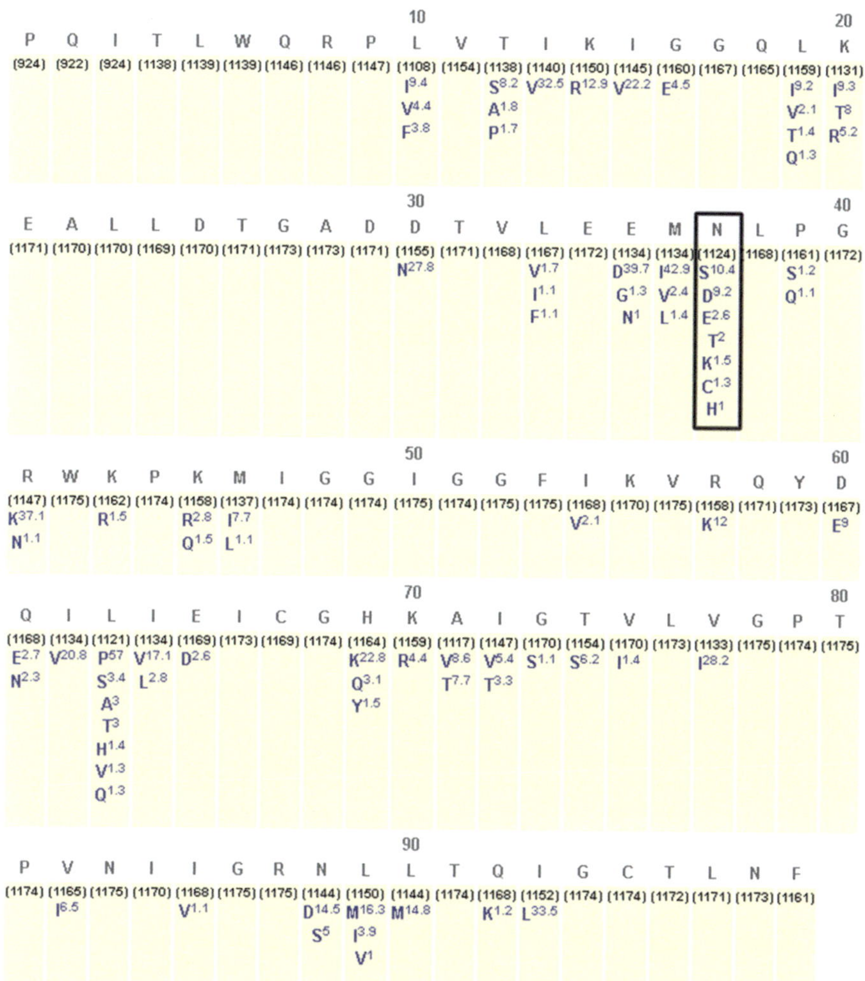

Fig. 16.3 Treatment profile for protease inhibitor nelfinavir (NFV), position 37

composite alignment (Fig. 16.3) indicates mutations as deviations from the consensus subtype B sequence. The total number of isolates analyzed at each position is shown beneath the consensus sequence; mutation superscripts indicate percentage of the total number of isolates with that mutation. Less than 1% occurrences are excluded. In close inspection of this alignment result, it was observed that 37N shows increased number of mutation (drug resistance) for nelfinavir drug.

16.4.2 Los Alamos HIV Sequence Database

This HIV databases (http://www.hiv.lanl.gov) contain data on HIV and SIV for sequence information, vaccine trials in nonhuman primates, and sequences for other viruses like hepatitis C virus (HCV) and hemorrhagic fever viruses (HFV). This database also contains many online free tools that can be used to analyze these data. This database was developed to Division of AIDS, the National Institute of Allergy and Infectious Diseases (NIAID). The main focus of this database includes collection and curation of HIV sequence data and makes it available to public access. This database is updated regularly to retrieve comprehensive data on HIV genetic sequences. This repository contains genome alignments and sequence analysis of HIV and SIV isolates. The alignments are updating annually for the newly added sequences and printed in the annual sequence compendium.

16.4.2.1 Search Tools in the HIV Databases

This database is used to update regularly with the newly developed number of interactive programs and also with newer versions of already existing programs to facilitate research in this field. Some of the important tools associated with HIV drug resistance predictions are discussed below.

HIV-BLAST

HIV-BLAST performs a sequence similarity search on the HIV sequence database alone. The output will be organized to facilitate the detection of contamination. One nucleotide sequence or one amino acid sequence may be submitted as input for single run. The sequence can be submitted by using any of three different options such as pasting sequence into text area in FASTA or raw sequence format, browsing sequence file from local computer, or specifying a GenBank accession number. The users can choose search of the query sequence against all sequences in the HIV database or sequences of subtypes A to P. The blast result contains annotated details such as subtype, sampling country, isolation year, and the genomic location of the sequence alignment. The results will also be sorted out based on the best score (Fig. 16.4).

HIV Sequence Locator Tool

This tool finds the positions of user given input sequence to the reference viral strain such as HIV or SIV sequence and gives annotations on gene and protein coding regions. This tool produces a graphical view for the prediction results. The tool

```
Z29296|HIV-1|HIVI1035|B|NL|-|HIV-1 DNA, V3 region...   541  e-154
L21486|HIV-1|HIVM21S2|B|US|-|Human immunodeficien...   454  e-128
U95417|HIV-1|HIVU95417|B|IT|-|HIV-1 clone 12 isol...   454  e-128
U95413|HIV-1|HIVU95413|B|IT|-|HIV-1 clone 8 isola...   454  e-128
```

Pairwise:

```
Score = 541 bits (273), Expect = e-154
Identities = 273/273 (100%), Positives = 273/273 (100%)
Query: 1 gtaattagatccgccaatttcacagacaatactaaaatcataatagtacagctgaatgaa 60
          ||||||||||||||||||||||||||||||||||||||||||||||||||||||||||||
Sbjct: 1 gtaattagatccgccaatttcacagacaatactaaaatcataatagtacagctgaatgaa 60

Query: 61 tctgtacaaattaattgtacaagacccaacaacaatacaagaaaaagtataaatatagga 120
           ||||||||||||||||||||||||||||||||||||||||||||||||||||||||||||
Sbjct: 61 tctgtacaaattaattgtacaagacccaacaacaatacaagaaaaagtataaatatagga 120
```

Fig. 16.4 HIV-BLAST result. It finds the database sequences most similar to query sequence

predicts the translation protein sequence of your input nucleotide. This tool produces the alignment results of both translated protein and nucleotide sequence against reference HXB2.

16.5 Conclusions

Computer-assisted anti-HIV therapy is an emerging field and has gained importance, as evidenced by the availability of many HIV-related databases and software tools. In future, use of bioinformatics software tools and databases will be very essential for decision-making of drug resistance prediction in HIV isolates. Once multiple drugs against the same target are in use and data on the relevant viral resistant are available, bioinformatics methods can be applied to selecting appropriate therapies. Hence, the development of HIV drug resistance databases and prediction tools must be freely accessible to broad community of researchers to promote discovery in the most efficient manner.

Acknowledgments J.M. is supported by BTIS scheme of Department of Biotechnology (DBT), Government of India, New Delhi, India.

References

Beerenwinkel N et al (2003) Geno2pheno: estimating phenotypic drug resistance from HIV-1 genotypes. Nucleic Acids Res 31(13):3850–3855

Beerenwinkel N et al (2005) Computational methods for the design of effective therapies against drug resistant HIV strains. Bioinformatics 21(21):3943–3950

Berger EA et al (1999) Chemokine receptors as HIV-1 coreceptors: roles in viral entry, tropism, and disease. Annu Rev Immunol 17:657–700

Brik A, Wong CH (2003) HIV-1 protease: mechanism and drug discovery. Org Biomol Chem 1(1):5–14
Davies DR (1990) The structure and function of the aspartic proteinases. Annu Rev Biophys Biophys Chem 19:189–215
Dimitrov DS et al (1998) HIV coreceptors. J Membr Biol 166(2):75–90
Dunn DT et al (2011) Genotypic resistance testing in routine clinical care. Curr Opin HIV AIDS 6(4):251–257
Flexner C (2007) HIV drug development: the next 25 years. Nat Rev Drug Discov 6(12):959–966
Gange SJ et al (2002) Effectiveness of highly active antiretroviral therapy among HIV-1 infected women. J Epidemiol Community Health 56(2):153–159
Kohl NE et al (1988) Active human immunodeficiency virus protease is required for viral infectivity. Proc Natl Acad Sci U S A 85(13):4686–4690
Kozal MJ (2009) Drug-resistant human immunodeficiency virus. Clin Microbiol Infect 15(Suppl 1):69–73
Kuiken C et al (2003) HIV sequence databases. AIDS Rev 5(1):52–61
Lambert DM et al (1992) Human immunodeficiency virus type 1 protease inhibitors irreversibly block infectivity of purified virions from chronically infected cells. Antimicrob Agents Chemother 36(5):982–988
Lengauer T, Sing T (2006) Bioinformatics-assisted anti-HIV therapy. Nat Rev Microbiol 4(10):790–797
MacArthur RD (2009) Understanding HIV phenotypic resistance testing: usefulness in managing treatment-experienced patients. AIDS Rev 11(4):223–230
Rhee SY et al (2003) Human immunodeficiency virus reverse transcriptase and protease sequence database. Nucleic Acids Res 31(1):298–303
Rose JR et al (1995) Defining the level of human immunodeficiency virus type 1 (HIV-1) protease activity required for HIV-1 particle maturation and infectivity. J Virol 69(5):2751–2758
Shafer RW (2006) Rationale and uses of a public HIV drug-resistance database. J Infect Dis 194(Suppl 1):S51–S58
Sluis-Cremer N et al (2015) Resistance to reverse transcriptase inhibitors used in the treatment and prevention of HIV-1 infection. Future Microbiol 10(11):1773–1782
Sterne JA et al (2005) Long-term effectiveness of potent antiretroviral therapy in preventing AIDS and death: a prospective cohort study. Lancet 366(9483):378–384
Terwilliger EF et al (1990) Mechanisms of infectivity and replication of HIV-1 and implications for therapy. Ann Emerg Med 19(3):233–241
Vandamme AM et al (1999) Managing resistance to anti-HIV drugs: an important consideration for effective disease management. Drugs 57(3):337–361
Zazzi M et al (2004) Comparative evaluation of three computerized algorithms for prediction of antiretroviral susceptibility from HIV type 1 genotype. J Antimicrob Chemother 53(2):356–360
Zeldin RK, Petruschke RA (2004) Pharmacological and therapeutic properties of ritonavir-boosted protease inhibitor therapy in HIV-infected patients. J Antimicrob Chemother 53(1):4–9

Bioinformatics Approaches for Animal Breeding and Genetics

17

Satendra Singh, Budhayash Gautam, Anjali Rao, Gitanjali Tandon, and Sukhdeep Kaur

Abstract

The main objective of animal genomics is to comprehend the genetic and molecular basis of all biological processes in animal. By understanding that, animals can be utilized as biological resources in the development of new breeds with improved quality and minimized costs. Animals with stress-resistant quality along with yield traits and reproductive traits are of major interest. This data, along with suitable technology, may help in designing predictive procedures for animal health and may also become part of future breeding decision management systems. Existing technologies generate a large amount of genomic data that requires proper processing, storage, and distribution. This data include sequence information as well as information on various markers, maps, functional discoveries, etc. In this chapter, we provide an insight on how different approaches, tools, and databases can be fruitfully utilized for the various animal breeding and genetics programs. Important objectives for animal bioinformatics comprises to encourage the submission of all sequence data into the public domain via various repositories; to make accessible the annotation of genes, proteins, and phenotypes; and to illustrate the relationships within the animal data and also between animal and other organisms.

Keywords

Animal · Genetics · Breeding · Bioinformatics · Databases

S. Singh (✉) · B. Gautam · A. Rao · G. Tandon · S. Kaur
Department of Computational Biology and Bioinformatics, Jacob Institute of Biotechnology and Bioengineering, Sam Higginbottom University of Agriculture, Technology and Sciences, Allahabad, Uttar Pradesh, India

G. Wadhwa et al. (eds.), *Current trends in Bioinformatics: An Insight*,
https://doi.org/10.1007/978-981-10-7483-7_17

17.1 Introduction

Bioinformatics is an implementation of computer technology for managing the biological information. The requirement for bioinformatics capabilities has been triggered by the explosion of genomic data available from wide range of the genome projects including humans, insects, microorganism, pathogens, plants, and animals. These projects aimed to decode DNA sequence of their entire genome. At the beginning of this genomic revolution, bioinformatics was meant to create and maintain the database for storing and retrieving the biological information such as nucleotide and amino acid sequences. Later these databases also included the interfaces whereby researchers could both access existing data as well as submit new or revised data. Presently the field of bioinformatics comprises the management, analysis, and interpretation of various types of data, including nucleotide and amino acid sequences, protein domains, and protein structures. Keeping all this in mind, we can easily understand that the animal domain is also not left untouched with the applications of bioinformatics (Fadiel et al. 2005).

Animal breeding started about 1000 years back, when they were captured and domesticated. Earlier phenotypic approaches were used for evaluations like different productivities and appearances. This approach altered extensively when more genetic information became available. Earlier genetic marker information was not used, but present analysis is highly dependent on them. Basic methods were developed after the Mendelian laws of inheritance were revived, and further advances were based on the knowledge on inheritance of genetic information via linkage and linkage disequilibrium. Theoretical approaches and progression on the field of quantitative genetics, especially the work from Fisher, Wright, and Lush, allowed multifactorial models and inclusion of complex pedigrees into defined breeding decisions (e.g., reviewed in Hill and Mackay, 2004; Hill, 2014). Blending of these theoretical approaches and various experimental achievements has played a vital role for the development of modern breeds. At present, structured breeding programs are available for most livestock species (Jonas and Koning 2015).

Several projects have been completed successfully in various areas such as gene identification, functional annotation, posttranslational modification prediction, protein-protein interactions, transcriptomics, and systems biology. Thus majority of these developments have focused on gathering the animal genome, transcriptome, proteome, and epigenome. This will allow how to use bioinformatics tools and databases for the analysis of large re-sequencing data sets and how to understand the functional sequence variants. This will also help on how to use software to identify various sequence variants (SNPs, indels, and CNVs) for discovering functionally relevant traits, making strong emphasis on its use for genetic improvement. Animal genetics research will contribute in understanding of basic genetics and livestock genetics and breeding, for example, analysis of genome-wide SNP data including association studies and classification problems for solving problems in animal genetics, as well as improving the knowledge of theoretical genetics to evaluate animal performance in production and welfare aspects (Daetwyler et al. 2007).

Various methods of bioinformatics are practiced to access various databases and to exchange information for comparison, storage, retrieval, and analysis of biological data. As on date, there are a number of databases on nucleotides and proteins pertaining to various animals and their breeds, bacteria, and other life forms that are available. For centuries, humans have selected animal varieties that best fit their purposes and developed animals that have many advantages compared to natural (wild) animals in quality, quantity, and farming practices. The revolution in life sciences signaled by genomics has changed the scale and scope of our experimental inquiry and application in animal breeding. There are several disciplines of bioinformatics which are directly applicable for animal breeding and genetics research. Some of these significant approaches and methods are discussed in this chapter.

17.2 Bioinformatics in Genetics

Genetics is a field of research where a number of varied investigative approaches are applied to determine the basis of heredity and the variation in the way these traits are differently expressed between individuals or populations of individuals. In more recent times, our fundamental understanding of the molecular basis of heredity and gene expression and the increased application of high-throughput technologies requiring new computer-based approaches to research has broadened the scope of the field of genetics research into a dynamic, multidisciplinary approach to scientific discovery. Population genetics is used to determine the genetic diversity among different groups of organisms as well as changes in gene frequencies over time and between geographical regions, leading to the definition of population structure and subdivision. Quantitative genetics is a highly focused form of population genetics. Now that the entire DNA sequence of many organisms (such as humans, mice, pigs, cows, buffalo, cattle, goat, etc.) is known, quantitative geneticists can include this available molecular information in their analyses. Molecular genetics, in contrast to quantitative genetics, is concerned with the characterization of specific genes whether that is the structure, expression, or evolutionary history of those genes. If a trait is governed by single gene, then understanding the underlying genetic mechanism might lead to the design of a genetic test to eliminate an undesirable condition. Genomic selection (GS) is a form of marker-assisted selection in which genetic markers covering the whole genome are used so that all quantitative trait loci (QTL) are in linkage disequilibrium with at least one marker. This approach has become feasible due to revolution in SNP discovery method like deep sequencing and throughput SNP genotyping on DNA chip. Such modern selection technology is heavily dependent on computational science or bioinformatics tools (Goddard and Hayes 2007). Following consortium of genomics and bioinformatics are working successfully with aim of better animal selection to increase productivity and health (cattle, buffalo, horse, poultry, sheep, goat, pig, camel, and rabbit).

17.3 Database Development and Maintenance

Databases are designed so that large amount of data can be stored, in a way that it can be easily accessed; data can be integrated and redundancy can be avoided. Modern breeding techniques, as well as disease control and quality assurance programs, are all on data. All three involve the same animals, and it thus makes sense to take an integrated approach to the databases supporting each of these (Wickham et al. 2013). The development and maintenance of database is one of the foundation steps. Database development and management on animal genetic resources is an essential task to characterize, utilize, and conserve these irreplaceable resources (Mitra and Acharya 2005). Generally a database on animal genetic resources is maintained on a regional/country basis. It stores information on breeds of various livestock and poultry species in the region. It also stores the data about the breeding tracts of breeds of various livestock species, breeding farms, and other information on breeds. Information on a breed includes physical, production, and reproduction traits of animals of the breed. Socioeconomic information about the farmers rearing the breed is an essential component of these databases. Database on animal genetic resources has a key role to play in documenting the breeds and highlighting their status. Various databases on animal genetic resources are available, for example, the Animal Genetic Resources - Information System (AGRI-IS) has been developed at NBAGR, Karnal, while the Domestic Animal Diversity Information System (DAD-IS) has been developed at the Food and Agriculture Organization (FAO- http://www.fao.org/). An information system for genebank management (ISGerm) has been developed for management of genebank. A database on genetic characterization of animal genetic resources is also being developed. This approach removes many technical barriers in analyzing animal genetic data and thus created a platform for converting results into predictive models for animal breeding studies (Baurley et al. 2013).

17.4 Databases on Animal Genetic Resources (Tables 17.1 and 17.2 and Fig. 17.1)

17.4.1 Information System on Animal Genetic Resources of India: AGRI-IS

This database supplies district-wise information on various animal resources with respect to population, production, farms, semen availability, vaccine production, import and export, and also breed description. It also covers data about various breeds of livestock and poultry species (general information, utility, geographical distribution, population, morphology, performance parameters, management practices). AGRI-IS can be accessed through http://www.nbagr.res.in/otherpub.html.

Table 17.1 Tools and databases in bioinformatics

S. No.	Tool/ database	Web address	Description
1	GenBank	www.ncbi.nlm.nih.gov	Nucleotide and protein sequences
	EMBL	www.ebi.ac.uk/embl.html	
	DDB	www.ddbj.nig.ac.jp	
2	TIGR	www.tigr.org/tdb/tgi.shtml	EST dB
3	Goat and sheep	http://www.itb.cnr.it/gosh/	GoSh dB database
4	ExInt	http://sege.ntu.edu.sg/wester/iekb/	Exon-intron structure of eukaryotic genes
5	TRANSFAC	http://www.gene-regulation.com	Transcription factors and binding sites
6	PIR	www.pir.georgetown.edu	A collection of protein sequence databases
7	SWISS-PROT	www.expasy.ch/sprot	Curated protein sequence databases
8	PROSITE	www.expasy.ch/prosite	Protein dB
9	Pfam	www.sanger.ac.uk/Software/Pfam/	Protein dB based on HMM
10	GO	www.geneontology.org	Gene ontology consortium database
11	CCSD	bssv01.lancs.ac.uk/gig/pages/gag/carbbank.htm	Complex carbohydrate structure databases (CarbBank)
12	PDB	www.rcsb.org/pdb/	Protein dB
13	KEGG	www.genome.ad.jp/kegg	Metabolic pathway dB
14	EcoCyc	www.ecocyc.org	*E. coli* K-12 genes, metabolic pathways, transporters, regulation
15	Ensembl	www.ensembl.org	Annotated information on eukaryotic genomes
16	Webcutter	http://users.unimi.it/~camelot/tools/cut2.html	Generate a map of enzyme sites
17	Nebcutter	http://tools.neb.com/NEBcutter2/	DNA sequence analysis
18	ORF finder	http://www.ncbi.nlm.nih.gov/gorf/gorf.html	For open reading frames prediction
19	CattleCyc	http://biocyc.org/CATTLE/organismsummary	Cattle-specific metabolic pathway
20	GENETPIG	http://www.infobiogen.fr/services/Genetpig/	Pig genome analysis
21	DbSNP	http://www.ncbi.nlm.nih.gov/SNP/	Database of single nucleotide polymorphism
22	Sheep QTLdb	http://www.animalgenome.org/cgibin/QTLdb/OA/index	Sheep QTL data published
23	Chicken	http://www.ncbi.nlm.nih.gov/mapview/map_search.cgi?taxid=903	Chicken genome assembly
24	AMOS	http://sourceforge.net/apps/mediawiki/amos/index.php?title=AMOS	Whole-genome assembly

Table 17.2 Various tools used in bioinformatics

TWAIN	A generalized pair HMM to predict genes simultaneously in two closely related eukaryotic organisms
ExAlt	A phylogenetic generalized Hidden Markov Model for finding alternatively spliced exons
JIGSAW	This program is use to predict gene models using the output from other annotation software
SIM4CC	An efficient program to align cDNA sequences (ESTs) to genomic sequences, specifically designed for cross-species alignment
PROFUNC	Identification of biochemical function of a protein from its three-dimensional structure
ABBA	Assembly Boosted By Amino acid sequence is a comparative gene assembler, which uses amino acid sequences from predicted proteins to help build a better assembly
AMOScmp	This is a comparative genome assembler, which uses one genome as a reference onto which assembles another genome of closely related species
MUMmer	*MUMmer* is an open-source software package for the rapid alignment of very large DNA (bacterial vs. bacterial and human vs. human genome) and amino acid sequences
TOPHAT	A short read aligner for RNA sequence experiments
Crossbow	Crossbow is scalable software for whole-genome resequencing analysis. It combines Bowtie, and *SoapSNP*, an accurate genotyper, within Hadoop to distribute and accelerate the computation with many nodes
FIGARO	A vector trimmer capable of accurately trimming vector from shotgun reads without prior knowledge of the vector sequence
ELPH	A motif finder that can find ribosome binding sites, exons splicing enhancers, or regulatory sites
MetaPath	MetaPath can identify differentially abundant pathways in metagenomics data sets, relying on a combination of metagenomics sequence data and prior metabolic pathway knowledge
MATLAB	It provides direct web connectivity and access to various databases like Genbank, PDB, EMBL, PIR, BLAST, etc., and can read files in FASTA formats
mfold	DNA/RNA secondary structure prediction using nearest neighbor thermodynamic rules
FoldX	Force field for energy calculations and protein designing, fast and quantitative estimation of the importance of the interactions contributing to the stability of proteins and protein complexes
Modeller	Used for homology and comparative modeling of protein 3D structure
Robetta	Protein structure prediction, protein-protein docking, and design of new proteins
HHpred	Template detection, alignment, 3D modeling
WHAT IF	Specialized on working with proteins and molecules in their environment like water, ligands, nucleic acids, etc.
Bhageerath	A computational protocol for modeling and predicting protein structures at the atomic level
ACEMD	NVIDIA GPUs. Heavily optimized with CUDA. ACEMD can read CHARMM/NAMD and AMBER input files
AMBER	MD tool
CHARMM	MD tool
GROMACS	MD tool
Discovery Studio	Discovery Studio is a comprehensive life science modeling and simulation suite of applications focused on optimizing the drug discovery process

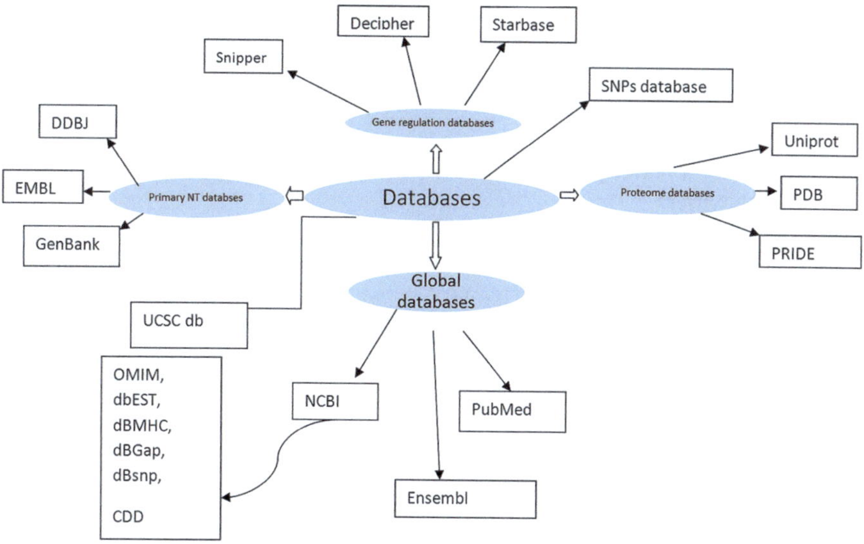

Fig. 17.1 Integrated databases system

17.4.2 DAD-IS (Domestic Animal Diversity Information System)

Food and Agriculture Organization (FAO) has developed Domestic Animal Diversity Information System (DAD-IS) for global management of animal genetic resources. It is accessible at http://dad.fao.org/. DAD-IS is a multi-language information system. It contains information on more than 14,877 breeds of around 38 domesticated animal species from 205 countries (Groeneveld et al. 2010). Various types of information stored in DAD-IS include general information (species, geographical location of a breed, local name), breed description (coat color, horn description, wool type, adult weight and size), genetic characters (marker genes, chromosomal aberrations), etc.

17.4.3 DAGRIS (Domestic Animal Genetic Resources Information System)

This system provides information on the amount of existing diversity, characteristics, and use of indigenous farm animal genetic resources in developing countries as the basis for their present as well as future sustainable utilization. Due to unavailability of a systematic database on this information, ILRI has been developing the Domestic Animal Genetic Resources Information System (DAGRIS- http://dagris.ilri.cgiar.org/default.asp) as a web-based electronic source of information on selected indigenous farm animal genetic resources (breeds/

ecotypes of cattle, sheep, goats, chicken, and pigs) with options to extend it further to cover geese, turkey, and ducks.

17.4.4 EFABIS (European Farm Animal Biodiversity Information System)

This is a platform to create structure of databases for monitoring farm animal biodiversity (FAB) in Europe. This project was developed to meet the specific requirements of the European continent as well as the need for full compatibility with the Global Information System (DAD-IS) of FAO. Historically, two databases were developed for the purpose of the management of animal genetic resources in the European perimeter. The European Association for Animal Production (EAAP) was one of the first organizations to develop a database to monitor the large variety of European breeds, making this information available through the Internet. The Animal Genetic Resources Group of FAO in Rome used this database, to set off a new development for its Internet-based dynamic Domestic Animal Diversity Information System (DAD-IS). As a result of the EFABIS project, a new system was developed (FABISnet) which was a merge of the EAAP and FAO database in terms of data structure with a strong emphasis on extended networking and communication capability. EFABIS is accessible through http://efabis.tzv.fal.de/.

17.4.5 National Animal Genome Research Program (NAGRP)

(USDA NRSP-8 Livestock Genome Research Projects)

This is the USDA NRSP-8 Program (National Animal Genome Research Program, NAGRP) Bioinformatics Coordination Project website. The project was supported by funding from the USDA-NRI (former CSREES) for the periods 2003–2008 and 2008–2013. The "precursor" of this site was known to some old-timers as "AnGenMap website."

17.4.6 Canadian Animal Genetic Resources (CAGR) Program

It's a joint program of Agriculture and Agri-Food Canada (AAFC) and the University of Saskatchewan (U of S). CAGR have three programs which include Genetic Diversity, Gamete and Embryo Biology, and Cryobiology. This information can be retrieved through http://www.agr.gc.ca/eng/?id=1297780434818.

17.4.7 Rare Breeds International (RBI)

It's a nongovernmental organization for conservation of animal genetic resources. This supports many project at time Enderby Island cattle, black pig, genebank, etc. All information found on this site (http://www.rarebreeds.co.nz/projects.html).

17.4.8 Econogene

This works on sustainable conservation of animal (sheep and goat in rural areas). This is funded by the EU with the quality of life five framework programs. Goat and sheep all related information are mentioned on this site (http://www.econogene.eu/).

17.4.9 NSA (National Sheep Association)

NSA works for sustainable development of sheep. NSA was promoted by Sheep Health and Welfare Group (SHAWG), and NSA can be accessed at http://www.nationalsheep.org.uk/.

17.4.10 Nucleotide Polymorphism Marker Database

Single nucleotide polymorphism (SNP) is the variation that occurs in DNA sequence, when a single nucleotide (among A, T, C, or G) in the genome differs between members of a species or paired chromosomes in an individual. SNP can occur either within coding sequence of a gene or noncoding sequence of DNA. SNPs within a coding sequence may not necessarily change the amino acid sequence of the produced protein, due to degeneracy feature of the genetic code. A SNP in which both forms lead to the same polypeptide sequence is termed *synonymous* (or silent mutation), and if a different polypeptide sequence is produced, they are *nonsynonymous*. SNPs are stable genetic markers and have low mutation rate in comparison with other genetic markers. The frequency of occurrence of single base difference in genomic DNA is very high when two equivalent chromosomes are compared. SNPs are evolutionary stable from generation to generation and thus, are useful in population studies and in tracking down the inheritance of any gene that contributes to a disease. SNPs are also useful in epidemiological studies, gene discovery, gene mapping, candidate gene polymorphism testing, diagnostics and risk profiling, and homogeneity testing (Sherry et al. 1999; Shah and Kusiak 2004). DoGSD (http://dogsd.big.ac.cn), all data of SNP detection for dogs, is such an example of nucleotide polymorphism marker database.

17.4.11 Animal Genetic Disease Database

Genetic diseases or inherited disorders occur in all animal species. These diseases lead to physical or functional anomalies with a harmful impact on health and productivity. For controlling various types of disease a database is developed – GDC (genetic disease control- http://www.gdcinstitute.org/) for dogs. Biotechnology helps in the possibilities of speedy and economical detection of carriers of genetic diseases. Similar methods are being developed to diagnose genotypes, such as normal, carrier, or affected, for conditions inherited by animals (Jovanović et al. 2009).

17.4.12 Database on Inherited Mendelian Traits and Disorders

Online Mendelian Inheritance in Animals (OMIA- http://www.ncbi.nlm.nih.gov/omia) is a comprehensive and annotated record of various inherited disorders and other familial traits in animals other than humans and mice. It is a complete resource of phenotypic information on heritable animal traits and genes in a strongly comparative context, relating traits to genes where possible (Lenffer et al. 2006). This database provides the total number of traits/disorders listed for each species, the Mendelian traits /disorders, the traits/disorders for which the causative mutation is known, and the traits which can serve as potential model for human diseases. Information can also be added in the database by the users. The database has been indexed such that queries can be constructed based on Entrez Gene ID, gene name, OMIM ID, Taxonomy ID, organism's scientific or common name, phenotype term, Mammalian Phenotype Ontology ID, PubMed ID, OMIA ID, or species-specific OMIA Phene ID (Nicholas 2003).

17.4.13 Links Between Databases

Links between different databases help in relating the information available in those databases. OMIA has links to OMIM for any disorder that is believed to be homologous to a human-inherited disorder. Since OMIM contains a wealth of information, this is an important source of invaluable information for veterinarians and others interested in animal disease. These links provide medical researchers with up-to-date information on animal models of inherited human disease (Nicholas 1998; Montaldo 2006).

17.4.14 Quantitative Trait Loci (QTL)

The use of genomic information (sequences or DNA marker polymorphisms) for the genetic improvement and selection of animals requires the knowledge of the effect of physically mapped genes with effects on economically important traits or quantitative trait loci (QTL). In marker-assisted selection (MAS) or transgenesis, the genomic information is combined with the classical performance records and genealogical information to increase selection accuracy, performing selection earlier in life and reducing costs (Bouchard and Mcgue 2003; Moreno et al. 2003). The traits, on which the application of marker-assisted selection can be more effective, are those that are expressed late in the life of the animal, have low heritability, are sex-limited, are expensive to measure, or are controlled by a few genes. Quantitative trait loci experiments, using crosses between breeds or lines with extreme genotypes for a trait, increase the power of detecting QTL for that trait, compared to within-family designs (Hu et al. 2010, 2013).

17.5 Bioinformatics Resources in Breeding Analysis

17.5.1 SIGENAE: Information System for Analysis of Breeding

SIGENAE is an Information System of AGENAE program. The AGENAE program (Analysis of Breeding Animals' Genome) is a national program with an ambition to develop research in the domain of breeding animal genomics – pig, chicken, trout, cattle, rabbit, and sheep.

17.5.2 Animal QTLdb

The Animal QTL database (QTLdb; http://www.animalgenome.org/QTLdb) is a repository for all publicly available QTL and single nucleotide polymorphism/gene association data on livestock animal species from which one can easily locate and compare QTL within species.

The Animal Quantitative Trait Loci (QTL) database (Animal QTLdb) attempts to collect all the publicly available trait mapping data, i.e., QTL (phenotype/expression, eQTL), candidate gene and associated data (GWAS), and copy number variations (CNV) mapped to livestock animal genomes, in order to enable finding and comparing discoveries within and between species. Various database tools are also added to link the QTL data to other types of genomic information, such as radiation hybrid (RH) maps, fingerprinted contigs (FPC) physical maps, linkage maps, comparative maps to the human genome, etc.

Currently, this database contains data on 13,030 pigs, 17,908 cattles, 1018 horses, 801 sheeps, 127 rainbow trouts, and 4525 chickens QTL (data as per release 26 dated 27th April, 2015), which are dynamically linked to respective RH, FPC, and human comparative maps. The Animal QTLdb provides a platform for comparative genomics studies across multiple species, with QTL information as a starting point.

17.5.3 AQUAFIRST

AQUAFIRST is to identify genes associated with stress and disease resistance in oyster, trout, sea bream, and sea bass in order to provide a physiological and genetic basis for marker-assisted selection.

17.5.4 EADGENE (European Animal Disease Genomics Network of Excellence)

EADGENE particularly focuses on the genomics of livestock disease caused by different pathogens. The major focus is on host (biology)-pathogen interactions relevant to animal and human health, including *Salmonella* in pigs and poultry, enterohemorrhagic *Escherichia coli* (EHEC) in cattle, and mastitis in cattle, goats, and sheep. This type of genomics research could assist the development of new or improved breeding of farm animals for disease resistance.

17.6 Application Areas of Bioinformatics in Animal Genetics

17.6.1 Genomics

Genomics helps in understanding of genes, noncoding regions, and regulatory regions of DNA sequences. Study of genomics can be categorized into structural genomics, functional genomics, and comparative genomics. Functional genomics is the analysis of gene expression and gene functions in the genome of a species or comparative analysis among genomes of various species, while comparative genomics is the analysis and the comparison of genomes of different species (Burt 2002; Cios et al. 2005). Most of the livestock species (cow, pig, horse, and chicken) have been sequenced, and others are in progress of completion; chicken was the first livestock to be sequenced and buffalo the latest (Andersson 2001; Javadi 2013). So genomics can help in finding genes and their functions.

17.6.2 Genome Annotation

Genome annotation is a process to identify gene location in a newly sequenced genome and to assign functions to identify genes and gene products. It involves two steps, gene prediction and functional assignment, e.g., gene annotation of human genome employs a combination of theoretical prediction and experimental verification. The predicted genes are compared with experimentally determined cDNA and EST sequences using pairwise sequence alignment programs such as *GeneWise*, *Spidey*, *SIM4*, and *EST2Genome*, and functional assignment of encoded protein is carried out by homology searching using BLAST searches against a protein database. Gene ontology is an annotation system for gene products using a set of structured, controlled vocabulary to indicate the biological process, molecular function, and cellular localization of a particular gene product (Burt 2002).

17.6.3 Proteomics

Proteomics includes separation, identification, and characterization of total proteins from a common source. The simplest method for identifying this separation is peptide mass fingerprinting (PMF) which identifies proteins by matching the masses of digest fragments with known masses from a nonredundant protein database (Table 17.3). Proteomics helps in analyzing thousands of proteins in a single experiment. This ability to analyze thousands of proteins gives the field of proteomics a unique capability to demonstrate how cells can dynamically respond to changes in their environment. Therefore, a goal of proteomics is to identify new and potentially unexpected changes in protein expression, interaction, or modification as a result of an experimental treatment (Lee et al. 2003; Daetwyler et al. 2013).

17.6.4 Genetic Diversity Analysis

The diverse livestock resources maintain a wealth of unique traits which have not been completely explored. This genetic diversity is useful for the identification and characterization of livestock breeds not only to prioritize them for conservation but

Table 17.3 Web resources for protein and microsatellite data analysis

S.No.	Tool	URL
1	Profound	http://prowl.rockefeller.edu
2	Mascot	www.matrixscience.com
3	PepSea	http://www.unb.br/cbsp/paginiciais/pe
4	MS-Tag	http://prospector.ucsf.edu
5	Peptident	http://us.expasy.org/tools/peptident.ht
6	Multident	http://us.expasy.org/tools/multiident
7	PopGen32	http://www.ualberta.ca/-fyeh/fyeh
8	GenAlEx	http://www.anu.edu.au/BoZo/GenAlEx
9	Arlequin	http://lgb.unige.ch/arlequin
10	GDA	http://hydrodictyon.eeb.uconn.edu/people/plewis/software.php
11	Phylip	http://evolution.genetics.washington.edu/phylip/getme.html
12	Microsatellite	http://oscar.gen.tcd.ie/~sdepark/ms-toolkit/
13	TreeView	http://taxonomy.zoology.gla.ac.uk/rod/treeview.html
14	Genome analysis	https://connect.innovateuk.org/web/
15	Genome analysis	http://www.mgrc.com.my/
16	Genome analysis	http://www.animalgenome.org/bioinfo/
17	Genome analysis	http://www.bioplanet.com/links.htm
18	Genome analysis	http://www.123genomics.com/databases.html

also for their improvement. Genetic diversity as well as relatedness within and among the population, parentage determination, possible bottlenecks, linkage disequilibrium, inbreeding coefficients are essential for analyzing complete population structure. These include allozyme loci, restriction fragment length polymorphisms (RFLPs), major histocompatibility complex loci, AFLP, microsatellites, mitochondrial DNA, etc. Of these the microsatellite markers have emerged as the most powerful DNA tools for genetic analysis owing to their several unique characteristics and are globally being exploited to establish genetic profiles of animal genetic resources. Since their discovery, microsatellites have been used in mapping programs and by population biologists for studies of population genetic structure and kinship investigations (Zhu et al. 2000a).

Microsatellites have been recommended by FAO as first priority molecular tools for the Measurement of Domestic Animal Diversity (MoDAD). Microsatellites are also known as simple sequence repeats (SSR), short tandem repeats (STR), and sequence-tagged microsatellite repeats (STMR). They occur at a frequency of one SSR per 10 Kb DNA and numbering to a total of 50,000–100,000 in the mammalian genome (Beckmann and Weber 1992; Zhu et al. 2000b). Genetic diversity tools are mentioned in Table 17.4.

17.6.5 Biodiversity Analysis and Parentage Testing

By analyzing the microsatellite, profiles for each individual across different loci inferences can be made about overall magnitude of genetic diversity within breeds. The priority breeds for conservation should be the ones with the largest within breed diversity. Microsatellites are most suitable to determine the relationships, expressed as genetic distances among breeds, possible levels of inbreeding in each breed, gene flow in livestock populations, most diverse and distinctive, i.e., "genetically unique" breeds/populations for higher priority in conservation programs, and relative contribution of each breed to the total (species) genetic diversity. These markers have been successfully used for differentiation of closely related breeds and assignment of individuals to specific breeds (Jarne and Lagoda 1996). A parentage test helps to confirm the potential parents of an individual. DNA similarity is valuable to know if ancestry is right or not via small DNA test tools such as DNA profile, pedigree verification, DNA match, Sire/DAM match (http://www.genomnz.co.nz/our-services/parentage-analysis/).

17.6.6 Next-Generation Sequencing

Genomics plays very crucial role in the field of life sciences. The advancement in genomics and the development of high-throughput techniques (HTT) facilitate to characterization of wide range of genes according to their functions like regulation of genes, metabolic pathways, and their reconstructions. In post-genomic era there serious challenges for to storage and analysis of these huge amount of important

Table 17.4 Genetic diversity tools and their references/resources

Tool name	Reference
TFPGA	Miller, M.P. 1997. Tools for Population Genetic Analysis (TFPGA), 1.3: A Windows Program for the Analysis of Allozyme and Molecular Population Genetic Data. Distributed by the author
Structure	Pritchard, J.K., M. Stephens and P. Donnelly. 2000. Inference of population structure using multilocus genotype data. Genetics 155:945–959
SITES	Hey, J and J. Wakeley. 1997. A coalescent estimator of the population recombination rate. Genetics 145:833–846
PowerMarker	Liu, J. 2003. PowerMarker: New Genetic Data Analysis Software, Version 3.0. Free program distributed by author over Internet at
POPGENE	Yeh, F.C., R.C. Yang, T.B.J. Boyle, Z.H. Ye and J.X. Mao. 1997. POPGENE, the User-Friendly Shareware for Population Genetic Analysis. Molecular Biology and Biotechnology Centre, University of Alberta, Canada
PHYLIP	Felsenstein, J. 1993. PHYLIP (Phylogeny Inference Package), Version 3.5c. Distributed by the author
PAUP	Swofford, D.L. 2002. PAUP*: Phylogenetic Analysis Using Parsimony (*and Other Methods), Version 4. Sinauer Associates, Sunderland, MA
NTSYSpc	Rohlf, F.J. 2002. NTSYS pc: Numerical Taxonomy System, Version 2.1. Exeter Publishing, Setauket, NY
MEGA2	Kumar, S., K. Tamura, I.B. Jakobsen and M. Nei. 2001. MEGA2: Molecular Evolutionary Genetics Analysis software. Bioinformatics 17(12):1244–1245
MALIGN	Janies, D. and W.C. Wheeler. 1998. MALIGN.pdf: Documentation for, software for multiple alignments of DNA sequences. Distributed by the authors over Internet at .
MacClade	Maddison, D.R. and W.P. Maddison. 2000. MacClade. Version 4. Sinauer Associates, Sunderland, MA
GeneStrut	Constantine, C.C., R.P. Hobbs and A.J. Lymbery. 1994. FORTRAN programs for analyzing population structure from multilocus genotype data. J. Hered. 85:336–337
GENEPOP	Raymond, M. and F. Rousset. 1995. GENEPOP (version 1.2): Population genetics software for exact tests and ecumenicism. J. Hered. 86:248–249
GDA	Lewis, P.O. and D. Zaykin. 1999. Genetic Data Analysis: Computer Program for the Analysis of Allelic Data, Version 1.0 (d12). Distributed by the authors
DnaSP	Rozas, J. and R. Rozas. 1995. DnaSP, DNA sequence polymorphism: an interactive program for estimating population genetics parameters from DNA sequence data. Comput. Appl. Biosci. 11:621–625
CLUSTAL W	Thompson, J.D., D.G. Higgins and T.J. Gibson. 1994. CLUSTAL W: improving the sensitivity of progressive multiple sequence alignment through sequence weighting, position-specific gap penalties and weight matrix choice. Nucleic Acids Res. 22:4673–4680
Arlequin	Schneider, S., D. Roessli and L. Excoffier. 2000. Arlequin: A Software for Population Genetics Data Analysis, Version 2.000. Genetics and Biometry Laboratory, Dept. of Anthropology, University of Geneva, Switzerland

data. Even that early stage of this era, so many commercial techniques and tools/ softwares are available to analyze NGS data. Many methods are available for NGS Illumina/Solexa, Roche454, Ion Torrent/PGM sequencing, and SOLiD.

17.6.7 Identification of Genes

When the contigs are obtained from assembly of DNA fragments, the next step is to identify the protein coding regions in them. This can be done in three ways: (1) using Hidden Markov Model (HMM)-based techniques, (2) searching the known database of genes to identify new genes, and (3) using algorithms based on decision trees that identify start and stop codons of the coding regions. Metabolic pathway analysis in livestock is a necessary step in understanding the evolutionary origins of metabolism and species-specific adaptive traits (Seo and Lewin 2009; Olvera et al. 2010). Tools for identification of genes are given in Table 17.5.

17.6.8 Expression Analysis Methods: RNA-Seq

RNA-Seq is a very influential technology for transcriptomics studies. It enables us to investigate the gene activities of organisms at different tissues, different stages, and/or under different conditions. RNA-Seq arrests almost all of the expressed transcripts for a snapshot of cells in theory, while microarrays rely on prior information that cannot detect novel splicing variants, novel genes, and novel transcripts. In addition, RNA-Seq has low background noise and high sensitivity, requires less RNA sample, and is becoming more cost-effective with the rapid advancements in the technology. Those advantages of RNA-Seq provide us the abilities to illustrate the complexity of transcriptome more comprehensively and generate an unprecedented global view of the transcriptome for various species. These aspects challenge the corresponding methods and algorithms to effectively process the large amount of RNA-Seq data (Cumbie et al. 2011; Garber et al. 2011). For those organisms that have relatively complete and have high quality reference genomes, we can directly map the RNA-Seq reads onto the reference and carry out

Table 17.5 Identification of genes

Tools	References/resources
AUGUSTUS	http://bioinf.unigreifswald.de/webaugustus/prediction/create
EUGENE	http://eugene.toulouse.inra.fr/
BGF	http://bgf.genomics.org.cn/
FGENESH	http://linux1.softberry.com/berry.phtml?topic=fgenesh&group=programs&subgroup=gfind
Geneid	http://genome.crg.es/software/geneid/geneid.html
GeneMark	http://topaz.gatech.edu/GeneMark/
Glimmer	http://ccb.jhu.edu/software/glimmerhmm/

diverse transcriptomics studies. However, for those organisms without reference genomes or their reference genomes are uncompleted, other methods are required to accomplish related research (Jiang and Wong 2009).

There are diverse applications for RNA-Seq, and for each application, there are usually a number of available software that can be chosen. Choosing suitable software to carry out related studies and selecting the optimal parameters for the software are both very important, and they both directly influence the results (Ozsolak and Milos 2011).

17.6.9 Microarray Analysis

Earlier microarray technology revolutionized the study of gene expression and has given rise to an unprecedented increase in the rate of data acquisition in identifying gene transcript regulation in complex eukaryotic genome. A microarray is a very powerful technology that allows large numbers of genes, up to the order of tens of thousands, to be evaluated simultaneously (Caetano et al. 2004). The core principle behind microarrays is the hybridization between two DNA strands, the property of complementary nucleic acid sequences to specifically pair with each other by forming hydrogen bonds between complementary nucleotide base pairs (Ushizawa et al. 2004).

17.6.10 Applications of Bayesian Statistics in Animal Bioinformatics

Bayesian methods due to recent developments in software are now used in many areas of science and technology. The emphasis in most published applications is on statistical inference rather than decision-making; although Bayesian theory tells us how to act under uncertainty, choose the decision which maximizes our expected utility (Bustamante et al. 2003). Bayesian insinuation provides an interesting general framework for phylogenetic analysis, which helps in incorporating a wide variety of modeling assumptions and provides a rational dealing of uncertainty (Corander et al. 2003; Hardcastle and Kelly 2010).

17.6.11 Whole-Genome Regression and Prediction Methods Applied to Animal Breeding

Genomic-enabled prediction is becoming increasingly important in animal and is also receiving attention in human genetics. For procuring accurate predictions of complex traits, implementation of whole-genome regression (WGR) models is required where phenotypes are regressed on thousands of markers concurrently. Methods exist that allow implementing these large-p with small-n regressions. This technique is implemented in many plant and animal breeding programs. (De Los

Campos et al. 2013). Many tools are available for statistical analysis such as rrBLUP, Synbreed, GEMMA, BGLR, and BigPSD.

17.7 Conclusion

The recent advances in bioinformatics and information technology have resulted in tremendous computational power available to understand or make better use of huge amount of molecular data generated in the field of animal breeding and genetics. Breeders can target traits related with growth rate, egg, meat, milk, wool production, and disease resistance or have other desirable traits that have revolutionized agricultural livestock production throughout the world. There are wide range of applications of bioinformatics in different domains like genetic diversity analysis using microsatellite and SNP data, genome annotation, genome-wide association studies, transcriptome analysis, microarray data analysis, protein structure prediction, metabolic pathway analysis, phylogeny, genetic diseases, etc. in the field of animal as well as veterinary science. Therefore we can conclude that the application of bioinformatics will allow the animal breeding and genetics for generating knowledge on the role and sustainable use of genetic variation in animals. It will also contribute to our quality of life by providing knowledge to support the adequate supply of safe and healthy food of animal origin and to enhance the health, welfare, and productivity of animals. It will also help in animal conservation of extinct endangered species of animals.

Acknowledgment The authors are grateful to the Sam Higginbottom Institute of Agriculture, Technology and Sciences (formerly Allahabad Agriculture Institute) (Deemed-to-be-University), Allahabad, for providing the facilities and support to complete the work.

References

Andersson L (2001) Genetic dissection of phenotypic diversity in farm animals. Nat Rev Genet 2:130–138

Baurley JW et al (2013) A web application and database for agriculture genetic diversity and association studies. Int J Bio-Sci Bio-Tech 5:33–42

Beckmann JS, Weber JL (1992) Survey of human and rat microsatellites. Genomics 12:627–631

Bouchard TJ, Mcgue M (2003) Genetic and environmental influences on human psychological differences. J Neurobiol 54:4–45

Burt DW (2002) Comparative mapping in farm animals. Brief Funct Genomics Proteomics 1:159–168

Bustamante CD et al (2003) Maximum likelihood and Bayesian methods for estimating the distribution of selective effects among classes of mutations using DNA polymorphism data. Theor Popul Biol 63:91–103

Caetano AR et al (2004) Microarray profiling for differential gene expression in ovaries and ovarian follicles of pigs selected for increased ovulation rate. Genetics 168:1529–1537

Cios KJ et al (2005) Computational intelligence in solving bioinformatics problems. Artif Intell Med 35:1–8

Corander J et al (2003) Bayesian analysis of genetic differentiation between populations. Genetics 163:367–374
Cumbie JS et al (2011) GENE-counter: a computational pipeline for the analysis of RNA-Seq data for gene expression differences. PLoS One 6:e25279
Daetwyler HD et al (2007) Inbreeding in genome-wide selection. J Anim Breed Genet 124:369–376
Daetwyler HD et al (2013) Genomic prediction in animals and plants: simulation of data, validation, reporting, and benchmarking. Genetics 193:347–365
De Los Campos G et al (2013) Whole-genome regression and prediction methods applied to plant and animal breeding. Genetics 193:327–345
Fadiel A et al (2005) Farm animal genomics and informatics: an update. Nucleic Acids Res 33:6308–6318
Garber M et al (2011) Computational methods for transcriptome annotation and quantification using RNA-seq. Nat Methods 8:469–477
Goddard ME, Hayes B (2007) Genomic selection. J Anim Breed Genet 124:323–330
Groeneveld L et al (2010) Genetic diversity in farm animals–a review. Anim Genet 41:6–31
Hardcastle TJ, Kelly KA (2010) baySeq: empirical Bayesian methods for identifying differential expression in sequence count data. BMC bioinformatics 11:422
Hu ZL, et al.(2010). QTLdb: a comprehensive database tool building bridges between genotypes and phenotypes. In: Proceedings of the 9th world congress on genetics applied to livestock production
Hu ZL et al (2013) Animal QTLdb: an improved database tool for livestock animal QTL/association data dissemination in the post-genome era. Nucleic Acids Res 41:D871–D879
Jarne P, Lagoda PJ (1996) Microsatellites, from molecules to populations and back. Trends Ecol Evol 11:424–429
Javadi FH (2013). Bioinformatics and molecular genetic studies of domestic and wild buffalo species: focus on evolutionary relationship of the DGAT1gene
Jiang H, Wong WH (2009) Statistical inferences for isoform expression in RNA-Seq. Bioinformatics 25:1026–1032
Jonas E, Koning DJD (2015) Genomic selection needs to be carefully assessed to meet specific requirements in livestock breeding programs. Front Genet 6:49
Jovanović S et al (2009) Genetic variation in disease resistance among farm animals. Biotechnol Anim Husb 25:339–347
Lee J et al (2003) Current status of comparative mapping in livestock. Asian Australas J Anim Sci 16:1411–1420
Lenffer J et al (2006) OMIA (Online Mendelian Inheritance in Animals): an enhanced platform and integration into the Entrez search interface at NCBI. Nucleic Acids Res 34:D599–D601
Mitra S Acharya T (2005). Data mining: multimedia, soft computing, and bioinformatics. Wiley, NewYork
Montaldo HH (2006). Genetic engineering applications in animal breeding. Electron J Biotechnol 9. doi:https://doi.org/10.2225/vol9-issue2-fulltext-7
Moreno CR et al (2003) Detection of new quantitative trait loci for susceptibility to transmissible spongiform encephalopathies in mice. Genetics 165:2085–2091
Nicholas F (1998) Genetic databases: online catalogues of inherited disorders. Rev Sci Tech (Int Off Epizootics) 17:346–350
Nicholas FW (2003) Online Mendelian Inheritance in Animals (OMIA): a comparative knowledgebase of genetic disorders and other familial traits in non-laboratory animals. Nucleic Acids Res 31:275–277
Olvera A et al (2010) Applying phylogenetic analysis to viral livestock diseases: moving beyond molecular typing. Vet J 184:130–137
Ozsolak F, Milos PM (2011) RNA sequencing: advances, challenges and opportunities. Nat Rev Genet 12:87–98

Seo S, Lewin HA (2009) Reconstruction of metabolic pathways for the cattle genome. BMC Syst Biol 3:1
Shah SC, Kusiak A (2004) Data mining and genetic algorithm based gene/SNP selection. Artif Intell Med 31:183–196
Sherry ST et al (1999) dbSNP-database for single nucleotide polymorphisms and other classes of minor genetic variation. Genome Res 9:677–679
Ushizawa K et al (2004) cDNA microarray analysis of bovine embryo gene expression profiles during the pre-implantation period. Reprod Biol Endocrinol 2:1
Wickham B, et al. (2013). Information system technology for integrated animal identification, traceability and performance recording: the example of the Irish cattle sector. ICAR Technical Series
Zhu Y et al (2000a) Genetic diversity and disease control in rice. Nature 406:718–722
Zhu Y et al (2000b) Insertions, substitutions, and the origin of microsatellites. Genet Res 76:227–236

α-Amylase Inhibitor's Performance in the Control of *Diabetes Mellitus*: An Application of Computational Biology

18

Jyoti Verma, C. Awasthi, Qazi Mohammad Sajid Jamal, Mohd. Haris Siddiqui, Gulshan Wadhwa, and Kavindra Kumar Kesari

Abstract

Diabetes mellitus is the most widespread disorders prevalent in current period. α-Amylase enzyme plays a key role in the onset of the abnormal condition by breaking starch into glucose; hence its inhibitors need to be studied thoroughly. Due to the various side effects posed by the existing commercial non-proteinaceous inhibitors, exploration of the natural plant-based inhibitors of the enzyme is the present-day demand. Ample of plants have been extensively studied and reported to exhibit hypoglycaemic properties. This article describes the mode of action of amylase enzyme, phytochemicals which behave as amylase inhibitors and classes of its inhibitors and summarizes various plants studied for their enzyme inhibitory properties including computational tools and techniques to analyse the binding pattern exploration of inhibitors using molecular interaction with enzymes of interest.

J. Verma · C. Awasthi
Department of Biotechnology, Gobind Ballabh Pant Engineering College, Pauri Garhwal, Uttarakhand, India

Q. M. S. Jamal
Department of Health Information Management, College of Applied Medical Sciences, East Qassim University, Al Qassim-Buraydah, Kingdom of Saudi Arabia

M. H. Siddiqui
Department of Bioengineering, Faculty of Engineering, Integral University, Lucknow, Uttar Pradesh, India

G. Wadhwa
Department of Biotechnology, Apex Bioinformatics Centre, Ministry of Science & Technology, New Delhi, India

K. K. Kesari (✉)
Department of Applied Physics, and Department of Bioproduct & Biosystem, Aalto University, Espoo 00076, Finland
e-mail: kavindra.kesari@aalto.fi

G. Wadhwa et al. (eds.), *Current trends in Bioinformatics: An Insight*,
https://doi.org/10.1007/978-981-10-7483-7_18

Keywords
Diabetes mellitus · α-Amylase inhibitors · Phytochemicals

18.1 Introduction

Amylase inhibitors play a defensive role in diabetic patients (Ali et al. 2006). Diabetes mellitus is a metabolic disruption with multiple aetiologies specifically marked by chronic hyperglycaemia along with disturbance of carbohydrate, protein and fat metabolism caused by flaws in insulin secretion or an action of insulin (WHO 1999). The World Health Organization (WHO) proposed that the global population is exposed to diabetes epidemic with Southeast Asian and Western Pacific people being at considerable high risk. Presently the count of diabetic instances is 171 million which is prognosticated to attain 366 million by 2030 (WHO 2014). Diabetes is primarily of two types: type I and type II. Type I diabetes is upshot from improper synthesis of insulin by pancreatic β-cells, whereas type II diabetes results due to insulin resistance (a state in which peripheral cells fail to respond normally to insulin) or cell malfunction. Non-insulin-dependent diabetes mellitus (NIDDM/type II) is the most common form of diabetes. Long-term ramifications of disorder include retinopathy, nephropathy, neuropathy, microangiopathy and elevated risk of cardiovascular complications (Laar 2008; Cheng and Fantus 2005; Aguiar et al. 2007). Type I diabetes can be treated by lifelong insulin therapy. Diminution of the insulin need, stimulation of endogenously secreted insulin, enhancement of the insulin work at target tissues and prevention of timely disruption of oligo- and disaccharides are curative remedies for the treatment of type II diabetes (Funke and Melzing 2006). Decreasing post-prandial glucose levels through the activity cease of the carbohydrate-hydrolysing enzymes, α-glucosidase and α-amylase, found in the small intestinal brush border that cause breakdown of oligosaccharides and disaccharides into monosaccharides suited for absorption is one of the therapeutic pathways for treating type II diabetes mellitus (Laar 2008; Inzucchi 2002; Goke and Herrmann-Rinke 1998; Lebowitz 1998). The management of blood glucose level may be accomplished by the use of oral hypoglycaemic drugs, namely, biguanides, insulin secretagogues and α-amylase inhibitors (Kazeem et al. 2013). Clinically used α-amylase inhibitors are acarbose, miglitol and voglibose (Bailey 2003). Acarbose is amylase inhibitor which lags carbohydrate digestion causing a step-down in glucose absorption rate in body (Laar 2008; Cheng and Fantus 2005).

Human salivary α-amylase possesses three varied functions in oral cavity (Ramasubbu et al. 1996). Primarily, its hydrolytic activity is accountable for breakdown of starch to oligosaccharides. Secondly, salivary α-amylase confined to tooth enamel or hydroxyapatite may have significance in the formation of dental plaque. Thirdly, soluble α-amylase has strong affinity for viridans oral *streptococci*, and bacteria-indentured α-amylase is capable of hydrolysing starch to produce glucose, which is an energy source and further metabolized to lactic acid. A vital gradation in dental caries progression is the dissolution of tooth enamel caused due to localized acid production by bacteria (Ramasubbu et al. 1996; Scannapieco et al. 1993). So there is a need to recognize natural amylase inhibitors having lesser or no fallouts. Plants provide an effective potential for inhibition of α-amylase and can be used as therapeutic sources for the intervention of disorders in carbohydrate ingestion, such as diabetes and obesity as emphasized in Ayurveda. An ample amount of research had been done to identify and characterize the compounds from plant sources that have been screened for α-amylase inhibitory activity. In diabetic patients, glycaemic index can be controlled by α-amylase inhibitors which are a class of compounds that helps in managing post-prandial hyperglycaemia by inhibiting the action of α-amylase enzyme preeminent to reduction in starch hydrolysis (Notkins 2002). The conventionally known amylase inhibitors which are acarbose and miglitol (a deoxynojirimycin derivative) competitively and reversibly inhibit α-amylase enzyme, but they are associated with gastrointestinal side effects like flatulence, abdominal pain and diarrhoea in the patients (Fujisawa et al. 2005; Singh et al. 2007). Also there are factors which conduce to lack of inhibition in mammals which are inactivation of the inhibitor by gastric juices (Carlson et al. 1983), difference in pH optimal activity for inhibition (pH 4.5–5.0) and pH in the duodenum (pH 6–7) (Le Berre-Anton et al. 1997) and excess production of α-amylase in the human gut (Fogel and Gray 1973).

18.1.1 Amylases

The α-amylase (α-1,4-glucan-4-glucanohydrolases; E.C. 3.2.1.1) is among the main secreted products of the pancreas (about 5–6%) and salivary glands (Whitcomb and Lowe 2007). They are commonly present in microbial, flora and fauna kingdoms. Amylase enzymes are able to hydrolyse α-1,4-glycosidic bonds of amylose, amylopectin, glycogen and their downstream products. Bonds between adjacent glucose units are hydrolysed to yield products characteristic of the particular enzyme involved.α-amylase action produces sugars of multiple lengths with an α-configuration and α-limit dextrins as degradation products (Whitcomb and Lowe 2007), which form a concoction of maltose, maltotriose and branched oligosaccharides containing both α-1,4 and α-1,6 associations of 6–8 glucose units (Maarel et al. 2002). There are six classes into which the enzymes, having prospective commercial grandness of microbial origin that split α-1,4 and/or α-1,6 bonds in these complex structures, may be divided:

1. Enzymes hydrolysing α-1,4 bonds and bypassing α-1,6 linkages, e.g. α-amylase (endo-acting amylases)
2. Enzymes hydrolysing α-1,4 bonds and not bypassing α-1,6 linkages, e.g. β-amylase (exo-acting amylases producing maltose as a major downstream product)
3. Enzymes hydrolysing α-1,4 bonds as well as α-1,6 bonds, e.g. amyloglucosidase (glucoamylase) and exo-acting amylase

4. Enzymes hydrolysing only α-1,6 bonds, e.g. pullulanase and other fragmenting enzymes
5. Enzymes hydrolysing α-1,4 bonds in short-chain oligosaccharides formed by activity of other enzymes on amylose and amylopectin, e.g. α-glucosidase
6. Enzymes hydrolysing starch to a chain of non-reducing cyclic D-glucosyl polymers called cyclodextrins or Schardinger dextrins, e.g. *Bacillus macerans* amylase which is a cyclodextrin-forming enzyme

High degree of homology is present between porcine, human, mouse and rat pancreatic α-amylase amino acid sequences (Pasero et al. 1986). The three-dimensional molecular exemplary of porcine pancreatic α-amylase (PPA) had been studied in detail (Qian et al. 1993). The human α-amylase is a calcium-bearing enzyme composed of 512 amino acids in a single oligosaccharide chain with a molecular weight of 57.6 kDa (Whitcomb and Lowe 2007). Humans have five α-amylase genes located on chromosome 1, at location 1q21. Three genes code for salivary amylase, AMY1A, AMY1B and AMY1C, and the other two genes code forAMY2A and AMY2B in the pancreas (Groot et al. 1988; Gumucio et al. 1988). High proximal similarity in amino acid sequence is found between human salivary and pancreatic α-amylases with 97% identical entities overall and 92% in the catalytic domains (Ramasubbu et al. 1996; Brayer et al. 1995).

Three-dimensional structure of the enzyme is capable of binding to substrate and promotes the breakage of the glycoside links by the activity of extremely specific catalytic groups (Iulek et al. 2000). The protein constitutes of three domains: A, B and C among which domain A has an (β/α) eight-barrel fold and constitutes the catalytic core domain containing around 280–300 residues along with the catalytic triad (Asp, Asp, Glu) (Vander et al. 2002; Janecek et al. 1997). Between A and C domains lies the B domain which is linked to A domain by disulphide bond. A polypeptide chain links A domain to C domain which is an independent domain presenting a β-sheet structure with unreported function. The active site (substrate-binding) of the α-amylase is present between the carboxyl end of both A and B domains. The calcium ion (Ca^{2+}) present between A and B domain is an allosteric activator and stabilizes the three-dimensional structure. The substrate-binding site is comprised of five subsites (−3, −2, −1, +1, +2) (Brayer et al. 2000). Double displacement reaction occurs for the hydrolysis of starch in which a covalent β-glycosyl enzyme intermediate is formed and hydrolysed by using active-site carboxylic acids (Rydberg et al. 2002). Asp197, Glu233 and Asp300 function as catalytic residues (Brayer et al. 2000; Rydberg et al. 2002). A covalently bound reaction intermediate is formed when nucleophile Asp197 attacks substrate at sugar anomeric centre cleaving off the sugar skeleton from reducing end of substrate. In second step, covalent bond between Asp197 and substrate is cleaved by water molecule that attacks the anomeric centre, attaching a hydroxyl group to it. Glu233 and Asp300 either independently or conjointly behave as acid/base catalysts in both the steps. Kinetic studies had revealed that active site of human α-amylase comprises of multiple major binding subsites and "−1", "−2" and "−3" pocket is the core of the catalytic reaction (Brayer et al. 2000). Several new enzymes linked with starch hydrolysis and related polysaccharides structures have been recognized and studied in the past few years.

18.1.2 Starch

Starch is a main storage carbohydrate in all higher plants and a natural substrate for the amylase enzyme. It is found in the form of water-insoluble granules of varied size and shape which are often the stature of plant species from which they are educed and form approximately 70% of the undried plant material in some cases. Hydrogen bonds hold the granules together in polysaccharide unit which weaken up when heated in water and allow the molecule to swell and gelatinize. Depending on the concentration of polysaccharide, they form paste or dispersion. Starch being a heterogeneous polysaccharide is composed of amylose and amylopectin having high molecular weight. Basic structure of amylose is linear, and it is unstable in aqueous solution with degree of polymerization and average chain length of $C10^3$. It is hydrolysed by β-amylase to 87% and at 650 nm, and it shows maximum interaction with iodine. However, the basic structure of amylopectin is branched, and it is highly stable in aqueous solution with degree of polymerization $C10^4$–$C10^5$ and average chain length of C20–25. It is hydrolysed by β-amylase to 54% and shows maximum interaction with iodine at 550 nm. Polar solvent like n-butanol can cause dispersion of starch into two components. Amylose is extensively degraded by α-amylase because it is composed of linear chains of α-1,4-linked D-glucose residues. Amylose can polymerize to several thousands of glucose units.

18.2 Inhibitors of α-Amylase from Plants

More than 800 plant species have been identified as potential treatments for diabetes mellitus (Perez et al. 1998). Grover et al. (2002) stated that more than 1123 plant species have been used ethno-pharmacologically or experimentally to cure diabetic patients. More than 200 pure compounds from plant sources are isolated that have been reported to show blood glucose-lowering property (Marles and Farnsworth 1994). The pharmacological mechanisms of the herbs can be classified as (1) reduction in carbohydrate absorption, (2) improvement of insulin sensitivity, (3) peripheral glucose uptake increment, (4) insulin secretion stimulation, (5) potentiating endogenous incretins, (6) exertion of antioxidant effects and reducing cell apoptosis and (7) the glycogenesis increment or hepatic glycogenolysis inhibition (Li et al. 2004; Prabhakar and Doble 2011; Bhat et al. 2011). Different formulation consists of multiple extracts and compounds, which exhibit multiple mechanisms. Various plant-derived compounds like alkaloids, steroids, glycosides, guanidine, galactomannan gum, polysaccharides, peptidoglycans, hypoglycans, terpenoids and glycopeptides are reported to have bioactivity against hyperglycaemia (Mentreddy 2007). *Syzygium cumini* leaves and *Psidium guajava* leaves are used in Indian traditional system of medicine to treat diabetes, exhibiting a dose-dependent inhibitory effect on α-amylase activity (Karthic et al. 2008). Conforti et al. (2005) demonstrated α-amylase inhibitory activity of two varieties of *Amaranthus caudatus* L. seeds (Oscar blanco and Victor red oil) which showed inhibition rate above 80% at 0.25–1 mg/mL. Various plant species, namely, *Balanites aegyptiaca* L., *Galega officinalis* L., *Holarrhena floribunda* L., *Melissa officinalis* L., *Mitragyna inermis* (Willd.), *Rosmarinus officinalis* L., *Camellia sinensis* L. Del., *Tamarindus indica* L., *Taraxacum officinale* L. and *Vaccinium myrtillus* L., possess considerable inhibitory activity (above 45% inhibition rate at 0.2 g/mL)[7]. Methanolic extracts of other Mongolian plant species, namely, *Rhodiola rosea* L., *Ribes pulchellum* L., *Vaccinium uliginosum* L., *Geranium pratense* L., *Paeonia anomala* L. and *Pentaphylloides fruticosa* L., showed greater than 30% enzyme inhibition (Kobayashi et al. 2003). Loizzo et al. (2008) demonstrated that methanolic extracts of *Salvia acetabulosa* L. and *Marrubium radiatum* L. exhibited the highest inhibitory activity against α-amylase among Lebanon medicinal plants. Use of traditional Ayurvedic antidiabetic plants is being promoted for over thousands of years due to their no or apparently lesser side effects (Bhat et al. 2011; Bhutani and Gohil 2010). Among the Indian medicinal plants, *Mangifera indica* L., *Embelia ribes* L., *Phyllanthus maderaspatensis* L. and *Punica granatum* L. showed exciting α-amylase inhibitory activity (Prashanth et al. 2001). Chloroform extracts of Ayurvedic plants, namely, *Azadirachta indica*, *Ocimum tenuiflorum* L., *Murraya koenigii* L., *Linum usitatissimum* L. and *Bougainvillea spectabilis* L., are reported to exert inhibitory activity on α-amylase (Bhat et al. 2011). Other plants screened for enzyme inhibitory activity are *Hibiscus sabdariffa* (Hansawasdi et al. 2000), *Amaranthus hypochondriacus* seeds (Martins et al. 2001), *Artocarpus heterophyllus* (Kotowaroo et al. 2006), *Arecae semen* and *Corni fructus* (Choi et al. 2000). Strawberry and raspberry fruit extracts are also reported to possess α-amylase inhibitory activity.

Other plants that intervene with digestion of carbohydrates, attaining better glycaemic control, are *Adhatoda vasica* Nees (Mc Dougall et al. 2005; Gao et al. 2008a, b), *Piper umbellatum* (Tabopda et al. 2008), *Tussilago farfara* (Gao et al. 2008a, b), *Terminalia chebula* (Gao et al. 2007), *Bergenia ciliata* (Bhandari et al. 2008), *Grateloupia elliptica* (Kurihara et al. 1999), *Syagrus romanzoffiana* (Lam et al. 2008), *Curcuma longa* (Du et al. 2006), *Fagara tessmannii* (Mbaze et al. 2007) and *Gypsophila oldhamiana* (Luo et al. 2008). In Europe, oleanolic acid, extracted from *Olea europaea* leaves, is widely identified as a folk medicine for diabetes and hypertension (Komaki et al. 2003). In north-western Argentina, ursolic acid and oleanolic acid derivatives derived from *Polylepis australis* plant are used for treatment of diabetes (De lampasona et al. 1988). Plants with already reported α-amylase inhibitory activity are shown in Table 18.1.

The proteinaceous inhibitor of α-amylase (αAI), which inhibits animal salivary and pancreatic α-amylase, has been recognized and extracted from different plant species (Wang et al. 2011). Among these plants, seeds of *Phaseolus vulgaris* L. that consist of proteinaceous inhibitors of the α-amylase and isoform inhibitor αAI-1 have been isolated and characterized (Yamada et al. 2001). The common bean αAI-1 has been reported to have comparatively higher potential as panoptic antiobesity and

Table 18.1 Plants with α-amylase inhibitory activity

Plant	Parts used	Type of extract	Activity (% inhibition) (concentration)	Control	Reference
Acalypha indica	Leaf	Aqueous	15 (25–75 μL)	Non-treated enzyme	Karthic et al. (2008)
Aconitum heterophyllum	Rhizome	Aqueous	6.5 (80 μg/mL)	Acarbose with 50% inhibition at 14.24 μg/mL	Loizzo et al. (2008)
Acorus calamus	Rhizome	Aqueous	30.80 (80 μg/mL)	Acarbose with 50% inhibition at 14.24 μg/mL	Loizzo et al. (2008)
Aegle marmelos	Leaf	Aqueous	6 (25–75 μL)	Non-treated enzyme	Karthic et al. (2008)
Aloe vera	Leaf gel	Aqueous	23.3 (2.5 mg/mL)	Acarbose with 50% inhibition at 10.2 μg/ml	Sudha et al. (2011)
		Cyclohexane	15.8 (2.4 mg/mL)		
Amaranthus caudatus var. Oscar blanco	Seed	Methanol	94.71(1 mg/mL)	Non-treated enzyme	Conforti et al. (2005)
		Ethyl acetate	93.82(0.5 mg/mL)		
		Hexane	90.64 (0.1 mg/mL)		
Amaranthus caudatus var. Victor red	Seed	Methanol	95.12 (1 mg/mL)	Non-treated enzyme	Conforti et al. (2005)
		Ethyl acetate	84.03 (0.25 mg/mL)		
		Hexane	91.63 (0.1 mg/mL)		
Andrographis paniculata (Burm. f.) Nees	Aerial part	Ethanol	50 (50.9 ± 0.17 mg/mL)	Acarbose with 50% inhibition at 14.9 ± 0.23 mg/mL	Rammohan et al. (2008)
Azadirachta indica	Leaf	Chloroform	50 (7.5 mg/mL)	Non-treated enzyme	Bhat et al. (2011) and Kazeem et al. (2013)
		Aqueous	50 (9.15 mg/mL)		
Balanites aegyptiaca L.	Bark	Aqueous	45–75 (200 mg/mL)	Acarbose with inhibition higher than	Funke and Melzing (2006)
				75% at 200 mg/mL	
Berberis aristata	Bark	Aqueous	11.40 (80 μg/mL)	Acarbose with 50% inhibition at 14.24 μg/mL	Rituparna et al. (2014)

(continued)

Table 18.1 (continued)

Plant	Parts used	Type of extract	Activity (% inhibition) (concentration)	Control	Reference
Bergenia ciliata Haw.	Rhizome	50% methanol	93.5 (150 mg/mL)	Non-treated enzyme	Bhandari et al. (2008)
			65.3(150 mg/mL)		
		Aqueous	84.3(150 mg/mL)		
		Ethyl acetate			
Bougainvillea spectabilis Wild.	Leaf	Chloroform	29.43 (25 mg/mL)	Acarbose with 50% inhibition at 1.22 mg/mL	Bhat et al. (2011)
Cajanus cajan L.	Seed	Aqueous	100 (2 mg protein)	Non-treated enzyme	Giri and Kachole (1998)
Camellia sinensis L. Del.	Leaf	Aqueous	45–75 (200 mg/mL)	Acarbose with inhibition higher than 75% at 200 mg/mL	Funke and Melzing (2006)
Cassia auriculata	Flower	Aqueous	60.30 (80 μg/mL)	Acarbose with 50% inhibition at 14.24 μg/mL	Rituparna et al. (2014)
Clerodendrum multiflorum L.	Stem	Methanol	50 (25.22 μg/mL)	Acarbose with 50% inhibition at 9.22 μg/mL	Sneha and Sanjay (2011)
		ethyl acetate	50 (36.86 μg/mL)		
Croton bonplandianum Baill.	Leaf	Ethanol	50 (17.22 ± 0.05 μg/mL)	Acarbose with 50% inhibition at 2.65 ± 0.03 μg/mL	Keerthana et al. (2013)
Cyperus rotundus	Tuber	Aqueous	40.40 (80 μg/mL)	Acarbose with 50% inhibition at 14.24 μg/mL	Rituparna et al. (2014)
Galega officinalis L.	Herb	Aqueous	35 (200 mg/mL)	Acarbose with inhibition higher than 75% at 200 mg/mL	Funke and Melzing (2006)
Geranium pratense L.	Aerial part	Methanol	43.9 (0.3 mg/mL)	Acarbose with 79.6% inhibition at 0.1 mg/mL	Kobayashi et al. (2003)
Gymnema sylvestre	Leaf	Aqueous	3 (25–75 μL)	Non-treated enzyme	Karthic et al. (2008)
Hibiscus sabdariffa L.	Flower	Methanol 50%	100 (10 mL/g of wt.)	Non-treated enzyme	Hansawasdi et al. (2000)
Holarrhena floribunda (Don) Durand & Schinz	Leaf	Aqueous	20–45 (200 mg/mL)	Acarbose with inhibition higher than 75% at 200 mg/mL	Funke and Melzing (2006)
Limonia acidissima	Seed	Aqueous	20 (25–75 μL)	Non-treated enzyme	Karthic et al. (2008)

Linum usitatissimum L.	Seed	Isopropanol acetone	55.7 (0.65 mg/mL)	Acarbose with 50% inhibition at 10.2 μg/mL	Sudha et al. (2011)
			39.1 (2.7 mg/mL)		
Mangifera indica L.	Bark	Ethanol	84.1 (1 mg/mL)	*Phaseolus vulgaris* with 59.4%	Prashanth et al. (2001)
				Inhibition at 0.0125 mg/mL	
Marrubium radiatum Delile ex. Benth.	Aerial part	Methanol	50 (0.0611 mg/mL)	Acarbose with 50% inhibition at 0.05 mg/mL	Loizzo et al. (2008)
		Ethanol	50 (3.33 mg/mL)		
				Non-treated enzyme	
Melissa officinalis L.	Leaf	Aqueous	50 (200 mg/mL)	Acarbose with inhibition higher than 75% at 200 mg/mL	Mc Cue and Shetty (2004)
Mesua ferrea	Dried buds	Aqueous	17.30 (80 μg/mL)	Acarbose with 50% inhibition at 14.24 μg/mL	Rituparna et al. (2014)
Mitragyna inermis Willd. O. Ktze.	Leaf	Aqueous	75 (200 mg/mL)	Acarbose with inhibition higher than 75% at 200 mg/mL	Funke and Melzing (2006)
Moringa oleifera	Leaf	Aqueous	16 (25–75 μL)	Non-treated enzyme	Karthic et al. (2008)
Murraya koenigii L.	Leaf	Chloroform	56.64 (25 mg/mL)	Acarbose with 50% inhibition at 1.22 mg/mL	Bhat et al. (2011)
Ocimum tenuiflorum L.	Leaf	Chloroform	24.57 (10 mg/mL)	Acarbose with 50% inhibition at 1.22 mg/mL	Bhat et al. (2011)
Paeonia anomala L.	Root	Methanol	33.1 (0.3 mg/mL)	Acarbose with 79.6% inhibition at 0.1 mg/mL	Kobayashi et al. (2003)
Pentaphylloides fruticosa L. O. Schwarz	Leaf and branch	Methanol	31.2 (0.3 mg/mL)	Acarbose with 79.6% inhibition at 0.1 mg/mL	Kobayashi et al. (2003)
Phaseolus vulgaris L.	Pericarp	Aqueous	45–75 (200 mg/mL)	Acarbose with inhibition higher than 75% at 200 mg/mL	Funke and Melzing (2006)
Phyllanthus amarus	Aerial part	Ethanol	50 (36.05 ± 4.01 μg/mL)	Acarbose with 50% inhibition at 83.33 ± 0.34 μg/mL	Iniyan et al. (2010)
		Hexane	50 (48.92 ± 3.43 μg/mL)		
Pithecellobium dulce	Seed	Methanol	50 (16.75 ± 1.81 mg/mL)	Non-treated enzyme	Dnyaneshwar and Archana (2013)

(continued)

Table 18.1 (continued)

Plant	Parts used	Type of extract	Activity (% inhibition) (concentration)	Control	Reference
Plumbago zeylanica	Root	Aqueous	52.90 (80 μg/mL)	Acarbose with 50% inhibition at 14.24 μg/mL	Rituparna et al. (2014)
Premna corymbosa rottl.	Leaf	Methanol	50 (145.72 μg/mL)	Non-treated enzyme	Radhika et al. (2013)
		Ethyl acetate	50 (176.60 μg/mL)		
		Hexane	50(183.95 μg/mL)		
Psidium guajava L.	Leaf	Ethanol	31.7(1.5 mg/mL)	Acarbose with 52.1% inhibition at 1.5 mg/mL	Wang et al. (2011)
Psidium guajava var. Pomiferum	Leaf	Aqueous	98 (200 mg/mL)	Non-treated enzyme	Karthic et al. (2008)
Punica granatum L.	Fruit rind	Ethanol	68.2 (1 mg/mL)	*Phaseolus vulgaris* with 59.4%	Prashanth et al. (2001)
				Inhibition at 0.0125 mg/mL	
Rhodiola rosea L.	Rhizome	Methanol	78 (0.3 mg/mL)	Acarbose with 79.6% inhibition at 0.1 mg/mL	Kobayashi et al. (2003)
Ribes pulchellum Turcz.	Aerial part	Methanol	78.9 (0.3 mg/mL)	Acarbose with 79.6% inhibition at 0.1 mg/mL	Kobayashi et al. (2003)
Rosmarinus officinalis L.	Leaf	Aqueous	60 (200 mg/mL)	Acarbose with inhibition higher than 75% at 200 mg/mL	Funke and Melzing (2006)
Salvia acetabulosa L.	Aerial part	Methanol	50 (0.0912 mg/mL)	Acarbose with 50% inhibition at 0.05 mg/mL	Loizzo et al. (2008)
Syzygium cumini L. Skeels	Leaf	Chloroform	22.31 (10 mg/mL)	Acarbose with 50% inhibition at 1.22 mg/mL	Bhat et al. (2011)
	Seed	Aqueous	98 (200 mg/mL)	Non-treated enzyme	Karthic et al. (2008)
Tamarindus indica L.	Leaf	Aqueous	90 (200 mg/mL)	Acarbose with inhibition higher than 75% at 200 mg/mL	Funke and Melzing (2006)
Taraxacum officinale Web. ex Wigg.	Herb	Aqueous	20–45 (200 mg/mL)	Acarbose with inhibition higher than 75% at 200 mg/mL	Funke and Melzing (2006)

Terminalia arjuna	Bark	Aqueous	62.50 (80 µg/mL)	Acarbose with 50% inhibition at 14.24 µg/mL	Rituparna et al. (2014)
Thespesia populnea L.	Leaf	Petroleum ether	50 (100 µg/mL)	Non-treated enzyme	Sangeetha and Vedasree (2012)
			50 (120 µg/mL)		
		Chloroform	98 (50 µg/mL)		
		Ethyl acetate	50 (20 µg/mL)		
		Methanol			
Tinospora cordifolia	Leaf	Aqueous	13 (25–75 µL)	Non-treated enzyme	Karthic et al. (2008)
Trigonella foenum-graecum	Seed	Aqueous	10 (25–75 µL)	Non-treated enzyme	Karthic et al. (2008)
Vaccinium myrtillus L.	Leaf	Aqueous	>75 (200 mg/mL)	Acarbose with inhibition higher than 75% at 200 mg/mL	Funke and Melzing (2006)
Vaccinium uliginosum L.	Leaf and wood	Methanol	80.7 (0.3 mg/mL)	Acarbose with 79.6% inhibition at 0.1 mg/mL	Kobayashi et al. (2003)
Ziziphus mauritiana	Seed	Aqueous	12 (25–75 µL)	Non-treated enzyme	Karthic et al. (2008)

anti-diabetes remedy (Wang et al. 2011).Two α-amylase inhibitors, called αAI-1 and αAI-2, that share 78% amino acid sequence identity and have a differential specificity towards mammalian and insect α-amylases occur in different accessions of the common bean (*Phaseolus vulgaris*). Seeds of *P. vulgaris* contain α-amylase inhibitor (αAI) which is synthesized as a pro-protein on the endoplasmic reticulum and is proteolytically processed post arrival in protein storage vacuoles to polypeptides of relative molecular weight (M_r) 15,000–18,000.

In barley, the endogenous inhibitor BASI (barley α-amylase/subtilisin inhibitor) is a 19.6 kDa bifunctional protein which binds with high affinity to the major α-amylase isozyme (AMY2) whereby the minor isozyme (AMY1) is not inhibited (Vallee et al. 1998). The affinity of AMY2 for BASI is vitiated with lowering pH levels from pH 8.0 to pH 6.0 (Abe et al. 1993) and elevating ionic strength up to 200 mM (Weselake et al. 1985). Vallée et al. (1998) determined the crystal structure of AMY2-BASI complex at 1.9 Å resolution which is the first reported structure of an endogenous protein-protein complex from a higher plant. The complex showed the occurrence of a completely solvated calcium ion at the centre of interface, which is absent in the structure of free AMY2 (Kadziola et al. 1994). The K_i of the inhibition has been reported to be 2.2×10^{-10} M at pH 8 and 37 °C (Abe et al. 1993). Specific arginine residue on the surface of BASI is vital for the inhibition of AMY2 (Abe et al. 1993). The crystal structures of AMY2 which is a 403-residue single-chain protein with three free cysteine residues have been determined by X-ray crystallography at 2.8 Å resolution (Kadziola et al. 1994) and in complex with the pseudo-tetrasaccharide acarbose inhibitor (Kadziola et al. 1998). Very high level of sequence identity had been observed between the bifunctional plant inhibitors of barley (BASI) and wheat (WASI) (92% identity) and to a lesser grade with rice (RASI) (58% identity) and the Kunitz family trypsin inhibitors STI (soybean trypsin inhibitor), WTI (wheat trypsin inhibitor) and ETI (Erythrina trypsin inhibitor) (around 25% identity) (Onesti et al. 1991). The three-dimensional structure of WASI has been reproduced in both its free state (Zemke et al. 1991) and in complex with a serine protease, proteinase K (Pal et al. 1994).

Ragi bifunctional inhibitor is reported to exhibit sequence homology with the α-amylase inhibitor from wheat and also with trypsin inhibitors from barley (Odani et al. 1982) and maize (Hochstrasser et al. 1970) and thus clearly establishes the existence in cereals of a new family of inhibitors that are active against exclusively unrelated enzymes (Campos and Richardson 1983). The Ragi bifunctional α-amylase/trypsin inhibitor (RBI) is only a member of cereal inhibitor superfamily that inhibits both trypsin and α-amylases, but the mode of inhibition of α-amylases by these cereal inhibitors is obscure. α-Amylase inhibition by commercially available inhibitors and natural plant-based inhibitors is shown in Fig. 18.1.

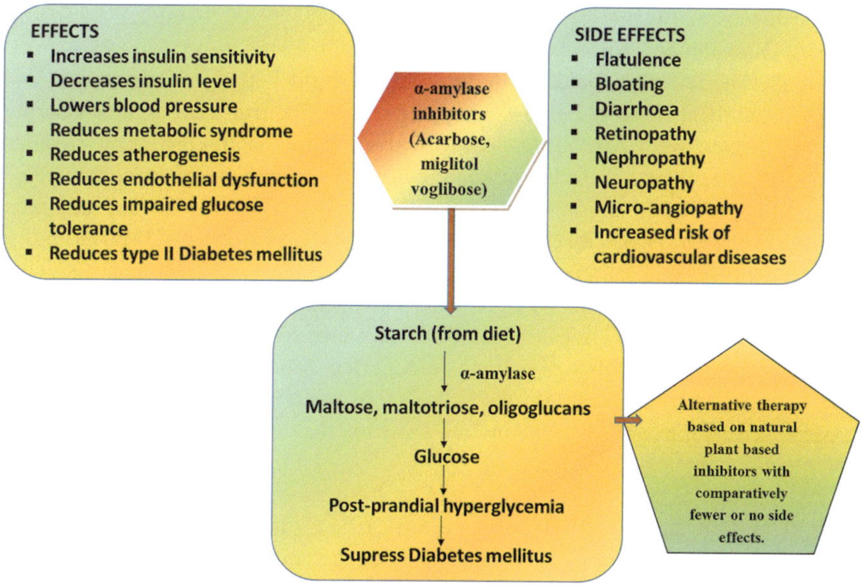

Fig. 18.1 α-amylase inhibition by commercially available inhibitors and natural plant based inhibitors. Double marks (=) indicate α-amylase inhibition

18.3 Phytochemicals with α-Amylase Inhibitory Activity

Oligosaccharide inhibitors of trestatin family, containing the acarviosine moiety (e.g. acarbose 1), are proteinaceous inhibitors purified from plant tissues and microbial sources (Svensson et al. 2004). Acarbose produced by *Actinoplanes* sp. fermentation is a pseudo-tetrasaccharide and a widely used drug for clinical treatment of diabetes mellitus. It comprises of a polyhydroxylated aminocyclohexane derivative (valienamine) associated via its nitrogen atom to a 6-deoxyglucose, which is itself α-1,4-linked to a maltose moiety. The mechanism of inhibition is of competitive type due to glycosidic nitrogen association and unsaturated cyclohexene ring that depicts the transition state for cleavage of glycosidic linkages (Yoon and Robyt 2003; Gyemant et al. 2003).

Phenolic compounds have at least a phenolic mediety in their structures and have high degrees of antioxidant properties as well as medicinal properties, hence used as drugs from quite long times. Flavonoids belong to the class of natural phenolic compounds and possess several biological properties. Their structures consist of a 15-carbon skeleton. Enzyme inhibition mechanism of flavonoids was investigated by Piparo et al. (2008). Phenolic compounds and flavonoid-rich plant species like *Hypericum triquetrifolium*, *Aloe vera*, *Osyris alba*, *Arbutus andrachne*, *Sarcopoterium spinosum*, *Geranium robertianum* and *Aesculus hippocastanum* had been screened using capillary electrophoresis for having α-amylase inhibitory activity (Hamdan and Afifi 2010).

Another widely distributed high molecular weight heterogeneous polyphenol group in the plant realm is tannins. They occur in unripe fruits but diminish during ripening. They are classified as hydrolysable tannins and condensed tannins. Hydrolysable tannins are further divided into gallotannins, derived from gallic acid units linked to a carbohydrate moiety, whereas condensed tannins are complex polymers derived from catechins and flavonoids (Kandra et al. 2004). Tannins being potential metal ion chelators and protein precipitation agents form insoluble complexes with proteins and biological oxidants, hence inducing distinct effects on the biological system (Quideau et al. 2003). Tannins inhibit α-amylase activity in situ although the mechanism of action is unreported. Kandra et al. (2004) suggested that interaction between tannins, like galloylated quinic acid, and human α-amylase is correlated with free OH groups in their structure that are capable of taking part in hydrogen bonding.

Terpenoids are compounds that originate from subsequent joining of isoprene units. They are classified as monoterpene (C10), sesquiterpene (C15), diterpene (C20), sesterterpene (C25), triterpene (C30) and tetraterpene (C40) on the basis of number of isoprene units present (55). They are derived from squalene or related acyclic 30-carbon precursors with many therapeutic uses (Connolly and Hill 1999). Sterols, steroids and saponins are triterpenoids with well-characterized biological activities (Xu et al. 2004). Inhibitory α-amylase activity is reported only for oleanane, ursane and lupane types of triterpenoids although the mechanism of action is unknown.

18.4 α-Amylase Inhibitor Classification

18.4.1 Non-proteinaceous Inhibitors

This class of inhibitors contains organic compounds such as hibiscus acid, acarviosine-glucose, acarbose, isoacarbose and cyclodextrins. The hibiscus acid forms, namely, the acarviosine-glucose, the isoacarbose and α-, β- and γ-cyclodextrins are purified from Roselle tea (*Hibiscus sabdariffa*), show inhibitory activity against porcine and human pancreatic α-amylase (PPA and HPA) (Kim et al. 1999; Nahoum et al. 2000; Qian et al. 2001; Hansawasdi et al. 2000). The inhibition is due to their cyclic structures, which resemble α-amylase substrates. Workers have studied the panoptic interactions between inhibitor and enzyme's active-site region that directly resemble protein-carbohydrate interactions observed in the PPA-acarbose complex. The structure of the PPA-acarbose complex at 2.2 Å resolution was determined, to study their interactions which revealed a number of structural variations between the unliganded and liganded enzyme (Qian et al. 1994). Bompard-Gilles et al. (1996) reported an extended protein-protein interface in the structure of the PPA-α-AI complex that accounts for large inhibition constant

($K_i = 3.5 \times 10^{-11}$), which was a similar attribute noticed in the structure of PPA complexed with the proteinaceous inhibitor tendamistat (Wiegand et al. 1995).

18.4.2 Proteinaceous Inhibitors

They are present in microorganisms, plants and animals (Ryan 1990; Silano 1987; Franco et al. 2000; Iulek et al. 2000). The structure of PPA in complex with microbial inhibitor tendamistat (74 amino acid residues), which is a class member of proteinaceous inhibitors from *Streptomyces* sp., had been reported (Wiegand et al. 1995). In plants, they occur primarily in cereals such as wheat (*Triticum aestivum*) (Franco et al. 2000; Petrucci et al. 1976; Feng et al. 1996), barley (*Hordeum vulgare*) (Abe et al. 1993), sorghum (*Sorghum bicolour*) (Bloch and Richardson 1991), rye (*Secale cereale*) (Iulek et al. 2000; Garcia-Casadoet al. 1994), rice (*Oryza sativa*) (Yamagata et al. 1998), cowpea (*Vigna unguiculata*) (Melo et al. 1999) and bean (*P. vulgaris*) (Grossi et al. 1997; Young et al. 1999). They have monomeric molecular masses of 5 kDa (Bloch and Richardson 1991), 9 kDa (Melo et al. 1999) and 13 kDa (Feng et al. 1996), homodimeric and heterodimeric masses of ≈26 kDa (Feng et al. 1996; Young et al. 1999) and tetrameric masses of 50 kDa (Kasahara et al. 1996). Giri and Kachole (1998) identified four AI isoforms in pigeon pea (*Cajanus cajan*) seeds that inhibited human salivary and bovine pancreatic amylase but failed to inhibit bacterial, fungal and endogenous amylase. These AIs were synthesized during late seed development and also degraded during late germination.

The active-site cleft in structure of barley α-amylase in complex with the proteinaceous inhibitor barley α-amylase/subtilisin inhibitor (BASI) was sterically blocked by the extended protein-protein interaction that might inhibit the enzyme (Vallee 1996). Bompard-Gilles et al. (1996) reported the structure of a porcine pancreatic α-amylase (PPA) in complex with the *P. vulgaris* inhibitor (α-AI1) which inhibits the activity of both insect and mammalian α-amylases. Molecular level analysis of the modality of action of these proteins, along with the general interest in inhibition of glycolytic enzymes, may prove to be an effective approach for the treatment of diabetes.

Richardson classified α-amylase inhibitors on the basis of their tertiary structure into classes like lectin-like, knottin-like, cereal-type, Kunitz-like, γ-purothionin-like and thaumatin-like (Richardson 1990).

18.4.3 Lectin-Like α-Amylase Inhibitors

The amylase inhibitor from beans has the same structure as lectin (agglutinin)-like proteins. α-AIs has been purified and characterized from white, red and black kidney beans *P. vulgaris* (Kasahara et al. 1996; Marshall and Lauda 1975; Wilcox and Whitaker 1984). α-AI-1, which is the best-characterized isoform, was cloned and identified as an α-amylase inhibitor homologous to phytohemagglutinin (PHA)

(Moreno and Chrispeels 1989). α-AI-1 shows inhibitory activity against PPA as well as the α-amylases of the *Callosobruchus maculatus* and *Cryptocarya chinensis*. A second allelic variant of α-AI, called α-AI-2 which has different inhibition specificity, is found in some wild accessions of common bean, and it inhibits the α-amylases of the *Zabrotes subfasciatus* (ZSA) (Grossi et al. 1997). A third isoform, α-AI-3 purified from *P. vulgaris*, is a single-chain α-amylase inhibitor-like protein which shows no activity towards any α-amylases (Finardi-Filho et al. 1996).

18.4.4 Knottin-Type α-Amylase Inhibitors

The smallest proteinaceous inhibitor of α-amylases till date is extracted from *Amaranthus hypochondriacus* seeds (AAI) with just 32 residues and 3 disulphide bonds (Chagolla-Lopez et al. 1994). The inhibitor structure was determined by NMR (Lu et al. 1999; Martins et al. 2001) and contains a knottin fold, three antiparallel β-strands and a characteristic disulphide topology. It was structurally similar to charybdotoxin and conotoxins (Park and Miller 1992) and the proteinase inhibitors from *Cucurbita maxima* (Bode et al. 1989). Gurmarin, a knottin related to the amaranth α-amylase inhibitor, has been used in traditional treatment of diabetes (Imoto et al. 1991).

18.4.5 Cereal-Type α-Amylase Inhibitors

α-Amylase inhibitors belonging to cereal family are composed of 120–160 amino acid residues forming 5 disulphide bonds (Franco et al. 2000; Buonocore et al. 1977; Lyons et al. 1987). Upon repeated exposure to humans, these inhibitors cause allergy, dermatitis and baker's asthma colligated with cereal flour, hence acting as sensitizing agents (Garcia-Casado et al. 1996; Kusaba-Nakayama et al. 2000). An inhibitor from rye which is the most reactive allergen is N-glycosylated for its reactivity (Garcia-Casado et al. 1995). The most focussed inhibitors from this family are exogenous wheat α-amylase inhibitor (Franco et al. 2000; Feng et al. 1996) and the bifunctional inhibitor from Indian finger millet (RBI) (Strobl et al. 1998; Alam et al. 2001).

18.4.6 Kunitz-Like α-Amylase Inhibitors

The Kunitz-like α-amylase inhibitors present in cereals, such as barley (Rodenburg et al. 1995), wheat (Gvozdeva et al. 1993) and rice (Ohtsubo and Richardson 1992), contain around 180 residues and 4 cysteine residues. The best studied inhibitor from this class is barley α-amylase/subtilisin inhibitor (BASI) which is a bifunctional double-headed inhibitor with strong inhibitory reaction, cereal α-amylase AMY2 and serine proteinases of subtilisin family (Mundy et al. 1983; Abe et al. 1993). Two disulphide bonds and a β-trefoil topology were revealed in the structure of BASI

(Vallee et al. 1998) which was homologous with wheat α-amylase subtilisin inhibitor (WASI) (Zemke et al. 1991), *Erythrina caffra* trypsin inhibitor (Onesti et al. 1991) and ricin B chain (Rutenber and Robertus 1991).

18.4.7 Thaumatin-Like α-Amylase Inhibitors

This class of inhibitors has molecular masses of ≈22 kDa which are similar in sequence to pathogenesis-related group 5 (PR-5) proteins and to an extremely sweet protein thaumatin from *Thaumatococcus daniellii* fruit (Cornelissen et al. 1986; Vigers et al. 1991). A bifunctional inhibitor zeamatin from *Zea mays* is the best studied inhibitor from this class.

18.4.8 γ-Purothionin-Like α-Amylase Inhibitors

The α-amylase inhibitors of γ-thionin superfamily are sulphur-rich proteins having 47 or 48 residues. Three isoforms from *Sorghum bicolor* called SIα-1, SIα-2 and SIα-3 inhibit insect α-amylases from guts of locust and cockroach, poorly inhibit α-amylases from *Aspergillus oryzae* and human saliva and do not inhibit the α-amylases from porcine pancreas, barley and *Bacillus* sp. (Bloch and Richardson 1991).The sorghum inhibitors of insect α-amylases were found similar in their amino acid sequences to gamma purothionins isolated from wheat endosperm with identities in the range 32–83% (Colilla et al. 1990). The wheat γ-purothionins which are a group of cysteine-rich thionins, toxic to insect larvae, show relative resemblance to α- and β-purothionins and hordothionins from wheat and barley (Ohtani et al. 1978; Hase et al. 1978; Jones and Mac 1977; Lecompte et al. 1982; Osaki et al. 1980) and the plant toxins crambin (Teeter et al. 1981), pyrularia (Vernon et al. 1985) and viscotoxin (Samuelsson and Petterson 1971). The modes of action as studied by workers are modification of membrane permeability (Aagerhofer et al. 1990), inhibition of cell-free protein synthesis (Garcia-Olmedo et al. 1983) and inhibition of ribonucleotide reductases by competing for reducing equivalents from thioredoxin (Johnson et al. 1987). Plants with α-amylase inhibitory activity are also discussed in Table 18.1. (Layer et al. 1985; Rituparna et al. 2014; Sudha et al. 2011; Rammohan et al. 2008; Sneha and Sanjay 2011; Keerthana et al. 2013; Mc Cue and Shetty 2004; Iniyan et al. 2010; Dnyaneshwar and Archana 2013; Radhika et al. 2013; Sangeetha and Vedasree 2012).

18.4.9 Computational Approaches to Predict Inhibition Patterns of Natural Compounds

The descried inhibitors from different resources could be proven as effective and beneficial in the treatment of several diseases. The current era of computational biology and bioinformatics developed many tools and techniques which can design, predict or analyse the functioning of any inhibitors in silico rather than a high expensive set of in vivo and in vitro approaches. Definitely, in silico technologies are saving cost and time of the experiments. So, why not we can implement computational approach towards analysis of different natural compounds?

18.5 Useful Tools, Software and Methodologies Which Are as Follows

18.5.1 Availability and In Silico Design of Inhibitors

In in silico analysis, we have to deal with only three-dimensional structure of inhibitors or natural compounds. Nowadays many databases and online resources (Irwin and Shoichet 2005; Mangal et al. 2013) are available which provide you 3D structures of inhibitors free of cost (i.e. not possible in wet lab experiments). So you are one step ahead. If you don't have 2D or 3D structure of natural compounds, then no need to worry. Many other tools are available through which you can draw 2D or generate 3D structure. The examples are available as Chembldb https://www.ebi.ac.uk/chembldb/, ChemSketch ChemDraw and ChemDraw Office 15. Once you have generated 2D structures, after that you can easily convert these structures into the 3D structures using other tools like CORINA (https://www.molecular-networks.com/products/corina), Open Babel (openbabel.org), Discovery Studio Visualizer, etc.

18.5.2 Preparation of 3D Structure of Receptor Molecules

Also we have to generate or obtain 3D structure of receptor molecules. The Protein Data Bank (www.rcsb.org) contains 3D structure of biological macromolecules and DNA (Vyas et al. 2012). The 3D data is obtained by X-ray crystallography, NMR spectroscopy or, increasingly, cryo-electron microscopy, and it is directly submitted by biologists worldwide, and we can access the information free of cost using the websites of its related member organizations PDBe (Cavasotto and Phatak 2009), PDBj (Chandonia and Brenner 2005) and RCSB (Vitkup et al. 2001). You can easily download 3D structure files for enzymes, proteins or DNA of interest. Unfortunately, if you did not find structure in the PDB, then don't worry; homology modelling approach is available to generate 3D structure of those enzymes of proteins. Homology modelling methods follow the relationship among the evolutionary-related proteins that have approximately similar structure (Chandonia and Brenner 2005; Vitkup et al. 2001; Bowie et al. 1991). I-TASSER is the best server for protein structure prediction according to the 2006–2012 CASP experiments (CASP7, CASP8, CASP9 and CASP10). It is freely available for the academic and research use. HHpred, MODELLER, SWISS-MODEL and Foldit are the popular software (Gullotto et al. 2013).

18.5.3 Molecular Interactions Analysis of Inhibitors with Enzymes

Once we prepared compound 3D file, i.e. *.pdb*, and receptor 3D file, i.e. *.pdb*, we can perform molecular interaction analysis of both using different computational approaches. Currently, many softwares and online resources are available to perform experiments computationally. The most popular freely available software is AutoDock 4.0 from MGL tools package. MGLTools is a software developed at the Molecular Graphics Laboratory (MGL) of the Scripps Research Institute for visualization and analysis of molecular structures. The researcher can freely download the software from http://mgltools.scripps.edu/downloads websites. The AutoDock 4.2 uses a semiempirical free energy force field to evaluate conformations during docking simulations. Using software we can analyse the binding pattern of these inhibitors with enzymes of interest. The major tools are GOLD, FlexX, Discovery Studio, etc., and most all the other docking programmes have the same scoring function and methodology, like AutoDock and FlexX. Some more useful software is available and documented on some online resources like:

Click2Drug (http://www.click2drug.org/directory_Docking.html);
SwissDock (http://www.swissdock.ch/);
Dock Blaster (http://blaster.docking.org/);
Docking Server (http://www.dockingserver.com/web);
VLS3D (http://www.dockingserver.com/web) and
PATCHDOCK (http://bioinfo3d.cs.tau.ac.il/PatchDock/) etc.

The computational biology filed is exponentially increasing, and new algorithm-based technologies are hitting the scientific experimentation to accelerate the outputs. The molecular dynamics is one of the most effective in silico techniques to validate the results obtained from initial computational study. Molecular dynamics simulations are important tools for understanding the physical basis of the structure and function of biological macromolecules. The early view of proteins as relatively rigid structures has been replaced by a dynamic model in which the internal motions and resulting conformational changes play an essential role in their function. This review presents a brief description of the origin and early uses of biomolecular simulations. It then outlines some recent studies that illustrate the utility of such simulations and closes with a discussion of their ever-increasing potential for contributing to biology (Karplus and Mccammon 2002).

Many software are available like:

VMD-Visual Molecular Dynamics (www.ks.uiuc.edu/Research/vmd/);
NAMD- Scalable Molecular Dynamics (www.ks.uiuc.edu/Research/namd/);
GROMACS (www.gromacs.org) and
AMBER (http://ambermd.org/) etc.

18.6 Conclusion

During the last four decades, the area of α-amylase inhibitors has gained considerable interest. A lot many reviews had stated presence, chemical attributes and reaction mechanisms of microbial and plant α-amylase inhibitors as well as their significance in nutrition. In nature, there are seven types of proteinaceous α-amylase inhibitors found as defined by sequence similarities and three-dimensional structures. These inhibitors in different forms are present in several organisms. The inhibitors have found application in obesity and diabetes therapy.

Intake of food or medicinal herbs rich in polyphenols has to be dose dependent, because we do not want to completely inhibit the enzyme but to reduce α-amylase activity to a limited extent. Complete inhibition in human system would result in unavailability of glucose to form ATP. On the other hand, evidences regarding the safe use of natural α-amylase inhibitor are required to prevent any serious side effects. For regulation of sugar uptake to control blood sugar level and prevention of oral diseases, designing of functional foods and exploring new therapeutic strategies are the need of the hour. To relate the variations among the obtained results, a standardized protocol should be developed to find potential therapeutic inhibitors. There is a need to explore new therapeutic agents to overcome the effects of the increasing trend in type II diabetes mellitus which has become a grievous medical concern worldwide. Effective approaches need to be accomplished to overcome the problem of drug resistance. These studies would be helpful to elucidate the pharmacological mechanism and also to develop medicinal formulas and nutraceutical or functional foods for diabetes and related symptoms.

Acknowledgement This study was supported by TEQIP-II (Technical Education Quality Improvement Programme, Government of India).

References

Aagerhofer CK et al (1990) Phospholipase activation in the cytotoxic mechanism of thionin purified from nuts of *Pyrularia pubera*. Toxicon 28:547–557

Abe JI et al (1993) Arginine is essential for the α-amylase inhibitory activity of the α-amylase/subtilisin inhibitor (BASI) from barley seeds. Biochem J 293:151–155

Aguiar LGK et al (2007) Microcirculação no Diabetes: Implicações nas Complicações Crônicas e Tratamento da Doença. Arq Bras Endocrinol Metab 51:204–211

Alam N et al (2001) Substrate–inhibitor interactions in the kinetics of α-amylase inhibition by Ragi α-amylase/ trypsin inhibitor (RATI) and its various N-terminal fragments. Biochemistry 40:4229–4233

Ali H et al (2006) α-amylase inhibitory activity of some Malaysian plants used to treat diabetes: with particular reference to *Phyllanthus amarus*. J Ethnopharmacol 107:449–455

Bailey CJ (2003) New approaches to the pharmacotherapy of diabetes. In: Pickup JC, William G (eds) Textbook of diabetes, vol 2, 3rd edn. Blackwell Science Ltd, UK, pp 73.1–73.21

Bhandari MR et al (2008) α-glucosidase and α-amylase inhibitory activitiesof Nepalese medicinal herb Pakhanbhed (*Bergenia ciliata*, Haw.) Food Chem 106:247–252

Bhat M et al (2011) Antidiabetic Indian plants: a good source of potent amylase inhibitors. Evid Based Complement Alternat Med 2011:810207

Bhutani KK, Gohil VM (2010) Natural products drug discovery research in India: status and appraisal. Indian J Exp Biol 48:199–207
Bloch JRC, Richardson M (1991) A new family of small (5kD) protein inhibitors of insect α-amylase from seeds of sorghum (Sorghum bicolor (L.) Moench) have sequence homologies with wheat δ -purothionins. FEBS Letter 279:101–104
Bode W et al (1989) The refined 2.0Å X-ray crystal structure of the complexformed between bovine β-trypsin and CMTI-I, a trypsin inhibitor from squash seeds (*Cucurbita maxima*). Topological similarity of the squash seed inhibitors with the carboxypeptidase A inhibitor from potatoes. FEBS Lett 242:285–292
Bompard-Gilles C et al (1996) Substrate mimicry in the active centre of a mammalian α-amylase: structural analysis of an enzyme-inhibitor complex. Structure 4:1441–1452
Bowie FU et al (1991) A method to identify protein sequences that fold into a known three-dimensional structure. Science 253:164–170
Brayer GD et al (1995) The structure of human pancreatic α-amylase at 1.8Å resolution and comparisons with related enzymes. Protein Sci 4:1730–1742
Brayer GD et al (2000) Subsite mapping of the human pancreatic alpha-amylase active site through structural, kinetic and mutagenesis techniques. Biochemistry-US 39:4778–4791
Buonocore V et al (1977) Wheat protein inhibitors of α-amylase. Phytochemistry 16:811–820
Campos FAP, Richardson M (1983) The complete amino acid sequence of the bifunctional α-amylase / trypsin inhibitor from seeds of ragi (Indian finger millet, *Eleusine coracana* Gaertn). FEBS Lett 152:2
Carlson GL et al (1983) A bean *α*-amylase inhibitor formulation (starch blocker) is ineffective in man. Science 219:393–395
Cavasotto CN, Phatak SS (2009) Homology modeling in drug discovery: current trends and applications. Drug Discov Today 14:676–683
Chagolla-Lopez A et al (1994) A novel α-amylase inhibitor from Amaranth (*Amaranthus hypocondriacus*) seeds. J Biol Chem 269:23675–23680
Chandonia JM, Brenner SE (2005) Implications of structural genomics target selection strategies: Pfam5000, whole genome, and random approaches. Proteins 58:166–179
Cheng AYY, Fantus IG (2005) Oral antihyperglycemic therapy for type 2 diabetes Mellitus. Can Med Assoc J 172:213–226
Choi HJ et al (2000) Inhibitory effects of crude drugs on alpha-glucosidase. Arch Pharm Res 23:261–266
Colilla FJ et al (1990) Gamma-Purothionins: amino acid sequence of two polypeptides of a new family of thionins from wheat endosperm. FEBS Lett 270:191–194
Conforti F et al (2005) In vitro antioxidant effect and inhibition of α-amylase of two varieties of *Amaranthus caudatus* seeds. Biol Pharm Bull 28:1098–1102
Connolly JD, Hill RA (1999) Triterpenoids. Nat Prod Rep 16:221–240
Cornelissen BJC et al (1986) A tobacco mosaic virus-induced tobacco protein is homologous to the sweet-tasting protein thaumatin. Nature 231:531–532
De Lampasona MEP et al (1988) Oleanolic acid and ursolic acid derivatives from *Polylepsis australis*. Phytochemistry 49:2061–2064
Dnyaneshwar MN, Archana RJ (2013) In vitro inhibitory effects of Pithecellobium dulce (Roxb.) Benth. seeds on intestinal α-glucosidase and pancreatic α-amylase. J Biochem Technol 4(3):616–621
Du ZY et al (2006) α-glucosidase inhibition of natural curcuminoids and curcumin analogs. Eur J Med Chem 14:213–218
Feng GH et al (1996) α-amylase inhibitors from wheat: sequences and patterns of inhibition of insect and human α-amylases. Insect Biochem Mol Biol 26:419–426
Finardi-Filho F et al (1996) A putative precursor protein in the evolution of the bean alpha-amylase inhibitor. Phytochemistry 43:57–62
Fogel MR, Gray GM (1973) Starch hydrolysis in man: an intraluminal process not requiring membrane digestion. J Appl Physiol 35(2):263–267

Franco OL et al (2000) Activity of wheat α-amylase inhibitors towards bruchid α-amylases and structural explanation of observed specificities. Eur J Biochem 267(8):1466–1473

Fujisawa T et al (2005) Effect of two alpha-glucosidase inhibitors, voglibose and acarbose on postprandial hyperglycemia correlates with subjective abdominal symptoms. Metabolism 54:387–390

Funke I, Melzing MF (2006) Traditionally used plants in diabetes therapy- phytotherapeutics as inhibitors of α-amylase activity. Rev Bras Farmacogn 16:1–5

Gao H et al (2007) Inhibitory effecton α-glucosidase by the fruits of *Terminalia chebula* Retz. Food Chem 105:628–634

Gao H et al (2008a) Inhibitory effect on α-glucosidase by *Adhatoda vasica* Nees. Food Chem 108:965–972

Gao H et al (2008b) α-glucosidase inhibitory effect by the flower buds of *Tussilago farfara* L. Food Chem 106:1195–1201

Garcia-Casado GL et al (1994) Rye inhibitors of animal α-amylases shown different specificities, aggregative properties and IgE-binding capacities than their homologues from wheat and barley. Eur J Biochem 224:525–531

Garcia-Casado G et al (1995) A major baker's asthma allergen from rye flour is considerably more active than its barley counterpart. FEBS Lett 364:36–40

Garcia-Casado G et al (1996) Role of complex asparagine-linked glycans in the allergenicity of plant glycoproteins. Glycobiology 6:471–477

Garcia-Olmedo F et al (1983) Biochem Biophys Acta 740:52–56

Giri AP, Kachole MS (1998) Amylase inhibitors of pigeonpea (*Cajanus cajan*) seeds. Phytochemistry 47(2):197–202

Goke B, Herrmann-Rinke C (1998) The evolving role of alpha-glucosidase inhibitors. Diabetes/Metab Res 14:S31–S38

Groot PC et al (1988) Human pancreatic amylase is encoded by two different genes. Nucleic Acids Res 16:4724

Grossi de Sá MF et al (1997) Molecular characterization of a bean α-amylase inhibitor that inhibits the α-amylase of the Mexican bean weevil *Zabrotes subfasciatus*. Planta 203:295–303

Grover JK et al (2002) Medicinal plants of India with anti-diabetic potential. J Ethnopharmacol 81:81–100

Gullotto D et al (2013) Probing the protein space for extending the detection of weak homology folds. J Theor Biol 320:152–158

Gumucio DL et al (1988) Concerted evolution of human amylase genes. Mol Cell Biol 8:1197–1205

Gvozdeva EL et al (1993) Enzymatic oxidation of the bifunctional wheat inhibitor of subtilisin and endogenous α-amylase. FEBS Lett 334:72–74

Gyémánt G et al (2003) Inhibition of human salivary α-amylase by glucopyranosylidene-spiro-thiohydantoin. Biochem Biophys Res Commun 312:334–339

Hamdan II, Afifi FU (2010) Capillary electrophoresis as a screening tool for alpha amylase inhibitors in plant extracts. Saudi Pharm J 18:91–95

Hansawasdi C et al (2000) α-amylase inhibitors from Roselle (*Hibiscus sabdariffa* Linn.) tea. Biosci Biotechnol Biochem 64:1041–1043

Hase T et al (1978) Disulfide bonds of purothionin, a lethal toxin for yeasts. J Biochem 83 (6):1671–1678

Hochstrasser K et al (1970) Physiol Chem 351:721–728

Imoto T et al (1991) A novel peptide isolated from the leaves of *Gymnema sylvestre:* characterization and its suppressive effect on the neural responses to sweet taste stimuli in the rat. Comp Biochem Physiol 100:309–314

Iniyan GT et al (2010) In vitro study on α-amylase inhibitory activity of an Indian medicinal plant, *Phyllanthus amarus*. Indian J Pharmacol 42(5):280–282

Inzucchi SE (2002) Oral anti-hyperglycemic therapy for type 2 diabetes. JAMA 287:360–372

Irwin JJ, Shoichet (2005) ZINC – a free database of commercially available compounds for virtual screening. J Chem Inf Model 45(1):177–182
Iulek J et al (2000) Purification, biochemical characterisation and partial primary structure of a new α-amylase inhibitor from *Secale cereale* (rye). Int J Biochem Cell Biol 32:1195–1204
Janecek S et al (1997) Domain evolution in the α-amylase family. J Mol Evol 45:322–331
Johnson TC et al (1987) Reduction of purothionin by the wheat seed thioredoxin system. Plant Physiol 85(2):446–451
Jones BL, Mac AS (1977) Amino acid sequences of the two α-purothionins of hexaploid wheat. Cereal Chem 54:511–523
Kadziola A et al (1994) Crystal and molecular structure of barley α-amylase. J Mol Biol 239:104–121
Kadziola A et al (1998) Molecular structure of a barley α-amylase-inhibitor complex: implications for starch binding and catalysis. J Mol Biol 278:205–217
Kandra L et al (2004) Inhibitory effects of tannin on human salivary α-amylase. Biochem Biophys Res Commun 319:1265–1271
Karplus M, McCammon JA (2002) Molecular dynamics simulations of biomolecules. Nat Struct Biol 9(9):646–652
Karthic K et al (2008) Identification of α- amylase inhibitors from *Syzygium cumini* Linn seeds. Indian J Exp Biol 46:677–680
Kasahara K et al (1996) Complete sequence, subunit structure and complexes with pancreatic α-amylase of an α-amylase inhibitor from *Phaseolus vulgaris* white kidney beans. J Biochem 120:177–183
Kazeem MI et al (2013) Inhibitory effect of *A. indica* Juss leaf extract on the activities of alpha amylase and alpha glucosidase. Pak J Biol Sci 16(21):1358–1362
Keerthana G et al (2013) In vitro alpha amylase inhibitory and anti-oxidant activities of ethanolic leaf extract of *Croton bonplandianum*. Asian J Pharm Clin Res 6(Suppl 4):32–36
Kim MJ et al (1999) Comparative study of the inhibition of α-glucosidase, α-amylase and cyclomaltodextrin glucanosyltransferase by acarbose, isoacarbose and acarviosine-glucose. Arch Biochem Biophys 371:277–283
Kobayashi K et al (2003) Screening of Mongolian plants for influence on amylase activity in mouse plasma and gastrointestinal tube. Biol Pharm Bull 26:1045–1048
Komaki E et al (2003) Identification of anti-amylase components from olive leaf extracts. Food Sci Technol Res 9:35–39
Kotowaroo MI et al (2006) Screening of traditional antidiabetic medicinal plants of Mauritius for possible alpha-amylase inhibitory effects in vitro. Phytother Res 20:228–231
Kurihara H et al (1999) Inhibitory potencies of bromophenols from Rhodomelaceae algae against α-glucosidase activity. Fish Sci 65:300–303
Kusaba-Nakayama M et al (2000) CM-3, one of the wheat α-amylase inhibitor subunits, and binding of IgE in sera from Japanese with atopic dermatitis related to wheat. Food Chem Toxicol 38:179–185
Laar FA (2008) Alpha-glucosidase inhibitors in the early treatment of type 2 diabetes. Vasc Health Risk Manag 4(6):1189–1195
Lam SH et al (2008) α-glucosidase inhibitors from the seeds of *Syagrus romanzoffiana*. Phytochemistry 69:1173–1178
Layer P et al (1985) Partially purified white bean amylase inhibitor reduces starch digestion in vitro and inactivates intraduodenal amylase in humans. Gastroenterology 88:1895–1902
Le Berre-Anton V et al (1997) Characterization and functional 149 properties of the [alpha]-amylase inhibitor ([alpha]-AI) from kidney bean (*Phaseolus vulgaris*) seeds. Biochim Biophys Acta 1343:31–40
Lebowitz HE (1998) Alpha-glucosidase inhibitors as agents in the treatment of diabetes. Diabetes Rev 6:132–145
Lecompte TJ et al (1982) Biochemistry 21:4843–4849

Li WL et al (2004) Natural medicines used in the traditional Chinese medical system for therapy of diabetes mellitus. J Ethnopharmacol 92(1):1–21
Loizzo MR et al (2008) In vitro inhibitory activities of plants used in Lebanon traditional medicine against angiotensin converting enzyme (ACE) and digestive enzymes related to diabetes. J Ethnopharmacol 119:109–116
Lo-Piparo E et al (2008) Flavonoids for controlling starch digestion: structural requirements for inhibiting human α-amylase. J Med Chem 51:3555–3561
Lu S et al (1999) Solution structure of the major α-amylase inhibitor of the crop plant amaranth. J Biol Chem 274:20473–20478
Luo JG et al (2008) New triterpenoid saponins with strong α- glucosidase inhibitory activity from the roots of Gypsophila oldhamiana. Bioorg Med Chem 16:2912–2920
Lyons A et al (1987) Characterization of homologous inhibitors of trypsin and α-amylase. Biochim Biophys Acta 915:305–313
Maarel MJEC et al (2002) Properties and applications of starch-converting enzymes of the α-amylase family. J Biotechnol 94:137–155
Mangal M et al (2013) NPACT: naturally occurring plant-based anti-cancer compound-activity-target database. Nucleic Acids Res 41(Database issue):D1124–D1129
Marles R, Farnsworth N (1994) Plants as sources of anti-diabetic agents. In: Wagner H, Farnsworth NR (eds) Economic and medicinal plant research, vol 6. Academic Press Ltd, UK, pp 149–187
Marshall JJ, Lauda CM (1975) Purification and properties of phaseolamin, an inhibitor of α-amylase from the kidney bean *Phaseolus vulgaris*. J Biol Chem 250:8030–8037
Martins JC et al (2001) Solution structure of the main alpha-amylase inhibitor from amaranth seeds. Eur J Biochem 268:2379–2389
Mbaze LM et al (2007) α-glucosidase inhibitory pentacyclic triterpenes from the stem bark of *Fagara tessmannii* (Rutaceae). Phytochemistry 68:591–595
Mc-Cue PP, Shetty K (2004) Inhibitory effects of rosmarinic acid extracts on porcine pancreatic amylase in vitro. Asia Pac J Clin Nutr 13:101–106
Mc-Dougall et al (2005) Different polyphenolic components of soft fruits inhibit alpha-amylase and alpha-glucosidase. J Agric Food Chem 53:2760–2766
Melo FR et al (1999) α-amylase from cowpea seeds. Prot Pept Lett 6:387–392
Mentreddy SR (2007) Medicinal plant species with potential antidiabetic properties. J Sci Food Agric 87:743–750
Moreno J, Chrispeels MJ (1989) A lectin gene encodes the α-amylase inhibitor of the common bean. Proc Natl Acad Sci U S A 86:7885–7889
Mundy J et al (1983) Barley α-amylase/ subtilisin inhibitor: isolation and characterization. Carlsb Res Commun 48:81–90
Nahoum V et al (2000) Crystal structures of human pancreatic α-amylase in complex with carbohydrate and proteinaceous inhibitors. Biochem J 346:201–208
Notkins AL (2002) Immunologic and genetic factors in type 1 diabetes. J Biol Chem 277 (46):43545–43548
Odani S et al (1982) Sequence homology between barley trypsin inhibitor and wheat alpha-amylase inhibitors. FEBS Lett 141(2):279–282
Ohtani S et al (1978) Complete primary structures of two subunits of purothionin A, a lethal protein for brewer's yeast from wheat flour. J Biochem 83:733–767
Ohtsubo K, Richardson M (1992) The amino acid sequence of a 20-kDa bifunctional subtilisin/ α-amylase inhibitor from grain of rice (*Oryza sativa* L.) seeds. FEBS Lett 309:68–72
Onesti S et al (1991) Crystal structure of a Kunitz-type trypsin inhibitor from *Erythrina caffra* seeds. J Mol Biol 217:153–176
Osaki Y et al (1980) Amino acid sequence of a purothionin homolog from barley flour. J Biochem 87:549–555
Pal GP et al (1994) The three-dimensional structure of the complex of proteinase K with its naturally occurring inhibitor PKI3. FEBS Lett 341:167–170

Park CS, Miller C (1992) Mapping function to structure in a channel-blocking peptide: electrostatic mutants of charybdotoxin. Biochemistry 31:7749–7755

Pasero L et al (1986) Complete amino acid sequence and location of the five disulfide bridges in protein pancreatic α-amylase. Biochem Biophys Acta 869:147–157

Perez RM et al (1998) Anti-diabetic effects of compounds isolated from plants. Phytomedicine 5:55–75

Petrucci T et al (1976) Further characterization studies of the alpha-amylase protein inhibitor of gel electrophoretic mobility 0.19 from the wheat kernel. Biochim Biophys Acta 420:288–297

Prabhakar PK, Doble M (2011) Mechanism of action of natural products used in the treatment of diabetes mellitus. Chin J Integr Med 17(8):563–574

Prashanth D et al (2001) Effect of certain plant extracts on [alpha]-amylase activity. Fitoterapia 72:179–181

Qian M et al (1993) Structure and molecular model refinement of pig pancreatic α-amylase at 2.1Å resolution. J Mol Biol 231:785–799

Qian M et al (1994) The active centre of a mammalian α-amylase: structure of the complex of a pancreatic α-amylase with a carbohydrate inhibitor refined to 2.2Å. Biochemistry 33:6284–6294

Qian M et al (2001) Enzyme-catalyzed condensation reaction in a mammalian alpha-amylase: high resolution structural analysis of an enzyme inhibitor complex. Biochemistry 40:7700–7709

Quideau S et al (2003) DNA topoisomerase inhibitor acutissimin A and other flavano-ellagitannins in red wine. Angew Chem Int Ed 42:6012–6014

Radhika S et al (2013) Phytochemical investigation and evaluation of anti-hyperglycemic potential of *Premna Corymbosa*. Int J Pharm Sci 5(4):352–356

Ramasubbu N et al (1996) Structure of human salivary alpha-amylase at 1.6Å resolution: implications for its role in the oral cavity. Acta Crystallogr Sect D Biol Crystallogr 52:435–446

Rammohan S et al (2008) In vitro α-glucosidase and α-amylase enzyme inhibitory effects of *Andrographis paniculata* extract and andrographolide. Acta Biochim Pol 55(2):391–398

Richardson M (1990) Seed storage proteins: the enzyme inhibitors. In: Rogers L (ed) Methods in plant biochemistry, 5th edn. Academic, London, pp 261–307

Rituparna C et al (2014) Screening of nine herbal plants for in vitro α-amylase inhibition. Asian J Pharm Clin Res 7:4

Rodenburg KW et al (1995) Arg-27, Arg-127 and Arg-155 in the β-trefoil protein barley α-amylase/subtilisin inhibitor are interface residues in the complex with barley α-amylase 2. Biochem J 309:969–976

Rutenber E, Robertus JD (1991) Structure of ricin B chain at 2.5Å resolution. Proteins 10:260–269

Ryan CA (1990) Protease inhibitors in plants: genes for improving defences against insects and pathogens. Annu Rev Phytopathol 28:425–449

Rydberg EH et al (2002) Mechanistic analyses of catalysis in human pancreatic α-amylase: detailed kinetic and structural studies of mutants of three conserved carboxylic acids. Biochemistry 41:4492–4502

Samuelsson G, Petterson BM (1971) The amino acid sequence of viscotoxin B from the European mistletoe (*Viscum album* L, Loranthaceae). Eur J Biochem 21:86–89

Sangeetha R, Vedasree N (2012) *In vitro* α-amylase inhibitory activity of the leaves of *Thespesia populnea*. ISRN Pharmacol 2012:1–4

Scannapieco FA et al (1993) Salivary α-amylase: role in dental plaque and caries formation. Crit Rev Oral Biol Med 4:301–307

Silano V (1987) α-amylase inhibitors. In: Kruger J, Lineback D (eds) Enzymes and their role in cereal technology. American Association of Cereal Chemists, St. Paul, pp 141–199

Singh SK et al (2007) Evidence-based critical evaluation of glycemic potential of *Cynodon dactylon*. Evid Based Complement Alternat Med 6(4):415–420

Sneha JA, Sanjay C (2011) Alpha-amylase inhibitory and hypoglycemic activity of Clerodendrone multiflorum Linn Stems. Asian J Pharm Clin Res 4(2):99–102

Strobl S et al (1998) A novel strategy for inhibition of α-amylases: yellow meal worm α-amylase in complex with the *Ragi* bifunctional inhibitor at 2.5Å resolution. Structure 6:911–921

Sudha P et al (2011) Potent α-amylase inhibitory activity of Indian Ayurvedic medicinal plants. BMC Complement Altern Med 11:5

Svensson B et al (2004) Review: proteinaceous α-amylase inhibitors. Biochim Biophys Acta 1696:145–156

Tabopda TK et al (2008) Bioactive aristolactams from Piper umbellatum. Phytochemistry 69:1726–1731

Teeter MM et al (1981) Primary structure of the hydrophobic plant protein crambin. Biochemistry 20:5437–5443

Vallée, F (1996). Structure cristalline à 1.9 Å de resolution d'un complexe protéine–protéine entreune α-amylase d'orge et un inhibiteur bifonctionnel. Ph.D. Thesis, CNRS-Marseille & Orsay, France

Vallée F et al (1998) Barley α-amylase bound to its endogenous protein inhibitor BASI: crystal structure of the complex at 1.9Å resolution. Structure 6:649–659

Vander MMJEC et al (2002) Properties and applications of starch-converting enzymes of the α-amylase family. J Biotechnol 94:137–155

Vernon LP et al (1985) A toxic thionin from *Pyrularia pubera*: purification, properties, and amino acid sequence. Arch Biochem Biophys 238:18–29

Vigers A et al (1991) A new family of plant antifungal proteins. Mol Plant Microb Interact 4:315–323

Vitkup D et al (2001) Completeness in structural genomics. Nat Struct Biol 8:559–566

Vyas VK et al (2012) Homology modeling a fast tool for drug discovery: current perspectives. Indian J Pharm Sci 74(1):1–17

Wang HH et al (2011) Comparisons of [alpha]-amylase inhibitors from seeds of common bean mutants extracted through three phase partitioning. Food Chem 128:1066–1071

Weselake RJ et al (1985) Effect of endogenous barley α-amylase inhibitor on hydrolysis of starch under various conditions. J Cereal Sci 3:249–259

Whitcomb DC, Lowe ME (2007) Human pancreatic digestive enzymes. Digest Dis Sci 52:1–17

WHO (World Health Organisation Consultation) (1999). Definition, diagnosis and classification of diabetes mellitus and its complications. Part 1: diagnosis and classification of diabetes mellitus. Report of a WHO Consultation, Geneva

World Health Organization (2014) Global status report on noncommunicable diseases 2014: attaining the nine global noncommunicable diseases targets; a shared responsibility. World Health Organization, Geneva

Wiegand G et al (1995) The crystal structure of porcine pancreatic α-amylase in complex with the microbial inhibitor Tendamistat. J Mol Biol 247:99–110

Wilcox ER, Whitaker JR (1984) Characterization of two amylase inhibitors from black bean (Phaseolus vulgaris). J Food Biochem 8:189–213

Xu R et al (2004) On the origins of triterpenoid skeletal diversity. Phytochemistry 65:261–291

Yamada T et al (2001) Purification and characterization of two [alpha]-amylase inhibitors from seeds of tepary bean (*Phaseolus acutifolius* A. Gray). Phytochemistry 58:59–66

Yamagata H et al (1998) Rice bifunctional α-amylase/subtilisin inhibitor: characterization, localization and changes in developing and germinating seeds. Biosci Biotechnol Biochem 62:978–985

Yoon SH, Robyt JF (2003) Study of the inhibition of four alpha amylases by acarbose and its 4IV-α-maltohexaosyl and 4IV- α-maltododecaosyl analogues. Carbohydr Res 338:1969–1980

Young NM et al (1999) Post-translational processing of two α-amylase inhibitors and an arcelin from the common bean, *Phaseolus vulgaris*. FEBS Lett 446:203–206

Zemke KJ et al (1991) The three dimensional structure of the bifunctional proteinase K/ α-amylase inhibitor from wheat (PKI3) at 2.5 Å resolution. FEBS Lett 279:240–242

MIX
Papier aus verantwortungsvollen Quellen
Paper from responsible sources
FSC® C105338

If you have any concerns about our products,
you can contact us on
ProductSafety@springernature.com

In case Publisher is established outside the EU,
the EU authorized representative is:
Springer Nature Customer Service Center GmbH
Europaplatz 3, 69115 Heidelberg, Germany

Printed by Libri Plureos GmbH
in Hamburg, Germany